ADVANCED ENVIRONMENTAL EXERCISE PHYSIOLOGY

ADVANCED EXERCISE PHYSIOLOGY SERIES

ADVANCED ENVIRONMENTAL EXERCISE PHYSIOLOGY

ADVANCED EXERCISE PHYSIOLOGY SERIES

Stephen S. Cheung, PhD
Brock University

Human Kinetics

Library of Congress Cataloging-in-Publication Data

Cheung, Stephen S., 1968-
   Advanced environmental exercise physiology / Stephen S. Cheung.
      p. ; cm. -- (Advanced exercise physiology series)
   Includes bibliographical references and index.
   ISBN-13: 978-0-7360-7468-1 (hard cover)
   ISBN-10: 0-7360-7468-6 (hard cover)
  1. Exercise--Physiological aspects. 2. Adaptation (Physiology) I. Title. II.
Series: Advanced exercise physiology series.
   [DNLM: 1. Exercise--physiology. 2. Adaptation, Physiological--physiology. 3.
Environment. 4. Stress, Physiological--physiology. WE 103 c526a 2010]
   QP301.c557 2010
   612'.044--dc22

                                    2009020458

ISBN-10: 0-7360-7468-6 (print)   ISBN-10: 0-7360-8549-1 (Adobe PDF)
ISBN-13: 978-0-7360-7468-1 (print)   ISBN-13: 978-0-7360-8549-6 (Adobe PDF)

The Web addresses cited in this text were current as of May 2009 unless otherwise noted.

Acquisitions Editor: Michael S. Bahrke, PhD; **Developmental Editor:** Jillian Evans; **Managing Editor:** Melissa J. Zavala; **Assistant Editors:** Katherine Maurer and Christine Bryant Cohen; **Copyeditor:** Joyce Sexton; **Indexer:** Gerry Lynn Shipe; **Permission Manager:** Dalene Reeder; **Graphic Designer:** Joe Buck; **Graphic Artist:** Dawn Sills; **Cover Designer:** Bob Reuther; **Photo Asset Manager:** Laura Fitch; **Photo Production Manager:** Jason Allen; **Art Manager:** Kelly Hendren; **Associate Art Manager:** Alan L. Wilborn; **Illustrator:** Gary Hunt; **Printer:** Edwards Brothers

Printed in the United States of America   10 9 8 7 6 5 4

The paper in this book is certified under a sustainable forestry program.

Human Kinetics
Web site: www.HumanKinetics.com

United States: Human Kinetics
P.O. Box 5076
Champaign, IL 61825-5076
800-747-4457
e-mail: humank@hkusa.com

Canada: Human Kinetics
475 Devonshire Road, Unit 100
Windsor, ON N8Y 2L5
800-465-7301 (in Canada only)
e-mail: info@hkcanada.com

Europe: Human Kinetics
107 Bradford Road
Stanningley
Leeds LS28 6AT, United Kingdom
+44 (0)113 255 5665
e-mail: hk@hkeurope.com

Australia: Human Kinetics
57A Price Avenue
Lower Mitcham, South Australia 5062
08 8372 0999
e-mail: info@hkaustralia.com

New Zealand: Human Kinetics
P.O. Box 80
Torrens Park, South Australia 5062
0800 222 062
e-mail: info@hknewzealand.com

Zachary and Jacob, may you feel the thrill of passion
in everything you choose in life.

Contents

Series Preface

Systematically detailing the effects of exercise on specific physiological systems and under various conditions is essential for advanced-level exercise physiology students. For example, what are the chronic effects of a systematic program of resistance training on cardiac structure and function, vascular structure and function, and hemostatic variables? How do different environments influence the ability to exercise, and what can pushing the body to its environmental limits tell us about how the body functions during exercise? When muscles are inactive, what happens to their sensitivity to insulin, and what role do inactive muscles play in the development of hyperinsulinemia and the type 2 diabetic state? These questions and many others are answered in the books in Human Kinetics' Advanced Exercise Physiology Series.

Beginning where most introductory exercise physiology textbooks end their discussions, each book in this series describes in detail the structure and function of a specific physiological system and the effects of exercise on that system or the effects of external conditions on exercise. Armed with this information, students will be better prepared to conduct the high-quality research required for advancing scientific knowledge, and they'll be able to apply the information to real-life scenarios such as health and fitness assessment and exercise guidelines and prescription.

Although many graduate programs and some undergraduate programs in exercise science and kinesiology offer specific courses on advanced topics in exercise physiology, there are few good options for textbooks to support these classes. Some instructors adopt general advanced physiology textbooks, but the limited number of textbooks available for advanced exercise physiology topics focus almost entirely on physiology without emphasizing *exercise* physiology.

Each book in the Advanced Exercise Physiology Series concisely describes a physiological system (e.g., cardiovascular and neuromuscular) or a particular topic (e.g., how the environment affects performance) and addresses the effects of exercise (acute and chronic, endurance and resistance types) on the specific system or in certain contexts. These textbooks are intended primarily for students, but researchers and practitioners will also benefit from the compilation of studies documenting the myriad effects of exercise and conditions on specific systems.

Preface

In my research and teaching career, most students entering environmental physiology typically are excellent students with a kinesiology or exercise science background, but with little actual experience in environmental physiology research or a good understanding of the breadth of the field. Therefore, the main drawback to their entry into the field is the enormous effort and time required to gain an advanced grounding in various research topics. My own immersion into the field was even more extreme, as my undergraduate degree was in oceanography and zoology, with not a single course in human physiology or anatomy! On the other hand, it was my passion for bicycling and desire to understand my body better that led to my switch to kinesiology, and it is that same desire to explore the limits of the human system that continues to this day.

Then as now, while excellent major volumes and classic texts are available on specific environments such as diving physiology or microgravity, students wishing to gain a survey of the scope and major ideas in environmental physiology have not had the option of one concise text. In my own undergraduate teaching, I have relied on cobbling together reading lists from highly detailed and specialized journal articles, supplemented by various review papers. For my graduate students, the process of orientation has involved reading major textbooks on the physiology and medicine of individual environments and at the same time jumping right into the specifics of a directed research project on a narrow and specialized topic. However, both methods generally result in an incomplete and somewhat random approach to each environment or the field as a whole, with some topics receiving major coverage while other topics are either absent or given only brief mention.

The present text does not strive to be a comprehensive document covering every physiological system in every environment. What I have aimed for instead is my own concise and personal survey of the field. For each environment, I have tried to provide a thorough overview of its major impact and also to highlight lively areas of current and future debate or investigation to stimulate further research and query. I am also a big proponent of the idea that the transition between fundamental, mechanistic research and applied research is seamless. Therefore, the direct application of fundamental laboratory research to athletic and occupational situations is highlighted throughout the text, while at the same time applied situations are used to illustrate how fundamental research ideas are developed and enhanced.

Overall, the environmental physiologist or ergonomist plays a central role in the design of safe systems and policies for athletes, workers, and the recreational public. This work requires a firm knowledge of basic exercise physiology, and then its application to specific environmental stressors. The latter thrust is the primary purpose of this volume, with each chapter demonstrating how different environments affect human physiology and performance. Each chapter also aims to highlight the major current debates within the given field, with the goal of stimulating potential avenues of inquiry for the reader.

Chapters 2 through 6 explore the thermal environment. As homeotherms, humans have an amazing capacity to maintain a relatively constant deep body temperature throughout a day and throughout their entire lives. This has been the case despite living in regions as diverse as the Saharan desert or the tropical jungles of New Guinea through to the high Arctic regions. What are the major physiological responses of humans to severe thermal stress that allow them to maintain such a strong homeothermy, and how does temperature affect exercise capacity and survival situations?

Chapters 7 through 9 explore the effects of environmental pressure differences on the human system. Unlike their ability to adapt to wide swings in temperature, humans are limited to a relatively narrow range of acceptable ambient pressures from sea level to approximately 5000 m (16,400 ft) for permanent habitats. It had long been scientifically theorized that humans were incapable of summiting Everest without supplemental oxygen until ultimately this idea was proven wrong by Reinhold Messner and Peter Habeler in 1978. The chronic adaptation of highland natives to altitude has also made the use of hypoxia very popular among athletes seeking to maximize their fitness. While humans are physiologically able to tolerate high ambient pressures, as evidenced by commercial saturation diving, the requirement for air breathing has necessitated recent technological advances to enable exploration of the hyperbaric environment.

Chapter 10 escapes Earth for the final frontier of space, with the microgravity environment a stressor for which the human body has had no opportunity to develop evolutionary adaptations. Space will also form the greatest physiological and technological challenge in the coming years with the proposed lunar base and the mission to Mars. Therefore, along with the undersea realm, the space environment is completely reliant on advances in technology and their successful incorporation with physiology and human factors research.

Chapters 11 and 12 deal with two nontraditional environmental stressors that partly originate from modern society. As discussed in chapter 11, respiratory problems from air pollution have become increasingly common in many major cities and even rural settings, primarily due to societal reliance on fossil fuels and industrial manufacturing. As highlighted by the 2008 Beijing Olympics, athletes are increasingly concerned about how exercise in polluted environments can affect both their performance and health, especially the ways in which multiple pollutants can synergistically affect long-term health issues. Chapter 12 explores the relationship between shift work and chronic sleep deprivation and risks of impaired cognitive performance and decision making, as well as the chronobiological impact on exercise capacity—a problem that is increasingly common. For athletes traveling long distances across multiple time zones for competitions, jet lag can provoke a host of physiological impairments that require countermeasures to either accommodate problems or to accelerate adaptation while not affecting fitness and training.

The exciting aspect of environmental physiology is that by its very nature it is multidisciplinary,

eBook

available at
HumanKinetics.com

requiring the practitioner to become a "specialized generalist." That is, scientists interested in the field must have a firm grounding in multiple physiological systems along with their interactions. Therefore, this text considers the capacity of the human to exercise in and tolerate different environments from an integrative approach, exploring how multiple systems interact during exposure and exercise in various environments. Overall, it is my hope that this text can serve as a gateway to the vibrant field of environmental physiology by providing a solid grounding in the key concepts and current debates in the field.

Acknowledgments

I started my career as a university scientist at Dalhousie University in 1998, a month after my marriage to my wonderful wife Debbie Hoffele. She has been there in every way imaginable through all phases and aspects of life as an academic. This has included the familiar joys of students graduating and succeeding, grants being successful, and papers getting published. It has also meant putting up with some long hours, occasional ranting, and my long-winded and convoluted explanations of why thermophysiology is more complex than simply people getting hot in the heat! She has also been there through the unexpected challenges of office and lab floods (sudden cold water immersion comes to life!). It's obvious that I could never have achieved a fraction of what I have without her love and support. To Debbie, I can't say more than that I love you in every sense of the word.

We are all a sum of our experiences, and I have had the fortune to have studied under four excellent scientists who have each taught me about life at the same time that they taught me the process and thrill of science. Paul J. Harrison took me out on the high seas during my B.Sc. oceanography studies and remained a friend and source of support even when my drastic switch to kinesiology took me far away from his own area of interest. Igor Mekjavic then literally threw me into cold water research for my M.Sc., as in my being a subject in a hypothermia experiment in my first week of grad studies! Igor also exposed me to the International Space University and to space research by asking if I was interested in a free summer in Spain. I was hooked! I worked with Tom McLellan as a summer research student during my M.Sc. studies, and ultimately when he became my doctoral advisor. Tom taught me how to organize big projects to investigate big ideas. Ron Maughan taught me the invaluable lesson that good ideas trump all, which served me in excellent stead as I developed my career at Dalhousie from a lab with four empty walls.

As all academics know, they may pretend to be the driving force in their labs and research programs, but it's really their lab members that do the brunt of the hard work and actually make the lab run. I've been blessed with a group of consistently outstanding undergraduate and graduate research students who are not just talented individually, but who have shared my same ethos of a lab based on honesty, teamwork, loyalty, and fun. They have pioneered many of the techniques and ideas from my lab, and I am honored to have played a small role in their development. One student in particular has been instrumental in numerous roles in my lab through the years. Robin Urquhart began as an undergraduate research student in my early years at Dalhousie. She later returned as a lab manager supreme, shepherding many techniques, students, and manuscripts through the academic process. Finally, she has also been instrumental in the coordination and preparation of this text. To all these students and my future students, I hope this text brings out the vibrancy and currency of environmental physiology.

Overview of Environmental Physiology

With increasing advocacy of the importance of physical activity and the continuing advances in technology, humans are becoming exposed to greater extremes of environmental conditions in recreational, athletic, and occupational settings. Much of the equipment previously limited to specialized occupations such as commercial diving, piloting, and professional mountaineering is now within the reach of keen amateurs or available for recreational pursuits. This is seen with the greater participation in endurance and ultra-endurance events among "weekend warriors," the increasing accessibility of guided mountaineering expeditions, and self-contained underwater breathing apparatus (SCUBA). At an extreme, the flights of the first "space tourists" in the first decade of the 21st century have opened up the Low Earth Orbit and potentially beyond to individuals outside of national astronaut corps. No longer must astronauts possess "the right stuff"; instead, this requirement can sometimes be circumvented by the right checkbook. Such developments expose humans of a wide range of ages, fitness levels, and backgrounds to environments never encountered in natural evolution. Therefore, research on environmental physiology has increasing relevance from scientific, clinical, and public health perspectives.

INTERDISCIPLINARY DESIGN ISSUES

What is exciting about environmental physiology is that by its very nature it is multidisciplinary, requiring the practitioner to become a "specialized generalist." That is, scientists interested in the field must have a firm grounding in multiple physiological systems along with their interactions. For example, the microgravity environment of space affects not just the cardiovascular system, but also the muscular, bone, and neurovestibular systems. Similarly, understanding manual function in the cold requires knowledge of microvascular blood flow dynamics to the fingers, neuromuscular characteristics, and also the neural control of grasping. Therefore, in addition to demanding knowledge of the unique biophysical nature of each environment, environmental physiology research typically requires a synthesizing and integration of environmental impacts on multiple but interrelated physiological systems.

In most extreme environmental conditions, workers and athletes may not be primarily concerned with individual physiological responses. Rather, the ultimate parameter of interest is performance, namely the ability to complete the required operational or survival task. Therefore, scientists must be cognizant of the practical applicability of their research on fundamental issues in environmental physiology. Such work may require close interactions with occupational health and safety specialists and policy makers, along with safety bodies such as the Red Cross or the Lifesaving Society.

Another layer of multidisciplinarity involved with environmental physiology is in the close collaboration with engineers and ergonomists to develop better and safer equipment. Unfortunately, physiological requirements and human factors are often the last design criteria to be considered or are else neglected altogether in the design of machines and work systems. For example, the nominal capacity of lifeboats adopted for abandoning offshore oil rigs is often highly optimistic, with seat widths designed to accommodate average humans rather than large individuals wearing bulky survival suits. This is exacerbated by inadequate ventilation that produces high carbon dioxide levels inside the lifeboat, which, along with motion sickness caused by constant tossing in rough oceans, may in turn affect thermoregulatory capacity and survival times.

Adding to the design constraints are inherent design dilemmas involved in protective clothing, namely the need for protection balanced against the needs of portability. While protective clothing is being continually improved and lightened, the requirement of adequate impact or environmental protection is generally contradictory to the desire for adequate ventilation or heat dissipation. The bulkiness and weight of clothing systems also add to the individual's metabolic costs, heat generation, and dehydration. Therefore, concurrent efforts must be made to design better materials and clothing systems and to understand the physiology of exercise in extreme environments.

COMMON TERMINOLOGY

Language and the correct use of terminology and definitions are important not just for such professionals as newspaper editors but also for scientists. Just as coaches, athletes, and exercise physiologists may mean completely different things when they talk about speed, strength, and power, so too can environmental physiologists create confusion when terms are used interchangeably. Without a common framework of definitions, it can become confusing and difficult to compare the findings of different research studies and across different disciplines. For the most part, no authoritative source for clear definitions exists; rather such definitions are derived from "best practices." To illustrate the importance of standardized terminology, the following section lists four commonly used terms that relate to individuals adapting to an acute or chronic environmental stress. In this particular case, the definitions have been standardized in the comprehensive "Glossary of Terms for Thermal Physiology" (2001), prepared and revised by the Commission for Thermal Physiology of the International Union of Physiological Sciences. One can contrast this standardization with the multiple uses and definitions for the term "fatigue."

Adaptation Terms

The following terms and definitions are fundamental to understanding the concept of adaptation to environmental stress:

- **Adaptation.** This is a broad umbrella term encompassing all of the concepts listed next; it can be defined simply as changes to the human system induced by exposure to an environmental stressor. Such changes can encompass both *genotypic* (changes in genetic factors) and *phenotypic* (changes in physiology) adaptations. Furthermore, adaptation can involve behavioral responses to environmental stressors.

- **Habituation.** Rather than the individual's adapting physiologically to a stress, habituation (also termed sensitization) refers to a reduction in the behavioral perception of a repeated stimulation. An example may be the reduction in severity of space motion sickness with several days in orbit as the brain behaviorally adapts to the disjointed feedback from the neurovestibular and visual systems. Another example is the reduced feeling of pain and discomfort with repeated cold immersion of the hand despite no change in actual finger temperatures.

- **Acclimation.** This term involves adaptive physiological responses to *experimentally induced changes in particular climatic factors.* An example is training for a competition in a hot environment by exercising inside a hot environmental chamber for an hour a day.

- **Acclimatization.** This refers to adaptive physiological or behavioral changes occurring in an individual in response to *changes in the natural climate.* In contrast to the heat acclimation process just described, there is the much more appealing acclimatization process of spending a week living and training in the Caribbean during the depths of a Canadian winter.

Fatigue

The concept of fatigue is possibly one of the most difficult terms on which to obtain a consensus definition. Fatigue is a multimodal phenomenon (Enoka and Duchateau 2008) and can be defined at different levels ranging from the cell to the entire organism, as illustrated by the effects of hyperthermia on muscle function. For example, at the level of individual motor units or muscle fibers, elevations in temperature can alter the nerve conduction velocity, the activity of different channels involved in excitation–contraction coupling, and the contractile mechanism itself. With isolated muscle groups during single-joint isometric movements, whole-body hyperthermia can reduce maximal voluntary contraction torque independent of local muscle temperature. At the systemic level of the entire organism, muscular endurance with large, multijoint movements can decrease following passive exposure to hot environments. Self-selected pace during time trial efforts also decreases in hot compared to thermoneutral environments, possibly reflective of a decreased neuromuscular capacity or central drive. Researchers from these perspectives may each use the term "fatigue" in describing the impaired muscular response.

COMMON RESEARCH THEMES

There are several major themes that are relevant to exercise in any environment; these themes are highlighted within each section and in relation to each environment throughout the text:

1. Acute exposure. What are the initial responses upon exposure to an environment for someone who is not adapted? For example, what happens when you jump into freezing water on New Year's Day as part of a polar bear event? What about when you first arrive at the base camp of Everest to find yourself tired, short of breath, and with a headache? Should you continue, or would that be dangerous?

2. Chronic exposure or adaptation. How does your body adapt to repeated exposures to an environment? Answering this question is fundamental to the concept of altitude training for athletes and to the use of hypoxic tents or intermittent hypoxia protocols. It is also of life-and-death importance to recreational and commercial divers breathing air or other gases at pressure for extended periods of time. What are the implications of chronic exposure to microgravity for astronauts and long-term space missions, especially upon return to Earth? What are the limits to physiological adaptation? Are there optimal exposures to maximize adaptation?

3. Performance issues. Not only are humans exposed to various challenging environments; the goal is typically to perform work or exercise in these demanding situations. How do different environments affect our ability to exercise, and what can pushing the body to its environmental limits tell us about how the body functions during exercise? For example, what can studies on the effects of hyperthermia on force generation tell us about central neuromuscular recruitment? At the same time, there is a growing realization in occupational settings that safe working levels should be defined not by physiological limits, but by task performance and the prevention of accidents resulting from impairments in mental function. Therefore, an emerging importance is being given to the integration of physiological and psychological influences on human performance in extreme environments.

4. Individual variability in response. Why is it that some people can seemingly tolerate exercise in stressful environments better than others? Are there individual characteristics that predispose one to greater risks of exertional heat illnesses or to heat waves? What specific physiological traits might be inherent in high-altitude natives such as the Sherpa that enable them to thrive and perform seemingly superhuman exercise at high altitudes? Can individual variability be used to predict exercise tolerance in different environments, or the rate at which adaptation occurs?

5. Countermeasures. To protect individuals exposed to extreme environments, physiological adaptations are only one layer of defense. Protection can also come from the design of technological, procedural, or pharmacological countermeasures to minimize the impact of environmental stressors. Maintaining adequate hydration status is one method of maximizing circulatory function and evaporative heat dissipation, as well as enhancing work capacity in the heat. At the same time, the continuing design and evolution of microclimate cooling systems underneath protective clothing are further enhancing work capacity in the heat, and the use of these systems to precool athletes prior to competition in the heat is increasingly common. What are the physiological mechanisms underlying these countermeasures, and where might future countermeasures take form?

SUMMARY

The field of environmental physiology is highly interdisciplinary, as it ultimately encompasses human response across a host of physiological systems to a multitude of environmental stressors. Therefore, the practitioner must keep abreast of relevant research and debates across a range of scientific disciplines, from epidemiology involving large data sets on heat stress mortality through to immunological markers and inflammatory response resulting from individual exposure to heat stress. In addition, environmental physiologists must interact with a variety of other professionals, including engineers, ergonomists, coaches, safety personnel, and manufacturers. Therefore, the environmental physiologist is superbly poised to perform a key role as the central hub of such multidisciplinary working teams, to ensure that scientific advances can make the transition to real-life improvements for workers and other individuals exposed to environmental stress.

Fundamentals of Temperature Regulation

Humans have proven to be one of the most adaptable vertebrate species in the history of the planet, as evidenced by their ability to tolerate and thrive in a multitude of extreme environments—ranging from the dry heat of the desert or the high humidity levels of the tropics through to the extreme cold found in the polar regions and at high altitudes. Lacking physical (e.g., blubber, fur) or strategic (e.g., hibernation, torpor) adaptations as evidenced by other homeotherms, humans have primarily relied on the physiological responses of sweating, shivering, and alterations in blood flow. However, what makes humans unique and quite remarkable is their highly advanced capacity to behaviorally alter the environment as necessary, by such means as the creation of clothing and shelter and the use of tools. Using this combination of behavioral and physiological responses, humans are able to maintain a reasonably constant core body temperature (T_c) of 37 ± 1 °C throughout their lives despite a wide range of ambient temperatures, with generally efficient thermoregulation occurring throughout a core temperature range of 35 to 40 °C (Parsons 1993). Thermal preferences can vary greatly across individuals based on physiological factors along with behavioral and cultural preferences. Preferred self-adjusted ambient temperature averages 24.9 ± 1.3 °C and 24.5 ± 1.5 °C for young and elderly Australians, respectively, though substantial interindividual and diurnal variability exists (Taylor et al. 1995). In turn, the perception of thermal stimuli and its subsequent influence on exercise performance and tolerance in the cold or heat remain a complex issue.

A challenge to the ability to maintain stable thermal balance is not just the stress imposed by the external environment but also the extreme variations in metabolic heat generation, as >70% of metabolic energy is converted to heat in the course of mechanical movement. As such, thermoregulation becomes an intricate balance between external and internal heat sources to ensure that the body does not become hyper- or hypothermic. Heat exchange is typically quantified in watts, where 1 W = 1 joule per second. This heat production can range from 70 to 100 W in resting adults and 280 to 350 W at mild walking pace to >1000 W during heavy exercise (Parsons 1993). Even with nontraditional exercise modalities like extravehicular activities during spaceflight, in which astronauts are not running but mainly straining isometrically, heat production can regularly exceed 600 W and can, in extreme cases, rise above 1000 W (Cowell et al. 2002).

In order to produce an appropriate behavioral or physiological response to environmental thermal stress, the body must be able to sense its thermal status. In keeping with the importance of behavioral responses in humans, the perception of the magnitude and the comfort quality of thermal stress plays a major role in human thermoregulation and possibly exercise capacity. This is accomplished by temperature sensors located throughout the body, which are then integrated centrally to achieve a coordinated response. While the basic biophysics of heat exchange can be laid out in a fairly simple heat balance equation, the actual modeling of thermal balance can be very difficult due to the complexity of the human sensory system. Indeed, the very nature of what the body is sensing and regulating in order to maintain homeothermy remains in active dispute. Similarly, the modeling of the thermal stress imposed on the body can vary across different environments, depending on the dominance of factors such as humidity and wind speed.

This chapter first explores the heat balance equation quantifying the rate of heat exchange between the human body and the environment. From there, we will survey the neural pathways by which the body senses, integrates, and then responds to thermal stimuli, along with the various models that have been proposed for how thermal homeostasis is achieved. We will also survey the scales and instruments developed for quantifying thermal stress, from temperature sensors to heat and cold stress indices. A detailed examination of the heat balance equation and thermal indices is provided in the textbook of Parsons (1993), and the anatomy and physiology of thermoreception are comprehensively covered in the text by Bligh and Voigt (1990).

HEAT BALANCE EQUATION AND FACTORS AFFECTING HEAT EXCHANGE

The basic heat balance equation modeling the rate of heat storage, $\dot{S}$, in $W \cdot m^{-2}$, is derived from the First Law of Thermodynamics and incorporates the four major pathways of heat exchange (radiation, conduction, convection, and evaporation) as follows:

$$\dot{S} = \dot{M} \pm \dot{W}_k \pm \dot{R} \pm \dot{C} \pm \dot{K} - \dot{E} \ (W \cdot m^{-2}) \ (Eqn. \ 2.1)$$

As presented here, a positive value for $\dot{S}$ represents a gain in heat storage by the body, eventually leading to hyperthermia, while a negative value represents net heat loss and eventual hypothermia. Breaking down this equation into its individual terms, we can see the following components of heat exchange and how they are typically calculated or measured in scientific studies. It is important to note that heat exchange can be either positive or negative for each individual component except for metabolism, depending on the gradient between the body and the environment. For example, when the ambient air temperature is greater than the body temperature, radiative, conductive, and convective heat exchange will be positive (heat stored by the body). The one exception is evaporation, which is defined as the change of water from liquid to vapor phase and can only be a source of heat dissipation in humans. Therefore, scalding burns are not a positive evaporative heat gain, but heat gain via convection.

$\dot{M}$ is the metabolic heat production. This is typically indirectly determined by the oxygen uptake, measured through analyzing the inspired and expired air from the subject.

$\dot{W}_k$ is the external work performed by the individual and is typically subtracted from the total heat production. External work can also be zero (e.g., as in going up an escalator) or positive in some situations in which limbs are moved without voluntary muscular contractions (e.g., as in having the legs moved by an externally powered cycle ergometer).

$\dot{R}$, the radiative heat exchange, refers to transfer of heat energy via electromagnetic waves between the environment and the human body. This is a combination of the sun's direct radiation, the ground's reflected radiation, and the diffuse radiation from the collision of atmospheric molecules. As such, the primary gradient of interest is the thermal gradient between the body and the environment. In environments with hot objects (e.g., workplaces exposed to the sun, or steel mills), the radiant temperature often exceeds the skin temperature, resulting in radiant heat transfer from the environment to the skin. The high reflection of ice and snow also contributes to the radiative heat load, which explains why mountaineers can often be in short sleeves even at high altitudes and low ambient temperatures. In addition, the radiation field surrounding an individual is dependent upon environmental conditions such as shade and cloud cover, time of day, and time of year, as well as altitude. Furthermore, radiative heat exchange is dependent upon individual factors, such as posture and the surface area exposed to the heat source and the amount and reflectivity or absorbance of the clothing worn.

$\dot{C}$ and $\dot{K}$ refer to conduction and convection, respectively, although they are often combined into the single term $\dot{C}$, encompassing both forms of heat exchange, due to their interconnectedness. These two forms of heat exchange involve the conveying of energy between media (solids, fluids, or gases) in direct contact with the body, and are primarily dependent on the thermal gradient between the body and the media in contact. Typically, conduction denotes direct contact with a solid surface such as ice or cold metal, while convection more commonly refers to contact with fluids or gases like water or air. Additionally, the heat conductivity of the surface or fluid in contact with the body has a major effect on heat exchange, with water having 27 times greater heat conductivity than air; this explains why 15 °C air temperature is comfortable but 15 °C water immersion is very cold and can quickly induce hypothermia. Similarly, direct contact of the bare hands with cold metal (high conductivity) can produce freeze burns due to high localized heat loss, while wood or rubber at the same temperature does not damage skin as rapidly.

An additional consideration with convective heat exchange is the speed of movement of the fluid against the body surface. We all know that convection ovens cook much more rapidly than conventional ovens, and that this is due to the addition of a fan that circulates the hot air in the oven and accelerates heat storage into the food. Conversely, while remaining still in a cold lake, the body loses heat but warms up the water layer immediately next to the skin, decreasing the thermal gradient for heat exchange. However, when the water is in a river or is moving in a current, the warm water layer is rapidly removed and replaced with cold water, maintaining the high thermal gradient and accelerating heat loss. As we see in chapter 6, a similar major increase in convective heat loss exists during swimming through cold water. A similar principle is responsible for the establishment of windchill scales for cold air temperatures, which will be discussed later.

$\dot{E}$ refers to evaporative heat exchange. Mechanisms of heat transfer are grouped into two general categories consisting of dry (radiative, conductive, convective) and wet (evaporative) pathways. Dry heat exchange is dependent on the temperature gradients within the organism (e.g., core to periphery) and also between the organism and the environment. In addition, the rate of cutaneous blood flow to transport heat from the core to the periphery influences the degree of conductive and convective heat exchange. Wet heat loss arises from the evaporation of water, typically secreted by the sweat glands within the skin, with full evaporation of each liter of sweat from the skin surface transferring approximately 2400 kJ of heat energy to the environment. Unlike what happens with the other methods of heat exchange, the potential for evaporative heat loss is determined primarily by the water vapor pressure gradient between the body surface and the environment. The water-carrying capacity of air is dependent on the air temperature in a roughly exponentially increasing fashion, with warmer air able to hold a greater amount of water vapor (see figure 2.1). Therefore, both air temperature and relative humidity need to be reported to accurately quantify water vapor pressure and environmental conditions in the heat.

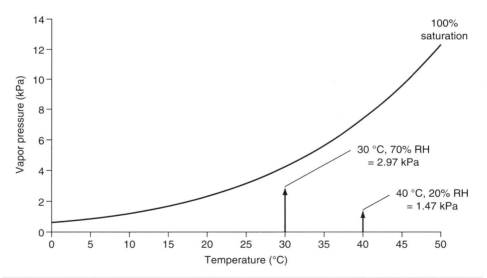

Figure 2.1 The amount of water required to saturate a volume of air increases with increasing temperature. The plotted line represents the water vapor pressure at 100% relative humidity. For each temperature, the water vapor pressure and therefore capacity for evaporative heat dissipation are greatly reduced at higher relative humidity levels. Heat stress scales such as the Humidex attempt to incorporate both the ambient temperature and relative humidity (RH) into a single numerical value.

When the temperature of the ambient environment is similar to that found at the skin, as is the case during exercise in the heat at approximately 35 to 40 °C, radiative, conductive, and convective heat exchange is minimized and the primary pathway for heat dissipation becomes the evaporation of secreted sweat. Maximal evaporative heat dissipation from the body occurs when secreted sweat is vaporized at the skin; sweat that either drips off the skin or is toweled off does not provide any evaporative cooling. The relative humidity of the environment plays a major role in determining the capacity for evaporative heat exchange due to its effect on water vapor pressure. For example, "dry" heat conditions, such as those found in a desert clime, feature high ambient temperatures but typically low humidity levels, thus still permitting significant evaporative heat dissipation. In contrast, the "wet" heat conditions of the tropics can have lower air temperatures but much higher humidity levels; therefore higher water vapor pressure in the air impairs both dry and wet heat dissipation, posing much greater hazards for hyperthermia. Referring to figure 2.1, we see that 100% water vapor saturation equates to 4.24 and 7.37 kPa at 30 and 40 °C, respectively. Therefore, environmental conditions of 30 °C and 70% relative humidity equal 2.97 kPa water vapor pressure, while hotter but drier conditions of 40 °C and 20% relative humidity equal a much lower vapor pressure of 1.47 kPa; thus the latter condition, while hotter, would still permit more evaporative heat dissipation. The section "Thermal Stress Scales" later in the chapter presents further discussion and examples of the quantification of these different heat exchange pathways into temperature indices.

Respiratory Heat Exchange

In addition to the skin, the respiratory system is a major avenue for conductive or convective and evaporative heat exchange, as the ventilation rate can exceed $150 \text{ L} \cdot \text{min}^{-1}$ during exercise. Inspired air is both warmed and humidified by the respiratory tract; and this is done very efficiently, such that at high ventilation rates, even air below 0 °C is near body temperature by the time it reaches the alveoli (Hartung et al. 1980). However, as exhaled air is at body temperature and fully saturated with water vapor, respiration represents a major site of heat and water loss during breathing of cold and dry air; in fact, respiratory heat masks are being designed for cold weather use in an effort to trap and filter the exhaled air to retain some of the heat and moisture. Therefore, to properly model heat exchange, it is necessary to measure ventilation rate along with the temperature and humidity of the inspired air.

In an attempt to use this pathway of heat exchange, researchers and industry have explored the breathing of warm and moist air as a method of rewarming from hypothermia. In theory, while the magnitude of absolute heat transfer may be relatively small compared to that from shivering heat production, the flow of heat through the respiratory tract close to the brain and hypothalamus may accelerate the rate of core rewarming. However, research in our laboratory and elsewhere has demonstrated that respiratory rewarming does not decrease postexposure core temperature afterdrop nor provide faster rewarming rates beyond that provided by passive rewarming via shivering (Wright and Cheung 2006). Chapter 6 presents more discussion on strategies for rewarming from hypothermia. In addition to heat and hydration effects, the breathing of cold and dry air can also be a trigger for exercise-induced asthma.

HUMAN CALORIMETERS

Calculation of the individual components of the heat balance equation remains an indirect measurement of total heat exchange, and the truest method of quantification is a direct air or water calorimeter, which measures the heat emitted by a mass. Bomb calorimeters have been used extensively in the fields of chemistry and physics to measure the heat given off during the combustion of a known mass, such as the energy contained in different food items. In general, calorimeters consist of a sealed chamber with a closely controlled rate of air or water flow around the chamber. Monitoring the volume of flow around the chamber and also the change in temperature of that gas or fluid provides an accurate measure of actual heat exchange. As applied to whole organisms including humans, direct calorimetry involves the measurement of the total heat dissipated by the body subsequent to anaerobic or aerobic metabolic oxidation (or both); thus, measurement of metabolic heat production is still required. While the principle may appear relatively straightforward, the actual technical requirements for establishing a calorimeter can be very demanding and difficult to execute. For example, the Snellen air calorimeter was recently constructed and developed by Dr. Glen Kenny in the Human and Environmental Physiology Research Unit at the University of Ottawa, requiring additional infrastructure including its location in a pressurized room, with control of operating temperature over a range of −15 to +35 °C; control of ambient relative humidity over a range of 20% to 65%; incorporation of an air mass flow measuring system to provide real-time measurement of air mass flow through the calorimeter; incorporation of a constant-load "eddy current" resistance ergometer; and an open-circuit, expired gas analysis calorimetry system (Reardon et al. 2006). Currently, it is the only fast-responding human air calorimeter in the world. Outside of such specialized facilities, indirect calculation of each component of the heat balance equation remains the more common method of quantifying heat exchange.

Photo courtesy of Dr. Glen Kenny, Faculty of Health Sciences, University of Ottawa.

Figure 2.2 The Snellen calorimeter. This facility requires highly specialized environmental control, but is capable of directly quantifying the heat exchange between a human and the environment during both rest and exercise. This removes the necessity of indirectly calculating the various components of radiative, conductive, convective, and evaporative heat exchange. A cycle ergometer is built into the calorimeter, and the heat from it is accounted for, permitting studies of both passive and active exposure to different temperatures and humidity levels. Metabolic heat production is calculated from the collection of expired air through the mouthpiece.

Effects of Clothing

The wearing of protective clothing can significantly alter evaporative heat transfer by impeding the movement of water vapor across the various layers, building a humid microenvironment and raising the water vapor pressure next to the skin. This results in a reduced vapor gradient between the skin and the environment, decreasing the rate of evaporative heat dissipation (Nunneley 1989). The size and nature of this microenvironment depend on the amount of surface area coverage (e.g., helmet and mask vs. bare head), the fit of the clothing, and the permeability of the clothing. Despite the increased perspiration, the majority of sweat drips from the body or is trapped within the fabric fibers of overlying protective clothing, with significant reductions in the capacity of evaporation to remove heat energy. The wetting or saturation of clothing by sweat may also affect the thermal characteristics of the clothing and further influence the rate of heat transfer. In addition, as diagrammed in figure 2.3, the site of phase change may be raised above the skin in a clothing microenvironment, where a portion of the heat energy may come from the environment rather than the body, thus decreasing the efficiency of evaporative heat loss. The modeling of heat exchange is further complicated by the dynamic nature of clothing use, as the quantification of insulation and clothing permeability is typically done on static models using new clothing rather than in real-life movement with clothing that is old or worn (Havenith et al. 1999). Depending on the fit of the clothing, the microenvironment can be up to 50 L in volume (Sullivan et al. 1987). With movement, this microenvironment experiences internal pumping and circulation, and the clothing may contact the skin differently than when stationary, wicking perspiration from the skin and altering heat exchange differently than in a static manikin or model.

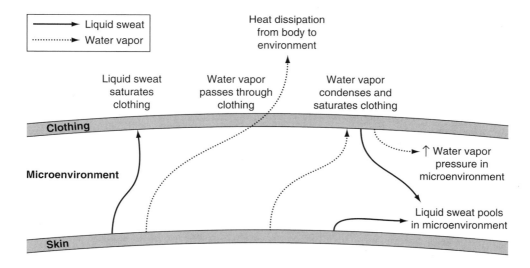

Figure 2.3 The wearing of clothing creates a microenvironment between the clothing and the skin, in turn affecting the ability of secreted sweat to dissipate heat from the body by evaporating into the environment. Any sweat that saturates the clothing or remains inside the microenvironment ultimately does not transfer heat away from the body. Such circumstances of impaired evaporation combined with additional heat generation from exercise can quickly lead to situations of uncompensable heat stress, in which the evaporative heat dissipation required to maintain thermoneutrality exceeds the capacity of the environment.

MODELS OF THERMOREGULATORY CONTROL

It may be somewhat surprising to find that even after centuries of research, the fundamental system of control of thermal balance in humans, as well as other mammals and homeotherms, remains controversial. One of the reasons may be that thermal balance is not confined to or dominated by a specific organ, such as the kidneys with renal control and fluid balance; accordingly, its physiological control is inherently integrative and open to theorization. Currently, three major models have been advanced to help explain thermal homeostasis in humans. The first two models propose that temperature is the regulated variable and share similar ideas concerning the underlying neural architecture. The third model is fundamentally different from the first two models, in that rather than temperature per se, body heat content is the regulated variable and body temperature a by-product of that regulation. In the following we consider these three models in some detail. As we will see, the models have individual strengths, but each remains imperfect and is difficult to accommodate to all situations, so that much work remains necessary to advance our understanding of thermal homeostasis.

Temperature Regulation: Adjustable Set Point

The traditional model of temperature regulation was proposed by Hammel and has been termed the set point model (Hammel et al. 1963). In this model, which explains thermal homeostasis using a set point signal as a comparator for body temperature, a corrective response is executed when body temperature differs from the comparator signal. A relevant basic analogy for this model is a household thermostat, which initiates either cooling or heating based on comparison of house temperature with an internally adjustable set temperature. This simple analogy breaks down in that, unlike a thermostat with its on/off response of maximal heating or cooling, humans are capable of graded thermoregulatory responses (e.g., higher-intensity shivering in more muscles) with greater deviations from normal temperatures. In the set point model, thermal afferents from throughout the body core and peripheries are integrated into an overall thermal signal at one or more central sites, generally assumed to reside within the central nervous system (CNS) and likely within the hypothalamus. This is informed in part by cold and warm receptors under human skin existing at an average depth of 0.15 to 0.17 mm and 0.3 to 0.6 mm (Boulant 2006), respectively. Further neural support for the model comes from the observation that the hypothalamus contains both cold- and heat-sensitive neurons, which increase their rate of firing based on either cooling or heating. This thermal signal is then compared to an internally stored thermal signal, and heat gain or heat loss responses are activated appropriately.

One common argument against this "set point" model is the false assumption that it cannot explain deviations from a presumably fixed and constant reference temperature stored within the CNS. Such deviations, for example a higher regulated body temperature, can occur normally over the course of the menstrual cycle, with baseline temperature during the luteal phase regulated at approximately 0.5 °C higher than during the follicular phase (see figure 2.4). Also, body temperature is generally defended appropriately, though at a higher absolute core temperature, during fever. In response, proponents for the set point model argue that there is no requirement for a fixed and unchanging reference temperature hard-wired into the CNS, but rather that the set point temperature may vary over a range based on the relative activity of heat- and cold-sensitive neurons throughout the body (Cabanac 2006).

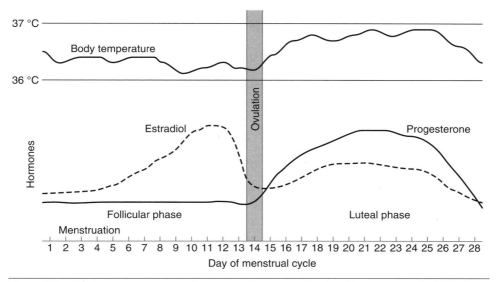

Figure 2.4   Baseline core temperature changes over the menstrual cycle, with an increase of ~0.5 °C during the luteal phase following ovulation. Note that these are average values; durations and values may vary among different females or different cycles.

Temperature Regulation: Reciprocal Inhibition

The reciprocal inhibition model of thermal homeostasis (Bligh 2006) utilizes the same neural architecture as Hammel's but explains temperature regulation slightly differently. Based on the vast array of interconnections between sensory receptors and thermal effectors for heat loss and heat gain, this model argues for a correspondingly complex integration and interconnections between sensors and receptors, as opposed to a more direct integration of a composite signal and comparison to a set signal as outlined by Hammel. This model assumes that warm and cold sensor inputs are integrated and summed to produce a net heat loss and heat gain response, respectively (see figure 2.5). One of the key empirical observations of this model is a "null zone" or "interthreshold zone" of thermal stability, wherein sweating (heat loss) and shivering (heat gain) effectors do not significantly act and vasomotor changes can maintain homeothermy. At the same time, very few normal situations (fever being an obvious exception) exist in which both occur simultaneously. Lastly, the model integrates homeothermy into a larger picture of whole-body physiological homeostasis by assuming the ability of a variety of nonthermal factors to affect temperature control (Mekjavic and Eiken 2006).

In building this model of homeothermy, Bligh adds a layer of complexity to the presence of warm-sensitive and cold-sensitive neurons within the body and especially the CNS. The added complexity comes in as the activity of one set of sensors interconnects with the pathway of the other sensors to provide an inhibitory stimulus, such that strong stimulation of warm-responsive neurons not only activates heat loss effector pathways but also inhibits cold-sensitive neurons and heat gain pathways. Thus, this reciprocal inhibitory process allows for only the net thermal signal (warm or cold) to evoke an effector response, or else only minor vasomotor changes if the net signal is not sufficiently strong. Similarly, nonthermal stimuli can produce either an

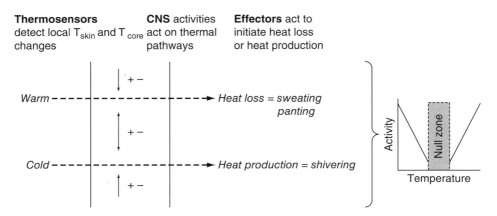

Thermosensors detect local T_{skin} and T_{core} changes

CNS activities act on thermal pathways

Effectors act to initiate heat loss or heat production

Figure 2.5 The reciprocal inhibition model of thermoregulation proposed by John Bligh. In this model, a heat stimulus activates heat loss effectors (sweating) and also simultaneously inhibits heat gain effectors (shivering). This cross-inhibition can result in a "null zone," a range of body temperatures where there is no active sweating or shivering and changes in vasomotor activity are the dominant means of thermoregulation.

HOMEO- VERSUS HOMOSTASIS

One commonly argued limitation of temperature regulation models, and especially of the set point theory, is the assumed constancy of the central reference signal, which at first glance appears unable to accommodate the altered sweating and shivering thresholds observed in cases such as fever. In argument, proponents of this model note that rather than a constant and invariable "reference" central temperature, the exact set point for the thermal signal is variable and adjustable by the body depending on individual situations such as disease and fatigue (Cabanac 2006). This raises an important semantic clarification that is required to avoid confusion in discussions of human thermal control. Often, temperature is assumed to be constant with a hard-wired invariable reference signal; this is a misconception. As Bligh (2006) noted, one of the earliest scientific articles on thermoregulation, by Claude Bernard in 1865, used the term "fixité," which may have contributed to the assumption of a constancy or invariability in temperature regulation. Rather, the distinction should be made away from the concept of "homo" (= same) to a more appropriate "homeo" (= similar) state of thermoregulation. Bligh (2006) further argues that "[t]his verbal distinction is crucial, for life depends on the sustained integration of myriad isolatable, but not isolated, functions; and their integration requires some degree of flexibility of them all. . . . Like all else of homeostasis, body temperatures vary to a greater or less extent between species, between environmental and organismic circumstances, with the time course of each day, and seasonally. Any hypothetical account of how homeothermy is 'managed' must incorporate tentative explanations for these variations" (p. 1332).

excitatory or inhibitory effect on either set of pathways, helping to explain the shifting of core temperature thresholds with individual changes such as hypoglycemia, dehydration, and inert gas narcosis. Another implication of this model is that there is no need for a central comparator signal as implied by the set point model, permitting an easier theoretical ability for core temperature shifts. Anatomical support for this idea can be seen in clinical tests in which the hypothalamus is impaired or blocked and thermoregulatory control, although coarser, is still possible. Thus, through evolution, more and more refined thermoregulatory interconnections and integration may have developed in different regions of the CNS.

Heat Regulation

The principal proponent of the heat regulation paradigm has been Paul Webb, who bases his position partly on his many years of research into human calorimetry in different conditions (Webb 1995). The heat regulation model is dramatically different in theoretical construct from both the set point and the reciprocal inhibition models of homeothermy. In those temperature regulation models, the body physiologically maintains a set point or narrow range of core temperature by activating blood flow changes (vasodilation, vasoconstriction), shivering, or sweating. Thereby, temperature is the controlled variable, and the amount of heat gained or lost by the body is the pathway toward that temperature homeostasis. In contrast, the underlying principle of the heat regulation model is that temperature is not the primary defended variable within the body. Rather, overall heat storage, or the homeostatic net balance of heat gained versus heat lost, is the regulated variable. So in this model, temperature is not a regulated variable but instead is the secondary result of the body's attempt to maintain a stable overall heat balance. Rather than sensing and integrating peripheral and core temperatures, the body uses heat flow and the temperature gradient across the skin surface as its primary afferent inputs. In turn, the central integration of heat balance operates with feedback from heat loss and possibly feedforward control from heat production, ultimately driving physiological responses that defend body heat content.

One advantage of the heat regulation model is that it is inherently more tolerant of deviations in body temperature. As a practical example (see general schematic in figure 2.6), over the course of a run, an athlete generally experiences a rapid initial rise in core temperature, followed by a plateau and stabilization at the higher level throughout the remainder of the run. The heat regulation model argues that, rather than the CNS readjusting its set point temperature to a higher value and then regulating body temperature, the initial rise in temperature occurs because heat production—which increases as soon as running begins—has a faster time course than the heat dissipation mechanisms (vasodilation and especially sweating). Sweating eventually becomes activated and matches the rate of heat production, resulting in the stable plateau of elevated core temperature. Furthermore, upon the cessation of exercise, temperature remains elevated for prolonged periods even though sweating and vasodilation drop rapidly (Kenny et al. 2008). This is difficult to accommodate with the temperature regulation models, which would predict that heat loss should remain elevated throughout recovery to return core temperature to baseline. In contrast, this observation is predicted by the heat regulation model, where homeostasis is primarily based on matching heat production with heat dissipation, and a slight postexercise increase in vasodilation and conductive heat exchange eventually dissipates the excess heat built up within the body.

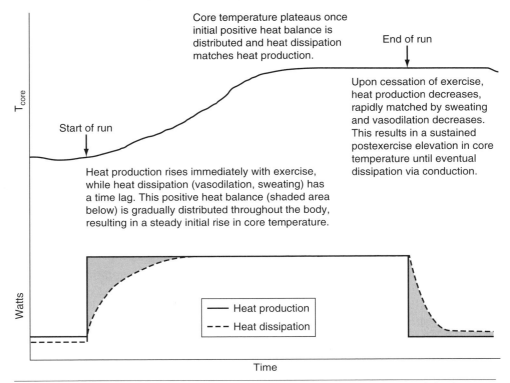

Core temperature plateaus once initial positive heat balance is distributed and heat dissipation matches heat production.

End of run

Start of run

Heat production rises immediately with exercise, while heat dissipation (vasodilation, sweating) has a time lag. This positive heat balance (shaded area below) is gradually distributed throughout the body, resulting in a steady initial rise in core temperature.

Upon cessation of exercise, heat production decreases, rapidly matched by sweating and vasodilation decreases. This results in a sustained postexercise elevation in core temperature until eventual dissipation via conduction.

T_{core}

Watts

—— Heat production

- - - Heat dissipation

Time

Figure 2.6   Schematic diagram of how heat regulation results in a stable body temperature during the course of a run.

Thus one strength of the heat regulation model is that inherently it does not presume a set point temperature preprogrammed into or located within the body. Rather, overall heat balance can vary and can occur at many different levels of internal heat production throughout the circadian rhythm or in the presence of other disturbances to overall physiological homeostasis, with heat content transiently variable and different states of heat balance corresponding to a fluctuating core temperature. The primary argument against this model is the anatomical necessity for a set of heat flow sensors that are anatomically difficult to demonstrate while at the same time accounting for the presence of temperature-sensitive neurons throughout the peripheries and the CNS. In response, peripheral thermal receptors at different depths throughout the skin and throughout different core tissues represent multiple thermal sensors, and the difference between these thermal signals can become integrated as rates of heat flow.

THERMAL STRESS SCALES

While an excellent representation of heat storage, the heat balance equation is extremely difficult to calculate outside of a laboratory and specialized facilities such as a direct air or water calorimeter (see the first sidebar). In order to estimate human thermal response and tolerance, the typical approach has been to model the general thermal characteristics of the environment, and then to develop exposure guidelines based on exercise intensity and duration of exposure. The quantification of the thermal environment can also be extremely complicated, as the dominant heat transfer pathway can differ depending on prevailing conditions. This section surveys some of the major heat and cold stress scales and discusses considerations in designing thermal scales and exposure limits.

Heat Stress

We have all had the experience of enjoying ourselves sitting outdoors on a café patio in short sleeves on one of the first sunny days of the spring, and then suddenly feeling a lot colder as soon as the sun goes behind a cloud and having to scramble to put on our sweaters. Similarly, there is a dramatic difference in perceived heat between a sauna with its dry heat and a steam room with its highly humid heat. These examples illustrate the respective importance of accurately modeling radiative and evaporative heat exchange when developing heat stress indices. In order to model and quantify the perceived heat stress imposed through all four heat exchange pathways, the United States Marine Corps aided in the development of the wet bulb globe temperature (WBGT) in the mid-1950s (Yaglou and Minard 1957). This scale requires the measurement of three separate temperature values (see figure 2.7):

- T_d or "dry bulb" temperature, measured with a thermometer shielded from moisture or radiation, to measure normal temperature.

- T_w or "wet bulb" temperature, measured with a thermometer that remains moist because it is wrapped in a fabric sock kept wet with water, to give an indication of humidity and evaporative heat exchange. At relative humidity levels <100%, evaporation will occur, such that the wet bulb temperature will be less than T_d.

- T_g or "globe" temperature, measured with a thermometer housed inside a black globe to ensure full absorbance of radiation and constant exposed surface area regardless of location of radiative heat source.

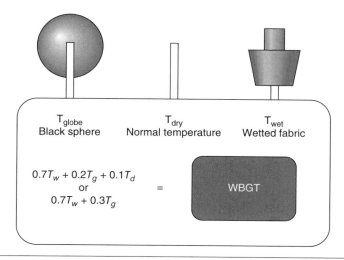

$$0.7T_w + 0.2T_g + 0.1T_d$$
$$\text{or}$$
$$0.7T_w + 0.3T_g$$

T_{globe} — Black sphere
T_{dry} — Normal temperature
T_{wet} — Wetted fabric

= WBGT

Figure 2.7   The wet bulb globe temperature (WBGT) is a common index used to assess thermal stress, incorporating three thermometers, that quantifies the four main pathways for heat exchange into a single value.

The three temperature readings are then converted to an overall WBGT value according to a weighting of the three into the equation:

$$WBGT = 0.7T_w + 0.2T_g + 0.1T_d \text{ (Eqn. 2.2)}$$

In situations without a significant radiant load, the equation is modified to

$$WBGT = 0.7T_w + 0.3T_g. \text{ (Eqn. 2.3)}$$

Since its initial development, the WBGT scale has been revised several times and has been adopted as the basis of many occupational heat exposure guidelines. Other composite temperature indices include the Heat Index used by the National Weather Service and the National Oceanic and Atmospheric Administration in the United States and the Humidex used by Environment Canada, which are similar in incorporating temperature and humidity but not solar radiation. Many other specific indices, such as the FITS (Fighter Index of Thermal Stress, intended for aircrew wearing light aircrew clothing but not for use when much more insulative marine immersion suits or chemical protective clothing is being worn, or in cold environments), have also been developed for specific applications (Nunneley and Stribley 1979). Most are derivations of concepts or temperature values employed by the WBGT. Common also to most indices are threshold temperatures for progressively stronger warnings of heat alerts, based on conservative estimates of anticipated tolerance.

While the use of such heat stress indices has undoubtedly greatly mitigated the risk for hyperthermia, some significant limitations are present with the WBGT and its derivative models of quantifying heat stress. From a practical perspective, an important limitation is the lengthy measurement time required for the temperature sensors to equilibrate and provide a stable and accurate reading. This makes it difficult to obtain accurate assessments of WBGT in dynamic situations or where environmental conditions change rapidly. From a public awareness and education perspective, the concepts of wet bulb and globe temperature are relatively obscure for many users of heat stress scales, making WBGT potentially difficult to understand and less likely to be adopted or widely used. Recently, in an attempt to replace the somewhat cumbersome and slow WBGT with terms and variables that are more commonly used and understood by the general public, a research group from Israel has proposed an Environmental Stress Index (ESI) based on rapidly equilibrating sensors for ambient temperature (T_a), relative humidity (RH), and solar radiation (SR) (Moran et al. 2003), according to the equation

$$ESI = 0.62T_a - 0.007RH + 0.002SR + 0.0043(T_a \cdot RH) - 0.078(0.1 + SR)^{-1}. \text{ (Eqn. 2.4)}$$

Comparing across three large sets of meteorological data, the ESI was found to correlate very closely to the WBGT (Moran et al. 2003). Ultimately, replacement of terms such as dry, wet, and globe temperature with simply temperature and humidity—terms already in common use in most public weather forecasts—gives the ESI a greater potential for understanding and use by the general public. In addition, the removal of globe and wet temperatures should make for a faster-responding index, increasing its utility in dynamic settings such as inside moving vehicles or for multiple sampling over the course of a marathon.

Cold Stress

As we have all noticed during winter months, there is a large difference in how cold we feel on a cold but calm day as opposed to a cold plus windy day. In cold weather, apart from dangerous drops in core temperature and hypothermia, the primary danger to the general and exercising population is from local tissue freezing and frostbite (see chapter 5). To educate the general population on the potential danger of exposure to the cold, a windchill index was first developed based on pioneering work by the polar explorers Paul Siple and Charles Passel (1999), who also first coined the term windchill in 1939. Siple and Passel measured the time required for plastic vials of water to freeze in different conditions during an Antarctic mission in 1945, and used these data to relate the rate of heat loss in watts per square meter at different temperatures and wind speeds. The original windchill indices therefore produced a windchill equivalent temperature, normalizing different temperatures and wind speeds to an equivalent temperature with zero wind. As with the heat indices discussed earlier, it is important to emphasize that the windchill indices are a perceptual scale or a reflection of how cold different conditions feel to humans and other animate beings, rather than quantification of an objective value. In contrast, inanimate objects (e.g., metal flagpoles, exposed water pipes) cannot cool below ambient temperature, and the concept of windchill does not apply.

During the later stages of the last century, both scientists and health officials began coming to grips with some of the limitations of the existing windchill models. Chief among these limitations, of course, is that neither water nor plastic vials accurately replicate the composition, conductivity, internal heat source, and complex physiological responses of human tissue. Furthermore, the original experiments were performed with the vials suspended at 10 m (33 ft) above the ground, where wind speed is much faster than at human heights due to lesser ground frictional forces. As the vials would freeze much faster than human skin, the result was an overestimation of the effects of convective cooling. Consequently the National Weather Service (USA) and Environment Canada joined forces with scientists and medical experts to reexamine the existing concept of windchill and to update both the underlying science and the practical calculation and public health messages, ultimately publishing a revised windchill index in late 2001.

The new windchill index (see figure 2.8) featured many updates and improvements, including the following:

1. The use of a facial cooling model rather than inert materials to develop cooling rates. This was first produced on a facial manikin with thermoconducting material simulating human skin. Testing was then done on actual humans in highly controlled laboratory settings inside a cold wind tunnel at Defence Research and Development Canada—Toronto. Testing consisted of monitoring the facial skin temperature of male and female subjects while they were exercising and exposed to a wide range of different environmental conditions with either dry or wet skin (Tikuisis et al. 2007).

2. The calculation of values for dangerous exposure, based on the 5% most susceptible individuals in order to be as conservative as possible. In the actual testing, cheek temperatures tended to exhibit the lowest values, and temperatures at this site were used to provide conservative estimates.

Temperature (°F)

Wind (mph) / Calm	40	35	30	25	20	15	10	5	0	–5	–10	–15	–20	–25	–30	–35	–40	–45
5	36	31	25	19	13	7	1	–5	–11	–16	–22	–28	–34	–40	–46	–52	–57	–63
10	34	27	21	15	9	3	–4	–10	–16	–22	–28	–35	–41	–47	–53	–59	–66	–72
15	32	25	19	13	6	0	–7	–13	–19	–26	–32	–39	–45	–51	–58	–64	–71	–77
20	30	24	17	11	4	–2	–9	–15	–22	–29	–35	–42	–48	–55	–61	–68	–74	–81
25	29	23	16	9	3	–4	–11	–17	–24	–31	–37	–44	–51	–58	–64	–71	–78	–84
30	28	22	15	8	1	–5	–12	–19	–26	–33	–39	–46	–53	–60	–67	–73	–80	–87
35	28	21	14	7	0	–7	–14	–21	–27	–34	–41	–48	–55	–62	–69	–76	–82	–89
40	27	20	13	6	–1	–8	–15	–22	–29	–36	–43	–50	–57	–64	–71	–78	–84	–91
45	26	19	12	5	–2	–9	–16	–23	–30	–37	–44	–51	–58	–65	–72	–79	–86	–93
50	26	19	12	4	–3	–10	–17	–24	–31	–38	–45	–52	–60	–67	–74	–81	–88	–95
55	25	18	11	4	–3	–11	–18	–25	–32	–39	–46	–54	–61	–68	–75	–82	–89	–97
60	25	17	10	3	–4	–11	–19	–26	–33	–40	–48	–55	–62	–69	–76	–84	–91	–98

Frostbite times: 30 min, 10 min, 5 min

Windchill (°F) = $35.74 + 0.6215T - 35.75 (V^{0.16}) + 0.4275T (V^{0.16})$

Where T = Air temperature (°F) V = Wind speed (mph)

a

Figure 2.8 *(a)* Windchill chart in °F.

Available: http://www.weather.gov/om/windchill/images/windchillchart3.pdf

T_{air} (°F)

V_{10}	Calm/5	0	−5	−10	−15	−20	−25	−30	−35	−40	−45	−50
5	4	−2	−7	−13	−19	−24	−30	−36	−41	−47	−53	−58
10	3	−3	−9	−15	−21	−27	−33	−39	−45	−51	−57	−63
15	2	−4	−11	−17	−23	−29	−35	−41	−48	−54	−60	−66
20	1	−5	−12	−18	−24	−30	−37	−43	−49	−56	−62	−68
25	1	−6	−12	−19	−25	−32	−38	−44	−51	−57	−64	−70
30	0	−6	−13	−20	−26	−33	−39	−46	−52	−59	−65	−72
35	0	−7	−14	−20	−27	−33	−40	−47	−53	−60	−66	−73
40	−1	−7	−14	−21	−27	−34	−41	−48	−54	−61	−68	−74
45	−1	−8	−15	−21	−28	−35	−42	−48	−55	−62	−69	−75
50	−1	−8	−15	−22	−29	−35	−42	−49	−56	−63	−69	−76
55	−2	−8	−15	−22	−29	−36	−43	−50	−57	−63	−70	−77
60	−2	−9	−16	−23	−30	−36	−43	−50	−57	−64	−71	−78
65	−2	−9	−16	−23	−30	−37	−44	−51	−58	−65	−72	−79
70	−2	−9	−16	−23	−30	−37	−44	−51	−58	−65	−72	−80
75	−3	−10	−17	−24	−31	−38	−45	−52	−59	−66	−73	−80
80	−3	−10	−17	−24	−31	−38	−45	−52	−60	−67	−74	−81

FROSTBITE GUIDE

- Low risk of frostbite for most people
- High risk for most people in 5 to 10 min of exposure
- High risk for most people in 2 to 5 min of exposure
- Increasing risk of frostbite for most people in 10 to 30 min of exposure
- High risk for most people in 2 min of exposure
- High risk for most people in 2 min of exposure or less

Windchill (°C) = $13.12 + 0.6215 \times T_{air} - 11.37 \times V^{0.16} + 0.3965 \times T_{air} \times V^{0.16}$

Where T_{air} = Actual air temperature (°C) V_{10} = Wind speed at 10 m in km/h (as reported in weather observations)

b

Figure 2.8 (b) Canadian windchill chart in °C.
Available: http://www.weather.gov/om/windchill/images/windchillchart3.pdf

3. The use of "temperature-like" numbers, without any specific units attached to them, to avoid confusion with actual temperatures. The identical scale was translated into both Celsius and Fahrenheit equivalents to provide consistency and usability throughout North America.

4. The addition of clear public health messages warning of frostbite risks. Especially relevant is the use of color-coded zones corresponding to low risk through to high risks of frostbite at 30, 10, 5, or 2 min of exposure.

Individual Strain Scales

While many scales quantifying thermal exposure are based upon ambient environmental conditions, one inherent limitation is the wide interindividual variability in response to a set environment and exercise load. Intraindividual response can also vary based on acclimatization, hydration, drugs (e.g., caffeine), and so on. This makes it very difficult to predict actual individual responses to a particular set of environmental conditions and exercise loads, and thus safety guidelines such as the Humidex, WBGT, and windchill index are generally highly conservative based on the most susceptible individuals within a population. Therefore, some researchers have proposed thermal exposure scales and safety policies based upon individual strain (response) rather than external stress (e.g., ambient temperature). After all, the primary concern in thermally stressful situations remains the actual response of any particular individual. As with modeling thermal stress, one challenge is selecting physiological variables that are scientifically relevant yet few in number and relatively simple to measure in the field to provide easier real-time monitoring of individuals.

Many examples of thermal strain models have been developed over the years. One example is the Physiological Strain Index (PSI) that was first proposed in the late 1990s. Taking data from a large number of experiments performed at the United States Army Research Institute of Environmental Medicine (USARIEM) in Natick, Massachusetts, Moran and colleagues (1998b) modeled exercise–heat stress as a combination of metabolic (represented by heart rate) and thermal (represented by rectal temperature) strain. Current measures of each variable are normalized to both resting values and assumed ceiling values to calculate a value ranging from 0 to 5 each. Specifically, these two variables were chosen for their relative simplicity in field monitoring and were designed to fit within a simple-to-understand 0 to 10 scale, from minimal strain to severe strain at approximately 7 and above:

$$PSI = 5(T_{ret} - T_{re0}) \cdot (39.5 - T_{re0})^{-1} + 5(HR_t - HR_0) \cdot (180 - HR_0)^{-1} \text{ (Eqn. 2.5)}$$

where T_{re} = rectal temperature; HR = heart rate; the subscripts t and 0 represent current and baseline values, respectively; and 39.5 °C and 180 beats · min^{-1} represent ceiling rectal temperature and heart rates, respectively.

While not necessarily intended to be precise scientific or lab-based measures of exercise–heat strain, such scales can be sufficiently sensitive for monitoring in the field, and appear sensitive to major individual changes such as hydration status (Moran et al. 1998a). Using a similar design philosophy of scientific relevance, ease of use, and a 10-point scale, the same group has advanced a Cold Strain Index (Moran et al. 1999) based on core and skin temperatures:

$$CSI = 6.67(T_{core} - T_{core-0}) \cdot (35 - T_{core-0})^{-1} + 3.33(\overline{T}_{sk-t} - \overline{T}_{sk-0}) \cdot (20 - \overline{T}_{sk-0})^{-1} \text{ (Eqn. 2.6)}$$

MODELING EXPOSURE LIMITS

Two basic approaches are possible for modeling exposure limits in hot or cold environments. The first is to estimate limits based on empirical data, or to examine existing experiments or case studies to establish "real-life" exposure limits. While it makes sense to utilize existing data whenever possible, this can be difficult because often only limited or even no data are available for the desired conditions. For example, survival times for victims of marine accidents in cold water require core temperature modeling well below the typical limit of 35.0 °C imposed by the majority of scientific ethics review boards, though some studies to 33.0 °C have been approved. As the clinical definition of hypothermia begins at 35.0 °C, scientists actually know very little about how the body responds in true hypothermia situations, and models are forced to extrapolate beyond scientific data limits. The results can be calibrated with isolated case studies of accident survivors, but each individual has his or her own unique situational factors that make generalization tricky.

The other approach to modeling human response is to take existing knowledge of how the body responds to thermal stress and then mathematically predict responses based on these principles. Examples of this can be seen in the calorimetric calculation of the heat balance equation itself and the modeling of heat exchange of individual body segments. The potential limitation of this approach lies in the inherent assumptions about how the different values are measured or calculated in real time, such as the surface area exposed to a radiant heat source. Another problem is the matching of laboratory-based assumptions with real-life use, as highlighted by the earlier discussion on the microenvironment caused by clothing use and the changes brought about by dynamic movement.

This discussion highlights the distinct extremes of each modeling approach. Of course, the vast majority of actual thermal models aim to combine the best available data and knowledge from both approaches into exposure models.

SUMMARY

Despite decades of research on how the human organism is able to regulate its thermal state to such a finely tuned balance, many questions remain to be definitively answered, including the very basic question of whether temperature itself is the regulated variable. While temperature-sensitive receptors have been located throughout the peripheries and the CNS, the fundamental architecture related to how these pathways are interconnected and integrated to produce coordinated heat loss and heat gain responses also remains open to exploration. As can been seen in the discussion of the heat balance equation and the various thermal stress indices, the development of simple yet accurate models of heat and cold strain, along with the education of end users, is an important component of occupational and also general population health and safety.

Heat Stress

Climatic warming, along with a growing participation in extreme or ultra-endurance sports and other outdoor activities, has contributed to the increased incidence of serious exertional heat illnesses worldwide. Hot environmental temperatures can severely impair exercise capacity and lead to risks of heat illnesses. A rise in body temperature from sustained heat storage happens through a continued imbalance in the heat balance equation (see chapter 2), and can arise through a combination of exercise-induced metabolic heat production, environmental conditions (temperature, humidity), and the wearing of clothing that impairs evaporative heat transfer.

The physiological responses of the body during exercise in the heat are not a new concern; a seminal work by Adolph (1947) summarized an extensive set of experiments on soldiers during World War II. Since that time, hyperthermia, heat exhaustion, and heatstroke have become well recognized as major risks during exercise in hot environments. The American College of Sports Medicine position stand on exertional heat illness (Armstrong et al. 2007) emphasizes that heat exhaustion and exertional heatstroke "occur world wide with prolonged intense activity in almost every venue (e.g., cycling, running races, American football, soccer)" (p. 556). While most frequent in hot and humid conditions, such problems can occur even in cool conditions with intense or prolonged exercise such as marathons or Ironman triathlons. In addition, heat exposure is recognized as a severe constraint on work capacity in settings such as the military, deep mining, firefighting, and hazardous waste disposal, increasing occupational interest in determining safe exposure guidelines and countermeasures to enable or prolong safe work. Finally, heat exposure is a significant clinical problem for even nonexercising populations, as evidenced by the history of heat wave–related deaths every summer, which were especially numerous across Europe in 2003 and in Chicago in 1995 (Bouchama 2004). Disturbingly, recent evidence suggests that transient temperature spikes can significantly raise mortality rates even in areas such as the southwestern United States, where summers are traditionally very hot and the population is presumably somewhat heat acclimatized (Yip et al. 2008).

The purpose of this chapter is to detail recent research on the major issues and the physiological responses experienced by the body during exercise in environments that are hot or humid or both, focusing on what we know about the mechanisms underlying our limits to exercise and potential cooling countermeasures to extend these limits to heat tolerance.

DIRECT EFFECTS OF HYPERTHERMIA

While it is generally accepted that exercise in the heat is more difficult than in other conditions, with studies demonstrating that exercise tolerance is lower even in 20 °C compared to 10 °C temperatures (Galloway and Maughan 1997), the actual mechanism or mechanisms by which impairment or exhaustion occurs remain unclear (Cheung and Sleivert 2004b). Some of the major mechanisms currently being explored are outlined in figure 3.1. One dominant view has been a systemic insufficiency in the cardiovascular system, due to competition for blood flow to both active muscles and the skin for conductive heat exchange, coupled with a loss of plasma volume through sweating and dehydration. However, scientific attention since the mid-1990s has also focused on the actual direct effect of body temperature on limiting exercise capacity. This interest arose from studies that, while investigating other issues, showed a striking consistency in the core temperature at which individuals voluntarily terminated exercise. One of these studies was from my own doctoral research at the University of

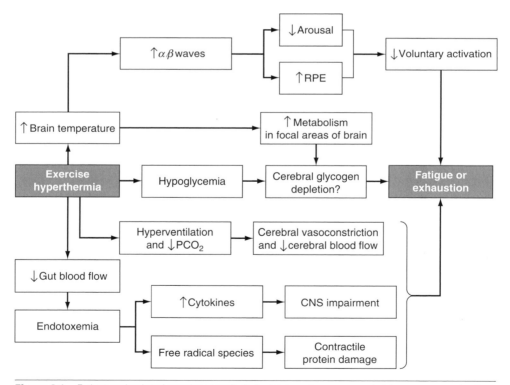

Figure 3.1   Fatigue and reduced exercise capacity in the heat can have many underlying causes, ranging from direct temperature effects on the brain's cognitive functioning and neurotransmitters through to thermal changes in the muscular, gastrointestinal, cardiovascular, and immune systems.

Reprinted, by permission, from American College of Sports Medicine, 2004, "Multiple triggers for hyperthermic fatigue and exhaustion," by S.S. Cheung and G.G. Sleivert, *Exercise and Sport Sciences Review* 32: 100-106.

Toronto and the Defence and Civil Institute of Environmental Medicine (now Defence Research and Development Canada—Toronto). In my particular case (Cheung and McLellan 1998), we were testing the separate and interactive effects of hydration status, physical fitness, and heat acclimation on exercise tolerance during exercise in the heat with nuclear, biological, and chemical (NBC) protective clothing. To do this, we had a highly fit and a moderately fit group walk on a treadmill in a hot–dry (40 °C, 30% relative humidity) chamber while either euhydrated or hypohydrated by 2.5% body mass, and before and after a two-week period of heat acclimation. We found that there was a very consistent core temperature cutoff point of ~38.7 and 39.3 °C for moderately and highly fit groups, respectively, which was resistant to changes in hydration or acclimation status. The majority of the moderately fit subjects all terminated exercise due to exhaustion, while most of the highly fit subjects stopped because they had reached ethical cutoff temperatures.

While interesting, the consistency of endpoint core temperature was not the primary focus in these initial studies and therefore may have simply been coincidental. At nearly the same time, three laboratories independently set out to determine whether this endpoint consistency was indeed valid using both animal and human models (Fuller et al. 1998; Gonzalez-Alonso et al. 1999; Walters et al. 2000). These studies altered the starting core temperatures by passive cooling or heating, which was followed by exercise performed at a set intensity to the point of voluntary exhaustion. In all cases, a very consistent terminal core temperature was reported despite differences in starting core temperature, rate of heat storage, or final skin temperature. Importantly, none of the human or animal subjects suffered any ill effects 24 h later, suggesting that exhaustion occurred well before health and system integrity were compromised. Table 3.1 details the consistency of core temperature endpoints found in human studies.

Taken together, these studies seem to firmly argue for a physiological "safety switch" in organisms to elicit voluntary cessation of exercise prior to catastrophic

Table 3.1   Endpoint Core Temperature for Voluntary Exhaustion in Moderately and Highly Fit Individuals During Exercise in Hot Environments

Study	Conditions	Moderately fit	Highly fit
Cheung and McLellan (1998)	Euhydrated, preacclimation	38.8 (0.3)	39.2 (0.2)
	Euhydrated, postacclimation	38.8 (0.3)	39.1 (0.2)
	Hypohydrated (2.5%), preacclimation	38.7 (0.3)	39.2 (0.1)
	Hypohydrated (2.5%), postacclimation	38.6 (0.3)	39.2 (0.1)
Gonzalez-Alonso et al. (1999)	Precooled		40.1 (0.1)
	Control		40.2 (0.1)
	Preheated		40.1 (0.1)
	Low rate of heat storage		40.1 (0.3)
	High rate of heat storage		40.3 (0.3)
Selkirk and McLellan (2001)	Trained subjects, low body fat		39.5 (0.0)
	Trained subjects, high body fat		39.2 (0.1)
	Untrained subjects, low body fat	38.6 (0.2)	
	Untrained subjects, high body fat	38.8 (0.2)	

Standard errors are in parentheses.

_____ UNCOMPENSABLE HEAT STRESS _____

In many occupational (e.g., firefighting, hazardous waste disposal) and athletic (e.g., American football, auto racing) settings, protective clothing is required to shield the individual from environmental hazards or from injury. Protective clothing, which is typically heavy, thick, multilayered, and bulky, exacerbates the challenge of thermoregulation by increasing the metabolic cost of exercise through adding weight and decreasing efficiency of movement. At the same time, the limited water vapor permeability across the clothing layers further decreases the rate of evaporative heat exchange (Nunneley 1989). Although protective clothing is being continually improved and lightened, the requirement of adequate impact or environmental protection is generally contradictory to the desire for adequate ventilation. Therefore, the evaporative heat loss required to maintain a thermal steady state (E_{req}) can exceed the maximal evaporative capacity of the environment (E_{max}) during exercise. In these uncompensable heat stress (UHS) situations, the body constantly stores heat, and hyperthermia onset is accelerated. In some situations, the environmental conditions can be severe enough that this can arise even at rest, making planning of countermeasures (removal from environment, active cooling) critical. It is also important to keep in mind that UHS situations are not limited to the use of protective clothing but readily occur with high rates of exercise in humid environments. Refer to figure 2.3 for a diagram outlining the pathways for water vapor dissipation through clothing.

systemic damage from hyperthermia. With a general consensus on the presence of a safety switch, research moved toward investigating different triggering mechanisms eliciting voluntary termination of exercise.

Neuromuscular Impairment

Temperature can have a variety of effects, potentially either positive or negative, on muscle function. Some studies show a higher power output during brief, high-intensity exercise with warm ambient temperatures, along with decreased maximal muscle strength with local muscle cooling. Partly based on this rationale of raising muscle temperature, warming up prior to athletic competitions or training sessions is ingrained in many sports. In laboratory settings, the ability of the brain to recruit and activate muscles has been one of the most widely studied potential mechanisms behind heat-related exercise fatigue.

Overall, while mild muscle warming or warm ambient temperatures may facilitate force generation, studies eliciting high levels of hyperthermia (above approximately 38.5 °C T_{core}) via exercise in the heat have generally supported an impaired level of maximal force generation and muscle activation. In order to further isolate core temperature as a factor in muscle function, our lab has developed a passive heating model. With this, we aimed to remove the potential confounding factors of high cardiovascular strain along with metabolic and muscular fatigue from exercise itself. At the same time, we systematically tested neuromuscular function at progressively higher core temperatures, providing a time course of temperature effects rather than just at baseline and the point of exhaustion. Lastly, as 39.5 °C was reached, we progressively cooled individuals and tested neuromuscular function again, providing a comparison of the same core temperature with hot and cool skin and helping to isolate the effects of core versus peripheral skin temperatures.

Using this model (Morrison et al. 2004; Thomas et al. 2006), hyperthermia was attained at a heart rate reserve of only 65%, and comparisons could be made at similar core temperatures with both warm and cool skin temperatures. As seen in figure 3.2, both maximal voluntary contraction (MVC) and central activation of the knee extensors, a measure of voluntary recruitment of the muscles, progressively decreased with increasing rectal temperature. Secondly, it appeared that core temperature was the primary thermal input causing hyperthermia-induced neuromuscular impairment, since when the skin was rapidly cooled (by ~8 °C) and core temperature held stable at ~39.5 °C, there was no recovery of MVC or voluntary activation. Furthermore, force and voluntary activation levels progressively returned to baseline values upon core cooling, indicating that the ability to activate the muscle and produce force was not depressed as a result of fatigue accumulating over the protocol, but likely directly influenced by body core temperature. The site of impairment along the neuromuscular pathway remains unclear, although transcranial magnetic stimulation demonstrates that the motor cortical excitability is not impaired with warm core temperatures, suggesting that it is beneath the level of the motor cortex.

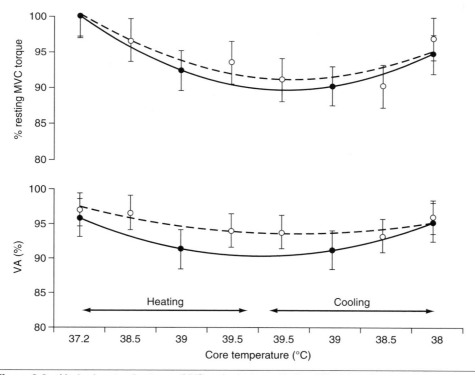

Figure 3.2 Maximal contraction torque (MVC) and voluntary activation (VA) of the soleus muscle during plantarflexion decreased progressively with increasing core temperature whether the soleus muscle temperature was maintained at thermoneutral values (unshaded circles and dashed lines) or heated up and then cooled (solid circles and lines), suggesting that central neuromuscular recruitment can be impaired with hyperthermia independent of peripheral influences.

Reprinted from M.M. Thomas et al., 2006, "Voluntary muscle activation is impaired by core temperature rather than local muscle temperature," *Journal of Applied Physiology* 100: 1361-1369. Used with permission.

Brain Arousal and Central Drive

A common feeling during heat waves is one of lethargy, and it appears logical that mental activity or acuity may be impaired with hyperthermia, in turn impairing the regulation of exercise by the central nervous system. Therefore, researchers have also examined the brain wave patterns and psychological responses to heat stress and hyperthermia. A reduction in electroencephalographic (EEG) frequency has been reported in primates with passive heating, and evoked brain signals from soldiers after exercise in the heat also demonstrated a slowing of responses to stimuli (Hocking et al. 2001). Brain activity, specifically the ratio of low-frequency (α = 8-13 Hz) and high-frequency (β = 13-30 Hz) brain waves as an indicator of arousal, during hyperthermia and exercise has been explored in humans cycling at 60% aerobic power in both hot (~40 °C) and cool (~19 °C) environments (Nielsen et al. 2001). A progressive reduction in β waves in the hot exercise condition was evident, such that the ratio of α to β waves was increased. This is similar to what happens during sleep, so it may reflect a reduced state of arousal in hyperthermic subjects. The functional significance of altered EEG activity remains to be determined, but it is worth noting that the altered brain activity was associated with changes in ratings of perceived exertion during exercise in humans. Subjects continually rated their effort higher during hyperthermic trials; the best predictor of the rate of perceived exertion was a reduction in EEG frequency in the frontal cortex of the brain (Nielsen et al. 2001).

While this is a very promising avenue of research, technical challenges are numerous in advancing the field. One primary issue is the difficulty of merging exercise with the technical requirements for EEG collection, which typically dictate that the subject be stationary and resting. This becomes especially difficult with the newest brain scanning technologies such as magnetic resonance imaging and positron emission tomography, which necessitate physical restraint of subjects. So, even with advances in miniaturization and telemetric data collection for many scientific instruments, a portable measuring system for brain signal analysis for field trials remains elusive.

Cerebral Blood Flow

While decreased mental arousal or activity may be an end result of hyperthermia, these changes in EEG activity may also be related to changes in blood flow dynamics within the brain. Hyperthermia results in a decrease in the cerebral blood flow velocity during prolonged submaximal exercise in humans (Nybo 2007); marked decreases in brain blood flow during exercise and hyperthermia have been measured in animal models, with decreases being particularly large and reaching critically low levels in unfit animals. These marked decreases in cerebral blood flow appear to be modulated by cerebral vasoconstriction, triggered by hyperventilation and a reduction in arterial carbon dioxide tension.

It is important to remember that decreases in cerebral blood flow are not necessarily associated with decreases in glucose utilization or overall metabolism. Furthermore, even with constant metabolic activity within the brain, heating may cause a selective redistribution of metabolic activity within different regions. Core heating of 1.5 °C is associated with a ~23% increase in resting metabolic rate and an increase in metabolic activity in such brain structures as the cerebellum as well as the hypothalamus—the thermoregulatory center of the brain (Nunneley et al. 2002)—along with concomitant decreases in the activity of other specific brain sites. It is possible that the increased cerebral metabolism in select areas during hyperthermia stresses the critical carbo-hydrate supply necessary for maintaining cerebral function. Given the increase of metabolic rate and energy demand in select regions of the brain and the hypoglycemia that accompanies hyperthermia, glycogen depletion of the brain could play a role in precipitating fatigue during hyperthermia in certain circumstances (Nybo 2007). Overall, the relationship between increased brain temperature, cerebral metabolism, and exercise performance remains unclear.

Heat Shock Protein Upregulation

One of the major technological advances in the past decades has been improved laboratory techniques for analyzing biochemical and immunological markers in blood and in tissues, permitting a much more detailed and integrated analysis of heat stress in multiple systems. Of these markers, one of the key immunological responses appears to be the production of various heat shock proteins (HSPs) with exposure to hyperthermia (Kregel 2002). Heat shock proteins are a class of proteins found across a wide range of organisms, are classified according to their molecular weight (e.g., HSP 70, HSP 90), and appear to perform multiple functions in many physiological systems. These proteins are also somewhat inappropriately labeled, as baseline levels are present even in the absence of overt stress, and their upregulation can be stimu-lated by many stressors apart from heat stress, including exercise, hypoxia, toxins, infections, and inflammations. As the name implies, however, a clear upregulation in the production of HSPs, especially HSP 72 and 90, occurs with prolonged heat stress and hyperthermia; and it is hypothesized that they play a significant role in heat adaptation (Kregel 2002). One potential pathway for enhanced thermotolerance may include reducing the risk of endotoxemia by maintaining permeability of the tight epithelial junctions within the gastrointestinal (GI) tract (Dokladny et al. 2006). As HSPs appear to provide an overall protective function to the body against stress, novel manipulations for heat therapy may also become feasible. For example, in a study utilizing hindlimb unloading as a model of muscle atrophy in rats, the applica-tion of whole-body hyperthermia both stimulated HSP 72 production and resulted in significantly less muscle atrophy than in control rats not exposed to hyperthermia (Naito et al. 2000). Thus, some forms of heat therapy may potentially become useful in countering the effects of muscle atrophy in patients exposed to prolonged bed rest and also during prolonged spaceflight.

Neurohumoral Factors

Mental function relies on proper signaling across neural pathways, and neurotransmit-ter activity and sensitivity may be altered by changes in temperature. Disturbances in cerebral neurotransmitter levels, especially serotonergic activity, have long been implicated in central fatigue (Meeusen et al. 2006), but few researchers have examined whether hyperthermia alters serotonin (5-hydroxytryptamine or 5-HT) levels in the brain. This neurotransmitter is of particular interest because it influences arousal levels: If 5-HT levels increase, that could contribute to the elevated perceived effort, as well as to the reduction in work rate or exhaustion (or both) often observed during hyperthermia. Certainly, animal work suggests that pharmaceutically altering 5-HT concentration in the rat brain, through the use of agonist or antagonist drugs, either impairs (5-HT agonist) or enhances (5-HT antagonist) endurance, while a 5-HT agonist also impairs endurance performance in humans (Davis and Bailey 1997). However, this work did not directly examine the influence of hyperthermia on fatigue, so it is difficult to extend this research and suggest a role for 5-HT in hyperthermic fatigue.

Dopamine is another neurotransmitter that is a candidate for modulating hyperthermic fatigue, since it plays a role in the control and initiation of movement and may also reduce 5-HT production. Levels of dopamine have been shown to increase during exercise, and precipitous falls in dopamine coincide with early fatigue (Davis and Bailey 1997). Nybo and colleagues (2003) reported that dopamine levels were elevated in both arterial and jugular venous samples during hyperthermia, but these elevated levels were most likely derived from tissues other than the brain and in fact were not related to an increase in cerebral dopamine release or increased uptake of the dopamine precursor tyrosine. More mechanistic work, perhaps using dopamine agonists or antagonists, is required to advance understanding of the influence of dopamine on hyperthermic fatigue and exhaustion.

Endotoxemia

The discussion of mental activity in the heat highlights the blurring and merging of thermal and cardiovascular mechanisms underlying exercise–heat stress. Recent evidence suggests that the well-documented redistribution of blood away from the GI tract during exercise–heat stress, to provide further cardiac output to the muscles, may cause a rebound problem of altered permeability and integrity of the GI tract lining. During severe exercise-induced hyperthermia above 40 °C, at levels associated with heat exhaustion or heatstroke, blood flow to the GI tract is markedly reduced, and this may compromise the integrity of the intestinal walls. As a result, the leakage of lipopolysaccharides (endotoxins) into the circulation can occur during severe exercise-induced hyperthermia. This endotoxemia can trigger a cascade of detrimental physiological responses including a cytokine-mediated rise in hypothalamic set point, which can induce a fever-like situation and accelerate heat storage and heatstroke (Lambert 2004). Of greater relevance to hyperthermic fatigue is the fact that endotoxemia triggers cytokine release, and cytokines have been implicated as factors that influence fatigue at the level of the central nervous system in infections like the Epstein-Barr virus or in diseases such as chronic fatigue syndrome. Thus, further research is required to explore whether endotoxemia influences the brain to increase the perception of effort, impair voluntary activation of muscle, or both.

In addition to causing an immune cascade and altering mental function, endotoxemia has another interesting consequence that has been only recently reported: an impairment of skeletal muscle intrinsic force-generating capacity, in the order of 20% to 40%, for skinned fibers (rats) treated in vitro with endotoxins (Supinski et al. 2000). Although few data have been reported, there is some evidence that contractile proteins are damaged by the endotoxemia-induced production of free radical species. Therefore, it is possible that the decrease in maximal voluntary contractile force and central neuromuscular drive observed with both exercise-induced and passive hyperthermia may be potentiated by an endotoxemic impairment of local muscular activity and that these factors may combine to induce fatigue and exhaustion.

An emerging avenue of investigation in endotoxemia induced by exercise–heat stress is nutritional countermeasures to maintain GI tract permeability. One example is the apparent ability of bovine colostrum, a dietary supplement often used to aid digestion and improve immune function, to decrease the changes in GI permeability. In essence, the colostrum may serve as a coating and sealant of the gut to minimize leakage into the bloodstream. This has been demonstrated in animal models with both colostrum and goat milk powder (Prosser et al. 2004), but quantifying gut leakage in humans remains difficult. Some initial unpublished work that has been done in humans to date suggests that bovine colostrum may enhance heat tolerance and reduce gastric distress in trained runners during prolonged exercise in the heat.

Central Versus Peripheral Influence

Core and peripheral thermal afferents can have differing effects on physiological responses; determining the relative role of these afferents on both perceptual and physiological responses to thermal stimuli is made difficult by methodological problems in manipulating one variable without altering the other. For example, Frank and colleagues (1999) used a combination of a thermally controlled mattress (at 14, 34, or 42 °C) to passively alter and maintain mean skin temperature to cool, neutral, and hot, followed by 4 °C intravenous saline infusion at 70 mL $\cdot$ min^{-1} to achieve rapid core cooling. A clear separation of thermal perception and physiological response was observed, with multiple linear regression analyses demonstrating that core and skin temperature contributed about equally to perceptions of perceived temperature but that core temperature dominated in driving vasomotor tone, metabolic heat production with core cooling, and epinephrine and norepinephrine responses. While such research is important for an understanding of the relative weighting of thermal afferents, care must be taken to avoid assuming a philosophical construct of central and peripheral afferents as two distinct entities with minimal interaction. Rather, to develop models of mean body temperature inputs into thermal stimuli integration, it would be interesting to extend such work to different levels and rates of core temperature alterations along with exploring regional skin temperature manipulations during exercise in the heat.

In addition to the interactions between central and peripheral thermoreceptors, thermal sensation may be determined in part by regional variability in the thermosensitivity of different skin surface regions to local heating or cooling (Nakamura et al. 2006). Local thermal stimulation of different skin surface regions plays a major role in thermal perception of heat stress independent of core temperature, and thermal sensation may be strongly tied to the comfort or discomfort of sweating rate and skin wettedness (Candas and Dufour 2005). The existence of thermosensitivity variability across different skin regions, notably the face, has been both supported and rejected. However, a systematic limitation of these studies is the closed-loop approach to thermal manipulation, such that one region is thermally stimulated while the temperatures of other skin regions are not controlled and are free to vary. Therefore, the actual overall neural input from the entire skin surface may vary across conditions, making firm conclusions on the relative importance of a single site difficult.

Recently, Cotter and Taylor (2005) employed a water-perfused suit enabling an open-loop approach, whereby skin regions could be stimulated while the skin temperature of nonstimulated regions could be maintained. With use of this design, cooling of the face was demonstrated to be two to five times more effective in suppressing sweating response and thermal discomfort than cooling an equivalent skin surface area elsewhere. The neurological origins of this increased sensitivity in humans are unclear, but as suggested by animal models may involve a higher facial thermoreceptor density. Overall, the behavioral and autonomic (e.g., sweating) sensitivity of the face suggests an effective potential site for targeted local cooling to lower perceived heat stress and possibly prolong exercise capacity during hyperthermia. The practicality of installing face cooling or misting during work or exercise can be logistically challenging. However, some potential avenues may involve the design of helmets that provide strong conductive cooling using air ventilation, misting, ice, or phase change materials. Alternatively, portable showers or misters and fans may be erected at worksites and athletic facilities to facilitate on-site cooling during or between work or exercise bouts.

Postexercise Thermoregulation

While thermoregulation during exercise in the heat has been extensively targeted in research settings, relatively little work has explored the dynamics of temperature regulation following exercise in the heat. However, it has been commonly reported that during postexercise passive recovery in warm environments, core temperature remains elevated by 0.5 to 1 °C above baseline values for a sustained period, 60 min or longer (Kenny et al. 2007). In contrast, heat production, as measured by oxygen consumption, decreased rapidly (<10 min) to baseline values. Furthermore, the primary conductive and evaporative heat dissipation avenues also rapidly returned to baseline; this included a drop in forearm blood flow, skin temperatures, and sweating rates.

Such post-hyperthermia thermoregulatory perturbations seem specific to the residual effects of exercise, as esophageal temperatures rapidly returned to baseline following passive warm water immersion. Therefore, nonthermal factors may also influence postexercise thermoregulatory response (Gagnon et al. 2008). One important mechanism may be changes in cardiovascular dynamics postexercise, with reports of sustained reductions in mean arterial pressure for several hours postexercise and of increases in the magnitude of this impairment with higher exercise intensities. This postexercise hypotension appears linked to an increase in systemic vascular conductance. In turn, this promotes the muscle mass as a major heat sink during the postexercise period, with a decrease in blood flow resulting in a pooling of blood and heat in the active limbs (Jay et al. 2007).

PACE SELECTION WITH HYPERTHERMIA

Concurrent with work on individual mechanisms of hyperthermic fatigue, other researchers began advancing a feedforward, anticipatory model of behavioral response during exercise in the heat. In this paradigm, well before exercise impairment via neuromuscular or other physiological mechanisms, the physiological impact of hyperthermia may be circumvented by feedforward and feedback voluntary control of effort at the commencement of and during heat stress. In other words, rather than permitting the body to reach a point of near collapse prior to preventive action, this model, as seen in the schematic in figure 3.3, proposes that prior to and throughout exercise, the body integrates the stress (both environmental and physical) imposed upon it, producing a work output that permits it to complete the task as efficiently as possible with the least risk of developing heat exhaustion.

Research to advance this model has encompassed both modeling and human studies. Modeling of the different components of the heat balance equation on subjects of different body masses running at different speeds demonstrates that the point of UHS is at much higher speeds for smaller runners (Dennis and Noakes 1999). Lower body mass and size provides a higher ratio of surface area to volume and also

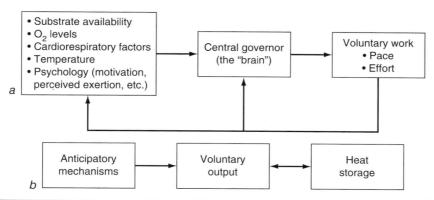

Figure 3.3   *(a)* The central governor receives input from a number of interrelated factors and integrates this input to regulate voluntary work output. This regulatory function ensures that an individual will maintain a work rate at an intensity that can be safely sustained by the cardiorespiratory and thermoregulatory systems for the duration of the exercise. *(b)* Feedforward anticipatory mechanisms reduce work output, thereby affecting (reducing) heat storage to levels ensuring that a catastrophic rise in core temperature does not occur. Continuous afferent feedback from temperature sensors and the rate of heat storage will modulate the work rate during exercise to further ensure that a critically high temperature is not reached.

results in lower absolute heat production at a given running speed; this higher possible pacing may be a contributing factor in the dominance of smaller individuals in distance running. Subsequent researchers have attempted to validate the modeling data with human studies, finding that body mass inversely and significantly correlated with self-selected running speed at 35 °C irrespective of aerobic fitness (Marino et al. 2004). The conclusion was that the larger body masses resulted in a higher rate of heat production and reduced heat dissipation capacity at a given speed, leading to a lower self-selected pace to maintain thermal balance during exercise in hot environments. When people were asked to maintain a constant perceived effort in cool, neutral, or hot environments, the rate of decline in voluntary power outputs was higher in hot environments than in the cool and neutral environments even though the rate of heat storage was similar in each condition, further supporting a model of some thermal signal influencing exercise capacity (Tucker et al. 2006).

Thermal Signals

Understanding or defining the signal or signals that become integrated to regulate self-paced effort is a difficult problem, as what is involved is most likely an amalgam of physiological and psychological sources and is further mediated by individual factors (e.g., training status, hydration, age). Core temperature itself would appear to be an obvious signal, and it has been argued that the progressive decline in central neuromuscular activation during isometric leg flexion with passive heating provides a neuromuscular basis for this idea. However, extrapolating this to whole-body exercise may be difficult, as the same passive heating and cooling model did not indicate any core temperature effects on maximal dynamic exercise capacity (Cheung and Sleivert 2004a). The role of absolute core temperature is also countered by self-paced studies in which, despite similar rectal temperatures, subjects in cooler environments were able to produce higher power outputs. The rate of core temperature increase is also unlikely to be a primary signal, as precooling seems to have a primary effect on reducing starting temperatures rather than the rate of heat storage. Clearly, further work is required to directly determine whether there is an anticipatory process in the brain that controls work output through feedback loops involving thermal or other inputs, along with interactions from perceptual and psychological parameters.

Overriding the Safety Valve

While hyperthermia clearly impairs physiological function and exercise capacity, it remains unclear how the physiological defense mechanisms against excessive hyperthermia may sometimes be overridden or, in cases of exertional heat illness, compromised to the point of catastrophe. Ultramarathoners can exercise at maximal capacity for 4 h in moderate ambient temperatures with only minor elevations in core temperature, suggesting a strong ability to thermoregulate even under conditions of high metabolic heat production. At the same time, some case studies show that endurance athletes can sustain elevated core temperatures much higher than those seen in laboratory studies (Kenefick et al. 2007). Furthermore, exertional heat illness can occur even in cool weather, with one case study of a well-trained male experiencing collapse near the finish of a marathon in 6 °C ambient conditions and presenting a rectal temperature of 40.7 °C approximately 30 min postcollapse (Roberts 2006). Studies on precooling that show a higher pace selection in trained runners in hot environments suggest a potential behavioral or perceptual override of hyperthermia

RESEARCHING EXERCISE AND FATIGUE

Something as nebulous as why someone stops exercising in the heat can be very difficult to examine in the laboratory, and is a good example of the challenge of research design. Imagine this to be your task (or research project!): How would you go about it? The straightforward approach may be simply to have individuals exercise with different thermal manipulations (e.g., precooling, different ambient temperatures) and observe their time to exhaustion or their voluntary pace selection. While this approach has the most ecological validity in terms of resembling real-life exercise, unless you are clever at its implementation, you may find it difficult to tease out the actual physiological mechanisms underlying changes in performance. Another possibility is to take a reductionist approach and try to isolate and investigate how individual physiological systems, such as cerebral blood flow or the brain's ability to neurally activate muscles, respond in both thermoneutral and hot conditions. This research paradigm has been used successfully to identify a wide range of physiological functions that are indeed impaired either in a hot environment or with an elevated body temperature. However, as in the task of rebuilding Humpty Dumpty, you may be left with a lot of pieces but unsure of which mechanisms are important or critical in something as multifactorial as impaired exercise capacity in the heat. The difficulty is further magnified by the strong behavioral thermoregulatory capacity of humans, such that psychological responses from perception of heat stress add a further layer of complexity to why humans perform differently in the heat. There are no correct approaches, and a comprehensive understanding of the field requires appreciation and integration of multiple disciplines and research paradigms.

signals. This is supported by reports of an attenuation of perceived exertion and thermal strain in highly fit individuals in UHS environments (Tikuisis et al. 2002). This may be further exacerbated by high inherent motivation in athletes, especially in the heat of competition, such that they may tune out physiological signals of discomfort or impairment. Ultimately, the mechanisms underlying the breakdown between voluntary exhaustion and exertional heat illness remain unclear and difficult to investigate in controlled laboratory situations.

INDIVIDUAL VARIABILITY IN HEAT TOLERANCE

The issue of individual variability in response to exercise in the heat is difficult to address. In an effort to minimize intersubject variability, most recent studies on hyperthermic fatigue with both constant and self-paced exercise have employed a fairly homogenous group of subjects, primarily males with high aerobic fitness, in order to ensure exercising to the point of hyperthermia or "true" voluntary exhaustion or both. This is coupled with the taxing nature of the studies, which generally self-select against individuals with low fitness or activity levels, children, and elderly persons, making it difficult to explore the role of sex, maturation, and fitness on heat stress and exercise capacity. Table 3.2 illustrates some of the major factors that may predispose individuals to heat intolerance.

Table 3.2   Predisposing Factors to Heat Illness During Exercise

Heat illness	Primary symptoms	Predisposing factors	Treatment
Exertional heatstroke[a]	• Central nervous system (CNS) dysfunction (disorientation, convulsions, coma) • Severe (>40 °C) hyperthermia • Nausea, vomiting, diarrhea • Severe dizziness and weakness • Hot and wet or dry skin • ↑ heart rate, respiratory rate • ↓ blood pressure • Extreme thirst and dehydration	• Obesity • Low physical fitness level • Dehydration • Lack of heat acclimatization • Previous history of heat illness • Sleep deprivation • Sweat gland dysfunction • Sunburn • Viral illness • Diarrhea • Medications (i.e., stimulants, anticholinergic and cardiovascular drugs, cocaine)[b] • Extremes of age[c] • Uncontrolled diabetes or hypertension, cardiac disease	• Institute aggressive and immediate whole-body cooling within minutes of diagnosis with water as cold as practical until temperature < 38.3 °C. • Remove equipment and monitor rectal temperature. • Monitor vital medical signs (airway, breathing, circulation, CNS status) continuously. • Perform IV saline infusion if feasible.

(continued)

Table 3.2   *(continued)*

Heat illness	Primary symptoms	Predisposing factors	Treatment
Exertional heat exhaustion[d]	• Fatigue and inability to continue exercise • Ataxia, dizziness, and coordination problems • Profuse sweating • Headache, nausea, vomiting, diarrhea	• Dehydration • High body mass	• Remove athlete from practice or competition and move to a shaded or air-conditioned area. • Remove excess clothing and equipment. • Lay victim down with legs above heart level. • Rehydrate using chilled fluids or normal saline IV. • Cool athlete to <38.3 °C—aggressive cold water immersion may not be required. • Monitor status and transport to emergency facility if needed.
Exertional heat cramps	• Intense pain • Persistent muscle contractions in working muscles during prolonged exercise	• Dehydration • Exercise-induced muscle fatigue • Large sweat sodium loss	• Rehydrate and provide sodium replenishment. • Prevent by sodium loading (e.g., 0.5 g in 1 L of sports drink) in cases of heavy or "salty" sweaters. • Use light stretching and massage.

Symptoms and treatment summarized from the 2003 National Athletic Trainers' Association "Inter-association task force on exertional heat illnesses consensus statement" (Casa et al. 2003).

[a]Exertional heatstroke = hyperthermia (core body temperature 40 °C) associated with central nervous system disturbances and multiple organ system failure.

[b]Stimulants (i.e., amphetamines, Ritalin, ephedra, alpha agonists) increase heat production; anticholinergic drugs (i.e., antidepressants, antipsychotics, and antihistamines) inhibit sweating; cardiovascular drugs (i.e., calcium channel blockers, beta blockers, diuretics, monoamine oxidase inhibitors) alter the cardiovascular response to heat storage; cocaine increases heat production and reduces heat loss by decreasing cutaneous blood flow.

[c]Children and elderly persons are particularly susceptible to heat accumulation due to decreased sweating ability, increased metabolic heat production, greater surface area–to–body mass ratio, decreased thirst response, decreased mobility, decreased vasodilatory response, and chronic medical conditions or medication effects (or both).

[d]Exertional heat exhaustion = inability to continue to exercise; may or may not be associated with physical collapse. Generally, this does not involve severe hyperthermia >40 °C or severe CNS dysfunction.

Aerobic Fitness

Besides increased exercise tolerance in thermoneutral environments, aerobic fitness is known to provide protective physiological responses to exercise similar to those observed with acclimatization to hot environments (Cheung et al. 2000). These benefits include greater evaporative heat dissipation through improved sweating response, with lower core temperature thresholds for the initiation of sweating as well as greater sensitivity of sweating response to increasing core temperatures. Improved aerobic capacity also leads to an elevated plasma volume and cardiac output, minimizing the competition for blood distribution between skeletal muscle and skin (heat dissipation and sweating output) during exercise and heat stress (Sawka et al. 1992). Other important benefits of aerobic fitness are a lowered resting core temperature coupled with an elevated endpoint core temperature that can be tolerated prior to voluntary exhaustion (Cheung and McLellan 1998). This latter response appears to be directly due to aerobic fitness rather than body composition, as differences across fitness groups remained evident when highly and moderately fit subjects were normalized for differences in body fatness (Selkirk and McLellan 2001). Therefore, fitness appears to benefit heat tolerance due to both a greater capacity for and a slower rate of heat storage. However, these benefits appear unique to long-term changes in aerobic fitness rather than associated with transient changes brought about by short-term training interventions (Cheung and McLellan 1998).

Heat Acclimation

Parallel to benefits in heat tolerance from improved aerobic fitness, heat acclimation generally elicits a lower resting core temperature, greater plasma volume, and an increased sweating rate (Cheung et al. 2000). Heat acclimation can be achieved by a variety of heat exposure protocols, with some variability in the frequency, duration, and intensity of heat stress required to attain and maintain adaptation. In healthy adults, the primary stimulus for heat acclimation appears to be a sustained elevation in core temperature of 1 to 2 °C for 60 to 90 min over 4 to 10 days, with aerobic fitness decreasing the required stimulus and also the amount of stimulus needed to maintain acclimation or reacclimate to heat (Pandolf et al. 1977).

The environment is also a significant component of a heat acclimation program. Wearing heavy sweat clothing in cool environments may elicit the same increased sweating response as with standard heat exposures, and this has been utilized as an alternative method of heat acclimation (Dawson 1994). The specificity and transferability of heat acclimation programs using protective clothing remain controversial, and the ideal heat acclimation program may be specific to both the environment that will be encountered and the clothing that will be worn (Fox et al. 1967). In contrast, Griefahn (1997) reported an equal degree of acclimation in subjects following exposure to three substantially different climates with a similar wet bulb globe temperature, 33.5 °C. Furthermore, this acclimation was transferable to other climates. Interestingly, subjects in this latter study wore minimal clothing for all experimental treatments, and it is possible that the different findings in the studies may be related to specific effects of the clothing.

Maturation

Maturation and age may influence an individual's ability to thermoregulate during rest or exercise in the heat (Falk and Dotan 2008). Children have a larger surface area–to–volume ratio than adults. While this is advantageous in temperate ambient environments by increasing the rate of radiative and conductive heat transfer, these benefits are negated with exposure to a hotter environment where the thermal gradient for exchange is smaller or toward positive heat gain by the body. At the same time, evaporative heat dissipation is also attenuated in children, with sweating response and sweating rate gradually increasing from childhood through adolescence and into adulthood, primarily due to changes in sweat gland activity rather than gland number (Falk et al. 1992). Coupled with these biophysical and physiological differences are an attenuated thirst perception and perceived effort of exercise, especially when one is "caught up in the game," which may lead to decreased awareness of heat stress and hyperthermia along with inadequate fluid replacement (Bar-Or and Wilk 1996). Countering this, however, children may be more sensitive to changes in core temperature than young adults, such that maturation may bring about different thermoregulatory strategies in the heat. This again suggests the potential importance of different monitoring and education strategies in work with children compared to adult populations.

While thermoregulatory changes with maturation appear to be clear during puberty and adolescence, findings on the capacity of people who are elderly to respond to heat stress are equivocal. Part of the challenge of research in this area involves a lack of consistency in defining "aging," which is especially relevant with the increase in life expectancy. Depending on the study, the minimum age threshold for the "elderly" subject group may range from as young as 55 to up to 75 years (Pandolf 1997). Another methodological issue in aging research is separating clinical issues (e.g., disease, prescription drugs), body composition changes (decrease in muscle mass and increase in body fat), and the concomitant tendency toward lower activity levels and aerobic fitness with increasing age. In the few studies that have included attempts to control these factors, core temperature thresholds for sweating and shivering were elevated and lowered, respectively, in older compared to young adults, along with a general attenuation of the magnitude of these responses. However, reviews generally conclude that there was minimal significant impairment in exercise thermoregulation due to aging per se when confounding factors such as fitness and body composition differences were removed (Pandolf 1997).

COOLING STRATEGIES

Regardless of the actual mechanism underlying exercise impairment from hyperthermia, the use of cooling protocols as an ergogenic aid or to counteract the risks of heat stress and hyperthermia has gained primacy in occupational and athletic settings. This section addresses work on cooling that is initiated prior to exercise, during exercise, or else during rest breaks or following exercise. In addition, we will explore the use of microclimate cooling garments, as well as postexercise thermoregulatory control.

Precooling

Where cooling during exercise is not possible, precooling—decreasing body temperature before exercise—has been popular among athletes prior to exercise in the heat. Some of the initial athletic proponents included the Australian rowing teams prior to the 1996 Atlanta Olympics, whose members used vests containing ice packs during warm-ups before competition. Since then, precooling has become popular for a variety of sports, including athletics, swimming, and cycling. In clinical settings, cooling helmets have been employed to induce brain hypothermia rapidly following cardiac arrest in order to minimize neural damage. Some athletes and coaches have turned to targeted precooling of the head, due to its high perceptual and sweating sensitivity, to improve subjective tolerance, physiological response to heat stress, or both.

As discussed earlier in this chapter, a lower body temperature may act to improve performance in events that rely on a high and sustained aerobic effort (e.g., running, rowing, long-distance cycling) via delaying heat buildup, upregulating voluntary pace selection, or both (Marino 2002). The consistent core temperatures at the point of voluntary fatigue with a constant workload in the heat would certainly suggest that the removal of heat before exercise could increase exercise tolerance by increasing possible heat storage prior to fatigue. So, although the rate of heat storage can remain unchanged with precooling, the lowered baseline temperature can enable both core temperature and heart rates to remain lower with prolonged exercise over time, delaying the attainment of a critical core temperature or cardiac insufficiency (Gonzalez-Alonso et al. 1999). Precooling also can decrease perceptions of heat stress and thereby possibly promote an upregulation of work intensity. This phenomenon has been reported in both elite runners and rowers with precooling of about 0.5 °C using cool water of about 20 °C or using ice vests (Arngrimsson et al. 2004; Booth et al. 1997). In turn, this improvement may occur because of either a higher consistent pace selection throughout the exercise or an ability to raise power outputs near the end of exercise due to a lower physiological strain during the initial phases. Less clear are the benefits of precooling for sports that are more reliant on short bursts of anaerobic power or strength, or for team sports (e.g., soccer) that place a high emphasis on intermittent sprints along with intricate decision making and strategy. Precooling prior to intermittent sprint protocols did not result in any improved power (Cheung and Robinson 2004; Duffield et al. 2003), and a higher ambient temperature actually may facilitate higher pedaling cadences and power outputs during anaerobic Wingate sprints (Ball et al. 1999).

The potential benefits of precooling extend beyond athletic and work situations to clinical settings and a variety of diseases. A primary example is multiple sclerosis, in which demyelination of the central motor nerves brings about symptoms including neuromuscular dysfunction and optic neuropathy. While exercise and improved muscle tone can delay the progression of the disease, the heat generated with exercise can lead to a slowing of nervous conduction and a worsening of symptoms. Precooling using liquid cooling garments (see figure 3.4) or cooling hoods, and the use of air conditioning, seem to attenuate symptoms and may improve some measures of muscle strength. Some of the latest work in our laboratory has focused on the effects of precooling with cooling hoods on a range of validated neuromuscular and ambulation tests. Especially interesting were our initial findings that head and neck precooling significantly decreased perceived fatigue and also increased distance covered during a 6 min walk test, suggesting that it may be a useful method of increasing exercise capacity and the abilities to perform activities of daily living in this population.

Microclimate Conditioning Systems

Where practical, the wearing of microclimate conditioning systems (MCS) during exercise can improve thermal perception and decrease the rate of heat storage. However, the capacity for MCS with protective clothing may be limited to only very light

From NASA: nasaimages.org.

Figure 3.4   Liquid cooling garments, consisting of tubing circulating coolant next to the skin, were first developed for use by the National Aeronautics and Space Administration in spacesuits. They have since been employed for a wide range of athletic and occupational uses, including precooling prior to competition in the heat and also under chemical protective clothing for hazardous waste disposal and military personnel.

workloads, as high rates of heat production and low rates of heat dissipation through the clothing quickly result in UHS situations (see chapter 2). At present, the majority of cooling garments rely on conductive heat exchange, using gas ventilation, phase change materials, or liquid mediums (Flouris and Cheung 2006). Gas-ventilated MCS have been developed since 1940 and during World War II were used as a heating countermeasure for pilots flying at high altitudes. At the same time, phase change materials, such as dry ice, have been frequently employed in MCS.

The potential for developing a MCS employing a liquid medium was first proposed in 1959 and implemented in 1964. The most common type of liquid MCS consists of a tight-fitting undergarment embedded with tubing through which coolant is run, with variations in the amount of skin coverage and sometimes in the number of different regions through which the amount of cooling can be varied. Such liquid MCS are now fairly common in a variety of occupational, medical, and athletic settings and continue to be the primary form of cooling underneath spacesuits for astronauts performing extravehicular activities. Currently, the basic liquid MCS design remains fairly crude, with minimal control of coolant distribution, flow rate, and temperature. For example, many versions consist of a whole-body (torso, hands, and legs) unit with full coverage and no adjustment capabilities. In addition, the cooling capacity of such suits remains limited by the small actual surface area of contact with the skin, due to both fit issues and the nonuniform distribution of the tubing throughout the suit. Along this research theme, the optimal fitting of the liquid MCS, the placement and design of tubing, and the control of cooling appear to be key avenues toward maximizing efficiency of heat exchange (Flouris and Cheung 2006). In response to the limitations and also aided by advances in materials technology, researchers are beginning to explore cooling garments based on heat exchange principles other than simply circulating cold water in tubes near the skin. For example, recent research in our lab and others has begun exploring the potential of membranes in clothing that dissipate heat from the skin by evaporation and then actively remove the water vapor from the clothing.

Most current cooling systems have only crude manual control that cools the entire body evenly rather than targeting specific zones of high thermal transfer. This is relatively inefficient in maximizing cooling efficiency and minimizing power requirements and is a major drawback to portability and the wider adoption of these systems in occupational settings. However, the design of an automatic MCS control is a challenging task, as the resulting system should be an automaton that provides cooling to specific sites and is adaptable to different individual characteristics and exercise types and intensities (Flouris and Cheung 2006). Potentially the most essential consideration is the identification of indicators of thermal state in order to guide cooling control. The feasibility of several indicators has been examined to date; these include coolant temperature, sweat rate, skin temperature, heart rate, and carbon dioxide production, as well as a combination of metabolic rate and body temperature. It seems reasonable that an optimum automatic MCS control should include practical indicators that can respond rapidly to changes in the wearer's condition and thermal comfort. These indicators will initiate instantaneous changes in MCS coolant temperature, maintaining the individual's thermal homeostasis despite alterations in heat production. In addition, indicators that respond more slowly may be used to provide more accurate and

_____ *MAXIMIZING CONDUCTIVE HEAT EXCHANGE* _____

One avenue for maximizing cooling garment efficiency has focused on maintaining peripheral skin blood flow despite cooling. In this way, vasoconstriction and the shunting of warm blood away from the peripheries are prevented and heat extraction can be maintained and maximized. Stephenson and colleagues (2007) tested the efficacy of skin temperature–based pulsed cooling of liquid cooling garments, whereby cooling was initiated when mean skin temperature increased above 34.5 °C but turned off when skin temperature dropped below roughly thermoneutral values of 33.5 °C. No difference was observed in core temperature or heart rate during exercise using protective cooling in the heat with any of the three protocols, but the T_{skin}-based pulsed cooling required 46% and 28% less power than constant cooling and time-based pulsed cooling, respectively. Another innovative example of ways to maintain peripheral blood flow during cooling is to combine conductive cooling of the hand on a metallic surface at the same time that the hand is encased in a slight vacuum, on the theory that the lower ambient pressure elicits local vasodilation even though the surface cooling of the hand stimulates vasoconstriction (Grahn et al. 1998). Such systems may be especially useful where workers or athletes repeatedly return to a particular spot for rest breaks following a bout of exercise (e.g., athlete benches).

stable long-term regulation of coolant temperature. Applicable examples of rapid indicators include heart rate and metabolic rate, while relevant examples of slower indicators include parameters of coolant temperature and ear canal temperature.

Ultimately, the design goal in microclimate cooling, from the physiological and engineering perspectives, is to maximize the efficiency of cooling in maintaining thermal comfort and neutrality with the least cooling possible in order to minimize coolant and power requirements. As many workers in heat stress situations are active and mobile, there is an acute need for power efficiency to ensure that cooling garments are portable and can last for prolonged periods without the need for recharging. Such advances will require a cross-disciplinary approach that merges engineering disciplines (e.g., biomedical, mechanical, electrical) in suit design with physiological advances in understanding human thermoregulation.

Intermittent Cooling During Exercise

The wearing of cooling garments may not be feasible or practical in some work settings, such as firefighting, due to the extreme heat and risk of scalding. Therefore, the use of rest breaks and cooling interventions may be required to prevent heat exhaustion. However, cooling advice for athletic and occupational settings must also balance scientific merit and practicality. Despite their efficacy, large and very cold air or water facilities are not readily available or portable. And while the periodic removal of headgear and impermeable rubber gloves can extend work tolerance in the heat (Montain et al. 1994), the practicality of such measures is low during actual work or possibly even rest breaks in occupational settings. Therefore, the onus is shifted to exploring potentially low-tech but practical measures that can be implemented during rest breaks. Within these constraints, the partial removal of clothing, in conjunction with immersion of the hands and forearms in cool to tepid (10-20 °C) water for 10 to 20 min, can significantly reduce core temperatures and prolong work tolerance in firefighters in hot and humid environments compared to partial encapsulation during rest breaks alone (Khomenok et al. 2008; Selkirk et al. 2004). As with work on liquid microclimate cooling systems, the tepid water appears to balance heat extraction with minimal vasoconstrictory stimulation and provides the added practical benefit of lower on-site cooling power requirements. An alternative to arm immersion may be face misting during rest breaks, focusing on the high heat extraction and effect on heat perception through the head and face (Selkirk et al. 2004). The logical extrapolation of this work on firefighters would suggest that similarly simple strategies may also prove effective in sports with intermission periods like soccer, football, and rugby.

Cooling From Heat Illnesses

In cases of exertional heat illness, priority must be given to the rapid lowering of core temperature, as the intensity of symptoms involves both the magnitude and the duration of hyperthermia. Due to its high heat conductivity, water immersion of the victim is typically recommended for rapid cooling. However, the ideal water temperature for this purpose has been a contentious issue, marked by arguments for balancing the need for rapid cooling with the possibility of cardiovascular shock from overly cold water temperatures. To resolve this issue, Proulx and colleagues (2003) investigated the effects of different water temperatures on core temperature recovery following exercise-induced hyperthermia of 39.5 °C, with subjects immersed to the neck for recovery in a range of temperatures from near-ice water to cool water (2, 8, 14, and 20

°C). Core cooling was most rapidly achieved in 2 °C water, requiring only 7 to 10 min to drop below 37.5 °C and eliciting minimal shivering. In contrast, cooling rates were significantly longer and were not different among the three other water temperatures, and the two intermediate water temperatures seemingly prolonged cooling by also eliciting shivering. At the same time, care must be taken to avoid leaving victims in cold baths for too long and overcooling them, as the rectal temperature typically used in clinical settings can greatly overestimate esophageal or aural canal temperature. In a follow-up study, the same research group reported esophageal temperature of 34.5 °C when rectal temperatures reached 37.5 °C following cooling with 2 °C water immersion, and concluded that T_{re} limits of 37.5 and 38.6 °C in water above and below 10 °C, respectively, were sufficient to safely remove the heat built up during exercise to 39.5 °C rectal temperature. However, the optimal cooling temperature may depend on situational factors such as whether the victim was exercising or not. Recent work on recovery from passive hyperthermia, more typical of the general population during heat waves, suggested that tepid water of ~26 °C was just as effective as colder temperatures (Taylor et al. 2008).

Another common use of cooling protocols for athletes is postexercise cold water immersion (CWI), on the premise that reduced core temperatures speed recovery and decrease exercise-induced trauma. However, the high degree of peripheral vasoconstriction from immersion may result in blood pooling in the limbs and decrease the rate of metabolic recovery, although the use of contrast baths alternating between cold and warm water may mitigate the blood pooling. What appears to be missing in the current literature are well-controlled studies examining the effects of passive CWI on subsequent aerobic or anaerobic performance, and especially compared to more traditional forms of postexercise recovery modalities such as mild active recovery, massage, and so on.

WIND SPEED IN THE LAB

A predisposing environmental factor leading to hyperthermia is low air circulation, leading to a decreased rate of heat exchange. Similarly, one methodological concern about the realism of laboratory settings is the typically minimal to low air movement or wind speed compared to that in actual field situations, such that the rate of convective and evaporative heat exchange and therefore heat storage could be dramatically misrepresented. For example, Saunders and colleagues (2005) reported that at higher air velocities simulating movement or wind in competitive cycling situations (33 and 50 km · h^{-1}, or 20.5 and 31 miles · h^{-1}), heat storage, body temperature, and perceived exertion during 2 h of cycling in 33 °C were lower than with 0 or 10 km · h^{-1} (6.2 miles · h^{-1}) wind speed. Furthermore, at the higher velocities, a higher rate of fluid ingestion did not influence heat storage, body temperature, or sweat rate. This suggests that the amount of air flow can be a critical factor in the realism of laboratory studies on heat stress, and that outdoor air movement and convective flow should be replicated for maximum transferability of lab findings. In our current studies of the effects of wind speed, we are validating this idea further by calorimetrically calculating the amount of heat extracted from the body during precooling and matching it to any extra work produced.

SUMMARY

Hyperthermia increases the physiological strain on the body and can result in a marked decrease in exercise capacity and potentially heat injury and death in the general, athletic, and occupational populations. Over the past decade, research interest has arisen concerning the direct effects of temperature per se on exercise capacity and tolerance in the heat. The effects of hyperthermia appear to pervade most physiological systems within the body, and these deleterious effects may begin at even moderate levels of temperature elevations. In turn, many factors contribute to individual tolerance to exercise in hot environments. Therefore, countermeasures for athletes and occupational applications revolve around maintaining body temperatures as close to baseline as possible during exercise. One strategy involves the use of microclimate cooling systems underneath protective clothing, but this is often countered by the increasing need for encapsulation and insulation (e.g., increasing personal armor for soldiers), along with technological limits on power and portability. Other countermeasure strategies include the use of cooling protocols before, between, or following work bouts.

Hydration Strategies for Exercise

A s we have seen in chapter 3, heat stress and heat illnesses are major consid-erations for both exercising individuals and the general population, and adequate countermeasures are critical in reducing the risk. That discussion also focused on various cooling protocols that can be performed in the lab and in practical applications. Beyond such technological interventions, an important comple-ment is to ensure that adequate hydration is maintained during exercise and in hot environments. Sweat rates exceeding $1 \, L \cdot h^{-1}$ are typical during moderate exercise, with a high of $3.7 \, L \cdot h^{-1}$ having been recorded by the American runner Alberto Salazar during preparation for the 1984 Olympic marathon (Armstrong et al. 1986). The thirst response is relatively slow, and dehydration of 2% or more of body weight may be undetected before a strong drinking response is observed. However, even with an adequate fluid supply and reminders to drink, the rate of ad libitum fluid intake rarely matches the rate at which fluid is lost through sweating, and an individual generally becomes gradually dehydrated over the course of prolonged exercise or heat exposure. If the fluid lost through sweat production is not adequately replaced, even minor decrements in fluid balance can impair cardiovascular and thermal regulation and exercise performance (Montain and Coyle 1992).

Walk into any gym, children's soccer league setting, or shopping mall, and chances are that you will see lots of people with water bottles slung over their hips or other-wise close at hand. Certainly within Canada and the United States, the message of the importance of adequate hydration has become entrenched in the general con-sciousness; this has come about as a result of both health educational initiatives and marketing by sports drink and beverage companies, making it potentially difficult to separate evidence-based knowledge from myths. While it is generally accepted that a lowered hydration status is detrimental to exercise capacity and heat tolerance, the definition of the word "adequate" remains unclear. Part of the problem is the difficulty in generating a health message that is simple to communicate to enhance acceptance and adoption; but in addition, a lively scientific debate continues on many apparently simple but fundamental questions concerning hydration. The following are some of these questions:

- How can we simply and reliably measure and monitor hydration status and changes both in the laboratory and in the field?
- Just how much fluid should an individual drink during exercise–heat stress to balance the opposing risks of dehydration and hyponatremia?
- Should individuals be drinking water or a special solution containing carbohydrates, electrolytes, protein, or some combination of these? Does this advice vary across different exercise durations and intensities?

The process of researching such questions has resulted in major changes in consensus among sport scientists, including a dramatic revision of the 2007 compared to the 1996 American College of Sports Medicine (ACSM) position stand on fluid replacement during exercise (Sawka et al. 2007). The purpose of this chapter is to review the current evidence concerning fluid replacement during exercise and heat stress and to sort through the scientific bases underlying current health messages, along with posing questions that continue to require sustained investigation.

PHYSIOLOGY OF FLUID BALANCE

The aim in this chapter is not to present a comprehensive survey on renal physiology, and the reader should refer to anatomy and physiology texts to obtain a grounding in kidney function and the interactions of various hormones (e.g., vasopressin, aldosterone) on fluid regulation (Eaton and Pooler 2009). However, before exploring hydration strategies, it is important to define the basic terminology used in the field to avoid confusion, as well as to discuss the physiological effects of hydration impairments on exercise capacity.

Hydration Terminology

A schematic of the various fluid compartments is presented in figure 4.1. Total body water (TBW) is generally categorized into the intracellular (ICF) and extracellular (ECF) fluid compartments, with the latter further divided into the interstitial fluid volume (ISF) and the plasma volume (PV). Total body water content is dependent on individual characteristics, such as fitness, body fatness, and glycogen storage. Assuming that TBW and ICF represent 60% and 40%, respectively, of the total body mass of a typical male, the body of a 75 kg (165 lb) individual would contain 45 L of water, with an ICF of 30 L and an ECF of 15 L. Blood volume (BV) in a healthy young male who weighs 75 kg is approximately 5 L, or 70 mL $\cdot$ kg^{-1}, with about 3 L of it PV. This can also be highly variable depending on individual characteristics; a high correlation has been reported between BV and aerobic fitness or lean body mass, with BV ranging up to 100 mL $\cdot$ kg^{-1} in trained and lean endurance athletes (Sawka et al. 1992).

Water balance during exercise is determined by a multitude of factors, including the environmental conditions, the nature and intensity of exercise, and the characteristics of the fluid replacement. Two terms related to hydration status that are often mistakenly interchanged are dehydration and hypohydration (see figure 4.2). In this book, dehydration is defined as the dynamic loss of body water due to sweating over the

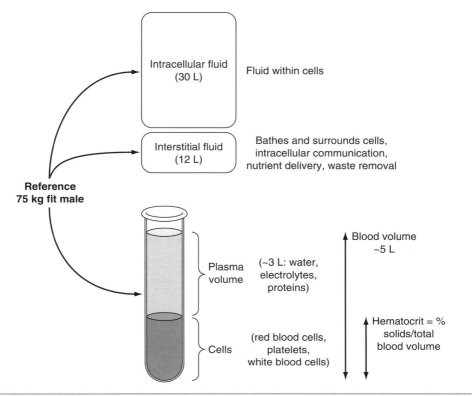

Figure 4.1   Schematic of typical body fluid compartments.

course of exercise without fluid replacement, or a process in which fluid replacement does not match the rate of fluid loss. In contrast, hypohydration refers to the state or level of hydration after the loss of a certain amount of body water from the body. As an example of the difference between the two terms, rowers may dehydrate themselves through exercise without fluid intake to make a certain weight category, then compete in the actual event in a hypohydrated state because they were not able to replace all of the lost fluid beforehand. Therefore, dehydration refers to a process, while hypohydration refers to a body state at a particular point in time. In turn, hyperhydration refers to a state of higher than normal body fluid balance, while rehydration refers to the dynamic act of increasing fluid balance before, during, or following exercise–heat stress to elevate hydration status. The nature of hypohydration is also dependent on the method used to achieve dehydration (Caldwell et al. 1984). Diuretics such as furosemide cause the even excretion of both electrolytes and fluid, resulting in iso-osmotic hypovolemia. However, with exercise or passive heat exposure, the production of a hypo-osmotic sweat results in a hyperosmotic hypovolemia within the intravascular space, raising plasma osmolality from a baseline value of approximately 283 mOsm $\cdot$ kg^{-1} to in excess of 300 mOsm $\cdot$ kg^{-1} with 7% hypohydration and producing a fluid shift from the ICF to the ECF to defend BV (Sawka et al. 1985).

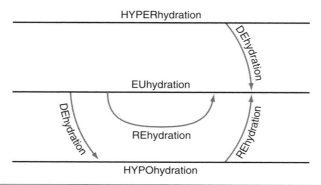

Figure 4.2 Various terms used in describing different states of hydration.

Physiological Responses to Dehydration

The most prominent physiological result of inadequate fluid balance in the body during exercise is a decrease in plasma volume, driven largely by the fact that the plasma volume is a major source of initial sweating losses (Costill et al. 1976). At the same time, the increased need for cardiac output and blood flow to the exercising muscles produces a host of adaptations in the cardiovascular system, affecting the ability of the body to maintain homeostasis. As an example, an initial cardiovascular response to dehydration is a reduction in cutaneous and systemic vascular conductance to maintain adequate venous return and blood pressure, along with a redistribution of blood flow away from the gastrointestinal tract, kidneys, and other visceral organs. This constrictor response is stressed during exercise in the heat, however, by increased demand for metabolic blood flow and peripheral vasodilation for heat dissipation (Rowell 1974).

Stemming from these challenges, one of the first observations during exercise is a progressively higher heart rate even with no changes in exercise intensity. This phenomenon, termed cardiovascular drift, is largely driven by both the decreased plasma volume and the redistribution of blood to the peripheral blood vessels for heat dissipation. Consequently, venous return to the heart is impaired, resulting in a decreased end-diastolic volume and stroke volume. Therefore, heart rate must rise to maintain a similar cardiac output and blood flow to the exercising muscles. The greater heat dissipation requirements and sweating rate in hot environments, or in those individuals with high natural rates of sweat loss (and in individuals acclimatized to heat; see chapter 3), may exacerbate the extent of cardiovascular drift. One implication for coaches and sport scientists is to be aware of such potential shifts in heart rate response when prescribing exercise intensities or monitoring training status and loads based on heart rate, in that the cardiovascular drift may lead to an overestimation of the actual training intensity.

Another notable result of dehydration and hypohydration is altered thermoregulatory capacity. Some studies show that thermosensitive neurons in the preoptic anterior hypothalamus are also sensitive to changes in osmolality and especially alterations in [Na^+], suggesting that hypohydration may influence thermoregulation centrally (Turlejska and Baker 1986). One possible result is an adjustment of the thermoregulatory set point and the resting core temperature, with reports of resting core temperature increasing linearly by approximately 0.15 °C with each percentage decrease in body weight (Sawka et al. 1985). However, while it is tempting to attribute physiological responses to a single variable, it is much more likely that changes in blood volume and osmolality produce overlapping effects and a synergistic response. Hypohydration also progressively impairs evaporative heat loss ability, with increasing levels of hypohydration both increasing the core temperature threshold for the initiation of sweating and decreasing the sensitivity of the sweating response, resulting in a lower sweating response overall and potentially less defense against continued heat storage (Sawka et al. 1985). Therefore, while core temperature may be able to stabilize at a level slightly higher than baseline when euhydration is maintained, the typical observation is a continued increase in core temperature when individuals exercise at a given intensity in a hypohydrated state.

A classic laboratory study on the effects of progressive dehydration on cardiovascular and thermal response addressed the effects of differing amounts of fluid replacement during prolonged exercise in warm conditions (Montain and Coyle 1992). Subjects exercised for 2 h in a 33 °C (50% relative humidity) environment while receiving either no fluid replacement or amounts equivalent to approximately 20%, 50%, and 80% of the estimated sweating rate. No changes in sweating rate were observed across the different rehydration rates, reflecting that the consistent exercise intensities across the four conditions elicited a similar level of heat production and therefore heat dissipation requirements. However, body weight loss and serum osmolality progressively decreased with increasing rehydration, resulting in body weight losses ranging from 1.1% to 4.2% and serum osmolality of 289 to 302 mOsm · kg^{-1} from the highest rate of rehydration to no fluid replacement, respectively. A greater thermal strain was also observed, as rectal temperature progressively increased (38.4-39.1 °C) with each lower rate of rehydration. The study clearly demonstrated that even mild levels of dehydration have the potential to cause significant physiological strain in the body and elevate the risk for exertional heat illnesses, and directly contributed to the consensus in the 1996 ACSM position stand on fluid replacement that recommended drinking fluid in sufficient amounts to offset the anticipated or measured sweat losses (Convertino et al. 1996).

The redistribution of blood with exercise has been theorized to affect metabolic function through a redistribution of blood away from the gastrointestinal tract (GI; Rowell 1974). This may affect the subsequent ability of the body to take in fluids and nutrients due to potential decrements in gastric emptying and also intestinal absorption. This may become especially important in sport nutrition planning, with a possible need to preload the body with nutrients early on during exercise while GI function is still optimal. However, the overall risk of impaired GI function with sustained hypohydration is equivocal. While exercise in a warm or hot environment and hypohydration can significantly impair the rate of gastric emptying irrespective of heat acclimation status (Neufer et al. 1989), the timing of fluid replacement of either a large bolus at the start of exercise or the same amount in even doses throughout exercise did not influence cardiovascular or thermal responses at the end of prolonged exercise in the heat (Montain and Coyle 1993). Acute plasma volume expansion of 12% to 13% via intravenous infusion also did not alter the rate or mix of substrate metabolism during submaximal exercise (Watt et al. 1999), nor did it enhance either thermoregulation or exercise tolerance time in a warm environment (Watt et al. 2000). Furthermore, hypohydration of 5% body weight did not impair the rate of muscle glycogen resynthesis postexercise (Neufer et al. 1991). However, as noted in chapter 3, in severe conditions of dehydration and heat stress, recent evidence suggests that GI ischemia can ultimately alter gut permeability to the point of eliciting endotoxemia and immune shock with hyperthermia (Lambert 2004).

Impact of Dehydration on Sport Performance

The previous discussion clearly demonstrated cardiovascular and thermoregulatory decrements from decreased hydration status, though the effect on metabolism does not appear as clear. The next question to address is whether dehydration and hypohydration affect actual exercise performance or capacity. Voluntary loss of body weight and water is often used by athletes to achieve a certain weight category prior to an athletic event. This practice of voluntary hypohydration to reach a weight category should obviously be balanced with the possible decrements in performance, although some time typically occurs between the weigh-in and competition during which athletes attempt to restore the lost body weight. Alternatively, in an industrial setting, inadequate time for rehydration and recovery may be an unintentional or unavoidable cause of hypohydration prior to a work session.

Conflicting conclusions have been advanced with regard to the influence of hypohydration on anaerobic performance or performance in sports of short duration and relatively high intensity. Using a supramaximal 30 s Wingate cycling sprint test, some authors have reported both decreased anaerobic power and capacity while others have found no performance effects with progressive hypohydration up to 5% body weight (Jacobs 1980); similar conflicting results have been reported with the use of other high-intensity exercise models (Judelson et al. 2007). However, one potential limitation of many studies on hypohydration is that passive heat exposure or exercise in the heat, along with diet and fluid restriction, was used to achieve the weight loss, with the actual hypohydration tests occurring almost immediately upon achievement of the weight loss. While realistic to many applied situations, such protocols make it difficult to isolate the direct effects of hypohydration from heat stress, exercise fatigue, and possibly lowered substrate levels and metabolic capacity; and future research designs should strive to remove these potential confounding factors.

With aerobic exercise of lighter intensity and longer duration in the heat, where the limit to exercise tolerance can be dominated by cardiovascular and thermoregulatory considerations, the detrimental effect of hypohydration is more clear-cut. One symptom of this impairment is a reduced maximal oxygen uptake, with $\dot{V}O_{2max}$ decreasing at hypohydration of approximately 5% body weight and beyond even in the absence of heat stress (Webster et al. 1990). In turn, this impairment may be exacerbated by heat stress and greater decreases in plasma volume, such that decreased $\dot{V}O_{2max}$ may occur with lower levels of hypohydration (Sawka et al. 2001). Subsequent to a decreased maximal aerobic capacity, submaximal exercise capacity should logically also be affected. In a hot environment or in situations of uncompensable heat stress, the additional stress imposed by hypohydration appears to potentiate physiological strain and severely impairs the ability of individuals to sustain exercise. In work done with military chemical protective clothing in the heat, our laboratory found a higher resting core temperature with 2% decrease in body mass, but the terminal core temperature at the point of voluntary exhaustion did not change, such that hypohydration prior to exercise negated any benefits in exercise capacity from heat acclimation or a short period of physical training. This emphasized the importance of ensuring adequate hydration prior to exercise as a preventive countermeasure (Cheung et al. 2000).

One project directly compared the effects of maintaining euhydration prior to exercise with the effects of fluid replacement during exercise in uncompensable heat stress. Whereas subjects were able to sustain a constant-paced treadmill walk for 106 min with both initial euhydration and 1 L · h^{-1} water rehydration, tolerance time at the same workload decreased to 93 min when they were euhydrated but not rehydrated, and equally to 87 min when they were initially hypohydrated by 2.2% body weight but given 1 L · h^{-1} fluid replacement (Cheung and McLellan 1998). Therefore, both inadequate fluid replacement and initial hypohydration independently impaired exercise capacity; importantly, fluid replacement was also unable to compensate for decreased fluid balance, making the maintenance of adequate hydration prior to the start of exercise critical.

Rather than setting a forced constant exercise intensity and then measuring time to exhaustion, which does not simulate real-life racing situations, many recent studies have utilized a cycling, rowing, or running time trial effort featuring self-selected pacing to investigate the effects of hydration status on aerobic exercise performance. Overall, findings from these studies of high-intensity but submaximal exercise generally concur with the decreased times to exhaustion observed with lighter constant-effort exercise (Cheuvront et al. 2005; Slater et al. 2005). After a 48 h weight loss protocol followed by an aggressive nutrition replacement for 2 h that elicited a 4% body weight loss, rowing times slightly but significantly increased compared to control values, and the decrement in performance time was greater when the trial was repeated in hot conditions (Slater et al. 2005). However, 48 h of energy restriction reduced the distance covered during a 30 min treadmill time trial by 10%, whereas 48 h of fluid restriction elicited similar body weight loss of ~3% but decreased time trial distance by a nonsignificant 2.8% (Oliver et al. 2007).

In summary, beginning exercise in a hypohydrated state or inadequate rehydration during exercise appears to significantly impair cardiovascular and thermoregulatory capacity, with a hot environment and its greater demand for heat dissipation potentiating the decrement from reduced fluid balance. While this altered hydration does not seem to have a major influence on activities heavily reliant on anaerobic capacity, the impact on aerobic capacity and submaximal exercise appears to be generally negative, with greater degrees of fluid loss and ambient temperatures magnifying the degree of impairment. As a result of such observations, current consensus is that the goal of fluid replacement during exercise should be to limit dehydration to <2% of body weight along with excessive alterations of electrolyte balance (Sawka et al. 2007). The remainder of the chapter addresses different factors to consider when one is attempting to achieve these hydration goals.

DEVELOPING A HYDRATION STRATEGY

Clearly, while a mild level of dehydration may be permissible or even inevitable throughout the course of exercise, long-term health and sustained performance depend upon maintaining adequate hydration levels. The degree of rehydration possible during exercise depends on a wide range of factors, from individual drinking behavior to environmental parameters, the nature of the activity or the rules of the sport, the exercise intensity, and the nature of the fluid itself. As it is impossible to develop a single guideline that can be applied across this wide range of individual and situational characteristics, the best advice for any athlete is to experiment with different hydration strategies and develop a plan based on his or her own needs (Maughan and Shirreffs 2008). In this section, we consider some of the major factors in rehydration planning and the current knowledge on fluid composition for recovery during and postexercise. A summary of the major recommendations from organizations such as the ACSM and the National Athletic Trainers' Association is presented in table 4.1.

Table 4.1   Summary of Major Considerations in the Development of an Individual Hydration Strategy for Exercising Individuals

Factors	Current consensus	Developing an individual hydration strategy
Sweat rates and composition	A large interindividual variability exists in sweat rates and composition. Women, children, and elderly persons generally have lower sweating rates than younger men. Increased fitness and heat acclimatization generally increase sweating rates. Hotter environments and nature of activity (e.g., may involve protective clothing) can increase sweating rate.	Athletes and workers should rely on individual determination of sweat rate and composition rather than general guidelines.
Hydration assessment		• Body weight, taken before and after exercise and over the course of days at consistent times, can give a general indication of sweating rates and fluid requirements. • Urine color (clear to straw color) and specific gravity <1.020 are simple complements to obtaining body weight. • Visual indicators (e.g., high level of salt encrusted on clothes) can be used as a general guide to individual saltiness of sweat and electrolyte requirements.
Hydration effects	• Dehydration results in progressive decreases in plasma volume, increasing cardiovascular and thermal strain. • Activities reliant on maximal strength, power, or anaerobic metabolism appear to be unaffected by 3% to 5% body weight dehydration. • Aerobic capacity and submaximal exercise may become progressively impaired beyond 2% body weight dehydration, especially with exercise in the heat.	Athletes and workers should strive to drink sufficient amounts to avoid dehydration beyond 2% body weight over prolonged exercise. Even with minimal anaerobic impairment, cognitive impairment may increase the risk of errors and accidents.

Factors	Current consensus	Developing an individual hydration strategy
Rehydration fluid composition	• CHO delivery of ~60 g per hour promotes carbohydrate availability and spares glycogen stores, enhancing submaximal exercise. CHO may have nonmetabolic benefits for high-intensity exercise <1 h. • CHO concentrations of 4% to 8% do not impair gastric emptying. Simple sugars (e.g., glucose, fructose) in combination appear to empty most rapidly, but fructose >3% can decrease GI absorption rate. • Small amounts (0.3-0.7 g · L^{-1}) of sodium can increase palatability and drinking via thirst stimulation, along with decreasing the risk for hyponatremia.	• Palatability of fluid should be a key consideration in promoting voluntary fluid consumption during exercise. This varies between individuals and also conditions (e.g., ↓ sweetness tolerance in the heat). • CHO fluids of 6% to 8% concentration should be consumed for prolonged exercise >1 h, and possibly for high-intensity exercise <1 h. • 0.3 to 0.7 g · L^{-1} of sodium can be added to fluid. • Individuals exercising >4 h should moderate their fluid intake to below sweating rates and avoid drinking solely water or low-solute fluids to minimize risk of hyponatremia.

CHO = carbohydrate; GI = gastrointestinal tract.

Summary collated from the American College of Sports Medicine 2007 Position Stand on Fluid Replacement (American College of Sports Medicine et al. 2007) and the National Athletic Trainers' Association Position Statement: Fluid Replacement for Athletes (Casa et al. 2000).

Assessing Hydration Status

In developing a hydration strategy, one of the first issues facing exercise physiologists and sport scientists becomes how to adequately monitor and quantify fluid balance in athletes and workers. Before tackling the multitude of potential methods, it is important to emphasize that measurements must be standardized to permit appropriate comparison; for example, body weight or urine samples should be taken each morning upon arising and voiding of the bladder.

The first and most obvious method for hydration monitoring would appear to be the simple measurement of body mass before and after exercise and on a daily basis. However, while useful in giving a gross measure, body weight changes do not necessarily equal changes in fluid balance in the body, as they provide incomplete information on the following (Maughan et al. 2007):

• Changes in water content of the bladder, GI tract, and the plasma volume itself. As dehydration progresses, fluid losses from the various body fluid compartments occur at different rates. With sweating throughout exercise, fluid is initially predominantly drawn from the plasma volume and interstitial fluid with subsequent compensation from the intracellular compartment, resulting in a rapid initial decrease and then defense in plasma volume with sustained body weight loss. Therefore, plasma volume can decrease by 10% with a 2% to 4% decrease in body weight (Costill et al. 1976).

• Electrolyte balance in both the blood and body tissues due to the dynamic balance between the composition of the sweat and also the fluid consumed. For example, for the same sweat rate and fluid consumed, a very "salty" sweater who drinks only water would be at greater risk for electrolyte depletion than a less salty sweater who drinks an electrolyte solution.

• Fluid stored within the body (e.g., due to muscle glycogen). This endogenous store of water would become available with carbohydrate metabolism, such that exercise intensity and the relative contribution of fat versus carbohydrate metabolism may affect overall fluid balance independent of body weight changes.

Given all this, body weight changes should form only one part of a multipronged approach to assessing overall hydration status. Table 4.2 broadly surveys the strengths and limitations of popular methods for quantifying hydration status. Of these options, urine color and specific gravity appear to be simple and practical complements to body weight when used consistently and with appropriate education of the athlete, along with periodic use of more invasive and technical blood measures such as plasma osmolality and hematocrit.

Table 4.2   Selected Options for Assessing Hydration Status

Measure	Technique	Threshold for determination of euhydration	Notes
Thirst perception	Subjective numerical ratings of thirst sensation	N/A; dependent on scale	Used primarily for educational purposes and as a very rough approximation, along with serving as a stimulus to drink.
Body weight	Weight taken in the morning or following exercise	<1% body weight from "normal" baseline	Often used to estimate acute sweat loss. High practicality but potentially low reliability even within an individual.
Bioelectrical impedance	Electrical conductivity of the body used in an algorithm to determine body fluid compartment sizes	N/A	• Not ideal for small or acute changes with fluid ingestion and sweat loss. • Can be reliable and accurate for chronic changes if protocol is standardized (e.g., machine, posture, time of day).
Urine color	Graded colored chart matching urine color to dehydration level	Scale ranges from 1 (very pale yellow) to 8 (brownish-green); 1 to 3 considered well hydrated	• High practicality. • Urine color can be influenced by diet (e.g., vitamin supplements) and also by large boluses of water or hypotonic fluid.
Urine specific gravity	Droplet of urine measured on a handheld refractometer	<1.020 g · mL^{-1}	High practicality. Main use for chronic rather than acute hydration assessment.
Urine osmolality	Measure of total solute concentration within urine	<700 mOsmol but no true consensus; highly variable across cultures	• Requires specialized equipment and training. Main use for chronic rather than acute hydration assessment. • Can be considered a measure of the kidney's concentrating ability.

Measure	Technique	Threshold for determination of euhydration	Notes
Plasma osmolality	Measure of total solute concentration within blood	<290 mOsmol	• Small changes highly correlated to thirst and hormonal changes. • Requires specialized equipment and training. • Must be measured shortly after blood sampling.
Hemoglobin and hematocrit	Changes in hemoglobin and hematocrit calculated as percentage change in blood and plasma volume	N/A	• Moderate practicality due to blood sampling, though centrifugation and measurement of hematocrit itself are relatively simple. • Hematocrit can be used by itself with a modified equation. • Can be affected by postural changes.
Tracer dilution	Double-labeled water with isotopes of hydrogen or oxygen is ingested and then sampled in blood	N/A; absolute values provided for various body fluid compartments	• Highly accurate and considered gold standard. • Expensive and requires highly specialized equipment and training. • Other tracers can be used to calculate different body fluid compartments. • Not practical for acute measurements or repetitive measurements due to time required for equilibration and clearance throughout the body.

Dehydration Versus Hyponatremia

The concern over hydration and heat stress during exercise, along with the scientific research on hydration summarized thus far, helped to promote a health education policy of raising as much awareness as possible about the dangers of dehydration and benefits of rehydration. This peaked in the mid-1990s with ACSM's 1996 position stand on fluid replacement, which advocated that athletes start drinking early and at regular intervals to the maximum that can be tolerated, ideally matching all the water lost through sweating (Convertino et al. 1996).

This particular portion of the ACSM's stand has come under heavy criticism from some circles, led by Dr. Tim Noakes at the University of Cape Town in South Africa, author of the seminal book *Lore of Running*. Noakes was concerned that the emphasis on high volumes of rehydration during exercise, especially over a prolonged ultra-endurance exercise, may lead to the onset of hyponatremia, marked by a reduction in serum sodium concentrations from norms of 135 to 145 mmoL $\cdot$ L^{-1} to levels below 130 mmoL $\cdot$ L^{-1}. In turn, this dilution of intravascular fluid results in a shifting of fluid into the ICF, which may produce dangerous swelling of the central nervous system and symptoms such as "confusion, disorientation, progressively worsening headache, nausea, vomiting, aphasia, impaired coordination, muscle cramps, and muscle weakness. Complications of severe hyponatremia include cerebral and pulmonary edema that can result in seizure, coma, and cardiorespiratory arrest" (Casa et al. 2005, p. 121). Unfortunately, these symptoms can be very similar to those seen with severe dehydration, making the correct diagnosis and medical response critical among first aid and other emergency personnel. "Water intoxication" has caused recent deaths, during the 2007 London Marathon and also during a radio promotional stunt in which contestants competed to drink large amounts of water and not use the bathroom. Such clinical problems can arise due to an abnormally high rate of water consumption (e.g., >2 L · h^{-1}) and are potentiated in individuals with very high sodium concentrations in their sweat. Another consideration for medical support at endurance events may be the slower participants (e.g., those requiring marathon completion times >5 h), who may be exercising at a lower intensity and consequently have lower sweating rates, exacerbated by greater rates of drinking and total time spent out on the course.

Rather than the message of drinking to replace sweat loss or drinking as much as tolerable, Noakes aims for a public awareness message encouraging individuals to be more attuned to their sense of thirst and to use this sense as a guide to ad libitum drinking during exercise. A few key observations form the primary foundation for Noakes' arguments (Noakes 2007):

• At issue is not a debate or scientific comparison between rehydration and not drinking at all, as the latter is rarely the chosen behavior in humans. Rather, the comparison in scientific studies should be between the control condition of ad libitum drinking versus other drinking protocols such as full replacement of sweat loss.

• Anecdotally, many world-class marathoners drink only very small amounts of water and well below either their anticipated sweat rate or the 1.2 L · h^{-1} recommendation. For example, it was estimated that Mizuki Noguchi, the winner of the women's marathon at the 2004 Athens Olympics, held in a starting temperature of 35 °C, spent only 30 s drinking during the entire race. Therefore, with such an apparent imbalance between fluid consumption and anticipated sweat loss, it is arguable whether moderate dehydration significantly impairs elite performance even in hot environments.

• Much of the research used to support the concept of drinking as much as tolerable is laboratory-based research with minimal realistic wind speed and therefore convective cooling, along with imposition of a set intensity of exercise on subjects. This research design may not adequately simulate real athletic situations, and some research has demonstrated that adequate convective cooling removes the differences observed in core temperatures or exercise performance between drinking as much as tolerable versus ad libitum drinking (Saunders et al. 2005).

Arguments such as these, along with continuing evolution of research into hydration and exercise, led to a significant revision of the fluid replacement guidelines, specifically regarding rehydration volumes, by ACSM in its 2007 position stand on exercise and fluid replacement (Sawka et al. 2007). One highlighted change in the revised document is recognition of the risks of hyponatremia from fluid consumption that exceeds the sweating rate. Specifically, the "goal of drinking during exercise is to prevent excessive (>2% body weight loss from water deficit) dehydration and excessive changes in electrolyte balance to avert compromised performance" (p. 377) while deliberately steering clear of a specific volume of recommended rehydration. Rather, the document provides information on various factors that may modify both sweating and electrolyte excretion rates, suggested methods of monitoring fluid loss, and finally individual and situational characteristics that may affect the ability to rehydrate during exercise. Overall, the position stand suggests a range of 0.4 to 0.8 $L \cdot h^{-1}$ as a starting point; large individuals who are heavy sweaters exercising in the heat may lean toward the higher end of this range, while small individuals exercising at low intensities in a cool environment may aim toward the lower rates.

Fluid Composition

Volume considerations aside, two further considerations are important in the design of rehydration fluids. First, even the best drink from a scientific standpoint is useless unless it is palatable and is something that individuals will want to drink sufficiently to ensure adequate hydration. Therefore, scientific research has also explored the effect of differing drink composition and characteristics on the rates of voluntary consumption in different populations. Secondly, the composition of the fluid and also the nature of the exercise can have significant influence on the ability of ingested fluids to rapidly affect the body compartments, as fluid must first be emptied from the stomach and absorbed from the intestines.

In keeping with the wide range of individual variability, people have different preferences for fluid taste and temperature. This can also vary within individuals based on the particular situation, with anecdotal reports of less tolerance for sweet drinks in the heat. In one of the few studies looking at voluntary drinking behavior in humans, a water temperature of 15 °C had the highest consumption rate in subjects allowed to choose the water temperature (Boulze et al. 1983). Importantly, this cool yet not cold temperature suggests that it is not critical to provide ice water at events, greatly aiding the ability of event organizers and occupational settings (e.g., military units) to logistically provide large volumes of fluids in the field. However, a recent study demonstrated that it is possible to achieve precooling of ~0.5 °C in rectal temperature by drinking 300 mL (~10 fl oz) of 4 °C water three times over a span of 30 min before exercise; this cooling effect was not observed with equivalent doses of 37 °C water. In turn, this cold water ingestion produced a decreased cardiovascular and thermal strain during submaximal exercise in the heat, and also prolonged constant-intensity exercise tolerance by >20% (Lee et al. 2008). Therefore, whenever available, cold water ingestion may serve both to ensure euhydration status and also as a practical precooling manipulation to enhance exercise performance.

Flavoring of drinks, irrespective of carbohydrate content, may also aid in voluntary consumption. For example, cooling and flavoring drinks had an additive effect on consumption rate, with a 120% greater consumption of 15 °C flavored water compared with 40 °C unflavored water (Hubbard et al. 1984). Similarly, work on the voluntary consumption rates in children during exercise in the heat showed that the addition of grape flavoring significantly increased the volume consumed compared to unflavored water. Interestingly, when the drink was grape flavored and also contained 6% carbohydrate and a small amount of sodium, the voluntary consumption rate increased again and exceeded the rate of fluid loss (Wilk and Bar-Or 1996). The difference in consumption rate may be explained either as an affinity for cool and flavored fluid or as a negative response to warm or unflavored fluid. One important thing to keep in mind when reviewing such research is that the number of subjects in these studies is small and that the samples are selected locally, which may introduce factors such as cultural preferences. Certainly, it appears that the emphasis in public health education should remain on the individual and on testing different fluids in practice situations to find a range of fluids that work best.

In addition to enhancing fluid palatability, the provision of carbohydrate (CHO) in a rehydration fluid can serve a strong ergogenic end by enhancing the availability of CHO for metabolic purposes while also maintaining fluid balance. The primary role of CHO feeding during exercise is to spare endogenous glycogen stores as a substrate for high-intensity exercise. Therefore, the exact amount of CHO fluid requirement for athletes is dependent principally on exercise intensity and duration and also on whether any solids or semisolids (e.g., sport gels) are consumed. Carbohydrate ingestion clearly benefits prolonged submaximal aerobic performance through metabolic pathways. Compared to water ingestion, CHO fluid ingestion did not have any effect on plasma volume or osmolality, nor on cardiovascular, thermal, or subjective response to 2 h of constant-intensity exercise. However, CHO dosages of 26 and 78 $g \cdot h^{-1}$ significantly improved subsequent 4.8 km (~3-mile) cycling time trial performance by approximately 30 s, while an intermediate dosage of 52 $g \cdot h^{-1}$ had a slight (~20 s) but nonsignificant effect (Murray et al. 1991). Therefore, it appears prudent to ensure that CHO forms part of a nutrition strategy for exercise of 2 h or more. Recent studies suggest that a CHO mouth rinse, with no actual ingestion of fluid, may also be of benefit to high-intensity exercise of much shorter duration, such as time trials of 1 h (Carter et al. 2004). As the amount of ingested CHO that would be metabolized in this short time is minor, the mechanism for improvement likely lies along nonmetabolic pathways, possibly through glucose receptors in the mouth and effects on central drive (Carter et al. 2004). However, no improvements were observed with 16 km (9.94-mile) cycling time trials of approximately 20 min duration with CHO fluid ingestion (Jeukendrup et al. 2008), so the true ergogenic benefit of CHO for short bouts of exercise remains unclear.

Protein is another macronutrient being experimented with in rehydration fluids, but its efficacy remains controversial. An initial study tested a 7.3% CHO drink and a 7.3/1.8% CHO–protein rehydration fluid during exercise and recovery in a temperate environment. Compared to the CHO drink condition, the CHO–protein solution during exercise elicited a 29% increase in time to exhaustion at 75% $\dot{V}O_{2peak}$, and its consumption following this test also improved time to exhaustion at 85% $\dot{V}O_{2peak}$ by 43% 12 to 15 h later (Saunders et al. 2004). However, when tested using a time trial protocol in which subjects could voluntarily determine exercise intensity, a 6% CHO and a 6/2% CHO–protein solution improved 80 km (48-mile) time trial performance equally by 4.4% compared with placebo (van Essen and Gibala 2006). One critical flaw in both studies was that the calories from protein were in addition to the similar CHO content of the two solutions, such that any benefits might have been primarily due to the approximate 20% increase in caloric content. Therefore, the ergogenic efficacy of protein solutions during exercise remains unclear, but these solutions may be promising as a recovery tool between training or exercise bouts.

Gastric Emptying and Intestinal Absorption

Due to both the delayed thirst response and the time required for fluid absorption, athletes have traditionally been advised to drink before exercise and also constantly throughout exercise, rather than wait until late in the exercise to begin rehydration. This appears to be sound advice. Gastric emptying is a dynamic function that is affected by many different factors, and these factors are highlighted in table 4.3. The

Table 4.3 Factors Affecting Gastric Emptying (GE) and Absorption of Fluid

Factor	Effect	Notes
Volume	• Increasing volumes of ingested fluid increase rate of GE, at least up to an ingested volume of 600 mL. • Increasing volume also increases net fluid absorption.	As gastric volume increases, distension of the stomach increases the activity of the stretch receptors in the gastric mucosa, thereby increasing gastric motility and speeding GE.
Temperature	Increasing fluid temperature (warmer beverages) decreases rate of GE.	Colder drinks can reduce gastric temperature, which may increase gastric motility.
Composition CHO Na$^+$ K$^+$	• Increasing CHO concentrations generally slow the rate of GE. • High osmolality of a CHO solution further delays GE, while low osmolality of a CHO solution leads to greater rate of fluid absorption. • Isotonic Na$^+$ solution increases GE compared to water. • Addition of NaCl to glucose solution increases intestinal absorption of glucose and water. • K$^+$ has no stimulatory effect on GE and may inhibit GE of CHO-containing fluids. • K$^+$ has an osmotic effect that decreases GE (similar to that of CHO).	Both CHO concentration and osmolality influence rate of GE; however, CHO content likely has a greater impact. To balance CHO needs with fluid replacement, CHO concentrations should not exceed 8%.
Exercise intensity	• Low-intensity exercise has little effect on GE. • Exercise at intensities >70% $\dot{V}O_{2max}$ decreases rate of GE. • Intermittent high-intensity exercise at overall intensities <70% $\dot{V}O_{2max}$ decreases GE relative to continuous exercise. • Exercise at moderate to high intensities may decrease rate of fluid absorption.	• Mechanisms behind exercise effects may include exercise-induced release of vasoactive hormones, which may inhibit gastric contractility and slow GE; reduced splanchnic blood flow, slowing GE or reducing rate of intestinal absorption (or both); and mechanical changes caused by upper body movements, increased respiration, or both, which influence gastric mobility. • Majority of studies examining GE and exercise use cycling as mode of exercise. Field studies in competitive arenas also show exercise effects: The rate of GE was reduced during 30 min of competitive soccer (low to moderate overall intensity with intermittent high-intensity sprinting) compared to 30 min of low-intensity walking.
Exercise mode	Treadmill exercise (walking, running) at low to moderate intensities (<70% $\dot{V}O_{2max}$) has been shown to enhance the rate of GE compared to resting conditions.	Treadmill exercise may increase intragastric pressure via the contractile activity of the abdominal muscles, thereby influencing GE.
Hypohydration	Exercising in a warm environment when hypohydrated may decrease the rate of GE.	

CHO = carbohydrate; Na$^+$ = sodium ion; K$^+$ = potassium ion.

rate of gastric emptying is influenced by the rate of fluid replacement and subsequently the volume of the stomach contents, with a direct relationship observed between the amount of fluid ingested and the rate of gastric emptying (Mitchell et al. 1994). However, it may be that as long as an adequate threshold volume of approximately 300 to 400 mL (~10-13 fl oz) is present in the stomach, further gastric volume does not enhance emptying rates but can potentially cause discomfort and impaired performance. With an increase in exercise intensity, the decreased splanchnic blood flow due to increased metabolic demands for muscle blood flow has been assumed to inhibit gastric emptying and intestinal absorption (Rowell 1974). The mode of exercise does not appear to affect gastric emptying rates, as a low value similar to that at rest was observed during exercise at 75% $\dot{V}O_{2max}$ with either cycling or running (Houmard et al. 1991). However, moderate walking and running at lower intensities may initially increase the rate of gastric emptying, possibly due to an increase in stomach motility from the walking motion, such that gastric emptying may become impaired only at high exercise intensities and greater rates of splanchnic ischemia. Some research has suggested that this threshold may occur at exercise intensities >70%.

In addition to exercise intensity, several other factors related to total blood volume and its redistribution during exercise–heat stress can influence gastric emptying. The first factor is the presence of heat stress itself during exercise, leading to an enhanced blood flow requirement to the skin for heat dissipation. This may be moderated somewhat by the enhanced total blood volume with both aerobic training and heat acclimatization. In contrast, hypohydration and the concomitant decrease of total blood volume magnify splanchnic ischemia during exercise–heat stress, and can reduce

HYPERHYDRATION

One common misconception in many aspects of nutrition and also exercise is that if a little of something is good, more must be better. In the case of hydration, the importance of promoting euhydration prior to exercise can lead to the notion that hyperhydration, or the attainment of a higher than normal body fluid balance, prior to exercise can prolong tolerance to exercise–heat stress. This may be achieved by drinking large amounts of water or other fluids in the days and hours prior to a competition. However, besides increasing the need to void during competition, this high level of water consumption can serve to dilute plasma sodium and enhance the risk of dilutional hyponatremia. Another method of achieving hyperhydration is through the ingestion of a glycerol solution, which aids in the retention of fluid in the intravascular spaces (Nelson and Robergs 2007). However, the total amount of hyperhydration possible is relatively small at about 1% to 2% above normal euhydrated body weight. This state is inherently unstable, as the body will rapidly strive to increase fluid excretion to return to euhydration, such that proper timing would be essential if this practice were to be employed. Furthermore, no scientific study has been able to demonstrate a clear thermoregulatory or ergogenic advantage from this practice in either a compensable or uncompensable heat stress environment (Latzka et al. 1997, 1998). Therefore, the best suggestion remains to focus on ensuring that athletes and workers are normally hydrated prior to exercise in the heat rather than seeking to use hyperhydration as an ergogenic aid.

the rate of gastric emptying and absorption. These factors again point toward the importance of ensuring euhydration prior to the start of exercise in hot environments rather than relying on full rehydration during exercise itself. Therefore, an optimal hydration plan should involve drinking prior to exercise to ensure both hydration and an adequate gastric volume to promote emptying and absorption, then maintaining the gastric volume and body fluid balance by drinking continuously throughout exercise.

Hydration Strategies Following Exercise

Many athletic and occupational situations require consecutive bouts of exercise with only a short recovery period. This can range from bouts of firefighting separated by short rest breaks of minutes to an hour to three-week events like the Tour de France with 18 h of recovery between race stages in summer weather. With the known efficacy of euhydration on exercise–heat tolerance, it therefore becomes critical to plan hydration strategies not just for the exercise itself, but also for the recovery period. Compared to work on hydration during exercise, though, relatively minimal research has been performed on recovery and hydration.

An excellent series of studies was performed by a research group in Aberdeen, Scotland, during the mid-1990s to isolate the effects of different drink characteristics on fluid balance recovery. A summary of the primary findings from these studies is presented in table 4.4. The general context of this series was the return of body weight to preexercise weight. The consensus from these studies suggests that adequate recovery rehydration requires the consumption of much more fluid than that lost through sweating (~150% of sweat loss) in order to compensate for fluid excretion through urination. Such work is especially relevant to weight-class athletes looking to recover following dehydration for a weigh-in. However, the work that is currently missing is research on the effect of different recovery rehydration fluids and protocols on actual aerobic and anaerobic exercise capacity. Such work, for example, would assist in planning for how much rehydration recovery is possible or required between weigh-in and competition. It could also aid in planning for when aggressive rehydration strategies, such as intravenous fluid infusion, are appropriate.

Table 4.4 Summary of a Series of Studies on Rehydration Following Exercise in the Heat to ~2% Body Weight Dehydration

Composition	Study	Conditions	Main findings	Best practice conclusions
Volume	Shirreffs et al. (1996)	50%, 100%, 150%, 200% sweat loss replacement of 23 or 61 mmol · L^{-1} [Na$^+$] solution	• ↑ urinary output with low [Na$^+$] fluid • ↑ urinary output with greater volumes • Net negative fluid balance with all low [Na$^+$] volumes • Net fluid recovery for high [Na$^+$] fluid with 150% and 200% volumes	• [Na$^+$] and volume interact in rehydration. • Rehydration volume must exceed sweat loss, placing great emphasis on palatability.

(continued)

Table 4.4 *(continued)*

Composition	Study	Conditions	Main findings	Best practice conclusions
Sodium [Na⁺]	Shirreffs and Maughan (1998)	0, 25, 50, 100 mmol · L⁻¹ at 150% of sweat loss	• Urine production inverse to [Na⁺] • Net even sodium balance with 50 mmol · L⁻¹ trial • Even volume balance with 100 mmol · L⁻¹ trial, but large urinary potassium loss	Na⁺ is essential for rehydration, but balance is needed in determining [Na⁺] and may be case specific based on individual sweat composition.
Potassium [K⁺]	Maughan et al. (1994)	Separate and combined effects of 60 mmol · L⁻¹ [Na⁺] and 25 mmol · L⁻¹ [K⁺] at 100% sweat loss	• NS in rate of PV recovery or urine output • ↑ net negative [Na⁺] balance with only K⁺ drink • ↑ net negative [K⁺] with only Na⁺ drink	May be useful in low concentrations to offset urinary K⁺ loss from Na⁺ ingestion.
Carbohydrate	Maughan et al. (1994)	90 mmol · L⁻¹ glucose alone versus combined with Na⁺ and K⁺	• ↑ urinary output and net negative fluid balance than with electrolyte solution • NS PV recovery • ↑ net negative [Na⁺] and [K⁺] balance	Carbohydrate alone is not sufficient for optimal rehydration. No additional benefit above that with electrolyte solution for fluid balance, but likely critical for substrate recovery.
Solid versus liquid	Maughan et al. (1996)	• Sports bar + flavored water • Equal total volume of sports drink	• ↓ urine output with sports bar, likely due to slightly greater electrolyte content • Similar plasma volume response	Food ingestion does not impair rehydration and may increase palatability and voluntary fluid consumption.
Alcohol	Shirreffs and Maughan (1997)	0%, 1%, 2%, 4% alcohol at 150% of sweat loss	• NS total urine volume • Peak urine flow rate later with 4% alcohol • ↓ PV and BV recovery with 4% alcohol	Alcohol <2% does not serve as a significant diuretic or impair rehydration.

Rehydration occurred over a 6 h recovery window of passive rest. NS = non-significant, PV = plasma volume, BV = blood volume.

SUMMARY

During exercise in both normothermic and hot environments, one of the primary countermeasures to both ensure safety and enhance exercise capacity and tolerance is the maintenance of adequate hydration. Maintenance of euhydration during the course of exercise or heat stress itself is rarely possible, due to both a behavioral tendency toward voluntary dehydration and also the physiological time lag required for ingested fluids to enter the body compartments. It is difficult to recommend a general guideline for either the appropriate amount of fluid replacement or its composition because of the wide range of mediating factors. This includes individual factors such as personal sweating rates and also situational ones such as practical constraints and exercise duration. The attainment of euhydration prior to exercise–heat stress appears to be a critical intervention, as exercising in a hypohydrated state seems to negate any advantages derived from heat acclimatization. Therefore, education about monitoring thirst and body fluid balance, along with the dangers of both dehydration and overconsumption of fluids, becomes an important issue in promoting health and safety among the general population and in athletic and occupational situations.

Cold Air Exposure

Polar explorers such as Peary, Scott, Amundsen, and Shackleton, along with mountaineers like Mallory, Hillary and Messner, have attained near-mythical status in Western culture; their exploits demonstrate the fascination that survival in the cold has in our collective consciousness. The cold environment places multiple stressors on our physiological systems, affecting work tolerance and also raising the risk of cold injuries. Each winter, news reports arise of individuals stranded in a car for hours or days, or hikers lost in the wilderness in winter conditions, who suffer from hypothermia and exhaustion. Indeed, the main life-threatening cold injury is a dangerous loss of body heat and whole-body hypothermia. However, in many occupational and survival situations, another critical factor determining optimal or safe work performance is the maintenance of manual function. As cold can rapidly cool the hands, the impaired function can greatly increase risks to the individual and precipitate dangerous situations. The feet are also highly susceptible to cold injuries because of their exposure to wet terrain and often constriction in footwear.

While cold injuries tend to be a less severe problem for civilian populations compared to conditions caused by heat waves and heat illnesses, the same cannot be said of specific populations such as the military. Cold environments and the loss of troops have plagued armies since the days of Alexander the Great. Perhaps the most famous examples of cold weather halting an advancing army and altering the course of history are Napoleon's and Hitler's futile attempts to invade Russia and the Soviet Union. The underprepared armies were literally frozen in their tracks by the onset of winter, and it can be argued that the winter climate did more to affect these conflicts than any military strategy. It has been estimated that nearly 10% of all U.S. military casualties in Europe during World War II and in the Korean conflict were due to cold injuries, with such losses in personnel bringing about a massive impact on military capacity and resources (Whayne and DeBuakey 1958). Therefore, it should come as no surprise that much of the research on hypothermia and cold injury, along with modeling of cold exposure guidelines, has been funded and developed with this military context in mind.

The purpose of this chapter is to focus on the effects of cold air environments on physical work capacity. We will explore the effects of cold temperatures on muscle function and especially on manual function. Chapter 6 on cold water immersion focuses more specifically on the effects of whole-body hypothermia and rewarming, and further discussion of the efficacy and limitations of precooling prior to exercise in the heat is presented in chapter 3.

MANUAL FUNCTION IN THE COLD

Whole-body hypothermia may be a significant problem during exposure or work in the cold, but a contributing factor or precursor to dangerous situations can often be the impairment or loss of manual function in cold environments. Most work or survival situations in cold environments are dependent on the proper functioning of the hands to manipulate objects. Consider the consequences for astronauts who need to work closely with their hands over the course of an 8 h spacewalk while encased in a spacesuit and subjected to extreme swings in external temperature, or mountaineers who lose a glove near the summit of Everest and become so cold that they cannot zip up their jacket. Therefore, research into the effects of cold exposure on human performance has centered on monitoring manual function during work in the cold, as well as on designing countermeasures to maintain hand temperature and capacity.

The fingers and toes are poorly designed for retaining heat, both in their anatomical arrangement and in their pattern of circulatory control (Burton and Edholm 1955). Long and narrow digits result in a relatively high ratio of surface area to volume for convective heat loss, with the hands composing approximately 9% of the total body surface area. At the same time, the low muscle mass and amount of fat provide minimal heat-generating capacity and insulation against heat loss, further increasing the potential for heat loss. The hands and feet also have a strong ability to regulate blood flow. Upon initial exposure to cold, vasoconstriction can very rapidly decrease distribution of blood flow to the extremities and divert blood and heat to the core and vital organs such as the heart and brain. Thus both nonfreezing cold injuries such as chilblains or trenchfoot, and freezing cold injuries like frostnip and frostbite, are predominant in the extremities. It should not be surprising, therefore, that a heavy research and design focus has been placed on the circulatory, thermal, muscular, and neural functioning of the hands during work in the cold.

Due to their high degree of vascularization, along with the large magnitude and rapid response time of changes in vasomotor tone, the hands and fingers have also been proposed as a site for monitoring both local and whole-body thermal status. This may prove especially useful in applications such as the development of thermal control garments that can regulate temperature separately over multiple zones of the body, as mean skin temperature calculations can become largely meaningless with such clothing.

Cold-Induced Vasodilation

Shortly upon exposure to cold environments, skin temperature of the extremities tends to decrease rapidly and exponentially to a level approaching that of the ambient environment. This vasoconstrictory response is a logical reaction within the peripheral tissues in order to maintain heat and temperature of the vital organs in the core. However, a common observation is that, after a brief period of lowered skin temperature, a seemingly paradoxical temporary warming occurs. The skin temperature can rise by as much as 10 °C during these rewarming bursts, and the fall and rise can occur repeatedly in a cyclic fashion. This pattern of periodic rewarming, generally measured as increased temperature or capillary blood flow, was first reported by Sir Thomas Lewis in 1930, who labeled it the "hunting response" for its apparent up and down pattern of seeking a stable temperature (Lewis 1930). This response has also been termed the "cold-induced vasodilation" or CIVD phenomenon; a typical response along with the main measurement parameters is presented in figure 5.1. Since Lewis's initial report of CIVD in the fingers, the response has been observed in various regions of the body, including the face, forearms, and feet (Fox and Wyatt 1962). Cold-induced vasodilation has generally been presumed to perform a cryoprotective function to maintain tissue integrity and minimize the risk of cold injuries. At the same time, the increased blood flow and temperature may help to maintain manual function in the cold, leading researchers to explore how CIVD may be stimulated or enhanced by both physiological and pharmacological means.

Mechanisms of CIVD

Despite more than 75 years of research, the actual mechanisms behind the CIVD phenomenon remain open to investigation, and there is little consensus. Anatomically, arteriovenous anastomoses (AVA) are thought to be the primary mechanical regulators of CIVD. Arteriovenous anastomoses are small blood vessels with thick muscular walls connecting the deeper tissues and blood vessels with the subcutaneous capillaries. Controlled by the sympathetic nervous system, AVAs can rapidly constrict, shunting blood away from the skin and thereby decreasing skin temperature and ultimately heat loss from the peripheral tissues. As discussed in chapter 3, the opening of large numbers of AVAs in the extremities may also be the mechanism behind the efficacy of hand and forearm immersion for decreasing body temperature

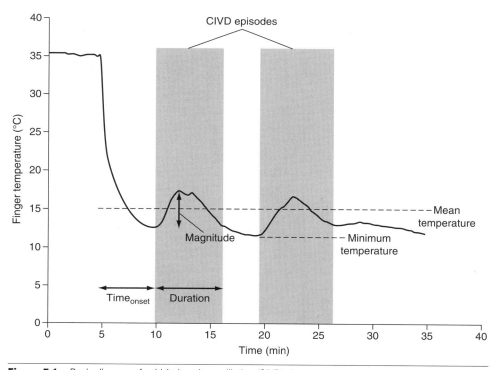

Figure 5.1   Basic diagram of cold-induced vasodilation (CIVD) showing different measurement components.

in heat-stressed individuals. A high concentration of AVAs in CIVD-capable regions would support this mechanism of action. However, even though a few attempts in the 1930s through the 1960s to quantify AVAs in the fingertips and several other regions turned up promising data supporting a high prevalence of AVAs in the extremities, it has been argued that these studies were erroneous or that they overcounted AVAs. The difficulty in isolating and quantifying these structures leads to questions about whether they are fixed or adaptable structures, and a systematic confirmation or refutation of their presence and function would appear to await improvements in imaging technology. For example, if AVAs are present and important to CIVD, do they increase in density or sensitivity to stimuli with prolonged cold acclimatization?

Apart from anatomical control, a variety of mechanisms have been proposed to be responsible for CIVD, as recently summarized by Daanen (2003). As with many issues regarding this phenomenon, the mechanism and subsequent potential manipulation remain unresolved and open to further systematic investigations:

• **Axon reflex.** The original mechanism proposed by Lewis was that the hunting response occurs as a result of an axon reflex, with noxious stimuli or electric stimulation causing a reflexive vasodilation. However, a series of more recent studies by Hein Daanen in the Netherlands, some of which I was a subject for, argues against this mechanism. With the body at normal temperatures, electrical stimulation of a hand immersed in warm water elicited vasodilation, but this could not be accomplished with stimulation of the other hand immersed in cold water (Daanen and Ducharme 2000).

• **Vasodilatory substances in the blood.** The identification of powerful local vasoactive substances, such as nitric oxide, which seems to act specifically on cutaneous blood vessels in response to cold stimulation in animals, has led to the suggestion that nitric oxide may contribute to CIVD. However, confirmation of the linkage between nitric oxide release and CIVD remains missing. In our laboratory, two weeks of repeated local cold exposure of the hand in 8 °C water did not alter either nitric oxide or endothelial-1 (local factors) or catecholamine (systemic factor) concentrations (Geurts et al. 2006). However, as the thermal responses of the fingers also did not change, firm conclusions are not possible.

• **Control via norepinephrine.** Activity of the sympathetic nervous system (SNS) has a vasoconstrictory effect on cutaneous blood vessels. In laboratory studies, the controlled increase of norepinephrine concentrations could greatly reduce or eliminate CIVD. This is further supported by a decrease in the prevalence and intensity of CIVD with stressors known to increase SNS activity, such as exercise, whole-body hypothermia, and emotional stress. However, while this model may help to explain the modulation of CIVD and possibly changes with adaptation, it does not actually provide a mechanism for CIVD itself, except potentially via periodic changes in SNS activity or local sensitivity to norepinephrine.

• **Direct effect of cold on cutaneous vasculature.** According to the model of CIVD control that proposes the most direct effect, continuous and prolonged cold causing sustained contraction of the AVAs of cutaneous blood vessels leads to eventual fatigue of the smooth muscle, resulting in a periodic release of vasoconstrictory tone that we observe as an increase in skin blood flow and CIVD. However, as this does not explain the sustained vasoconstriction and lack of CIVD observed with norepinephrine infusion or whole-body hypothermia, this mechanism remains an incomplete explanation for CIVD.

Trainability of CIVD

In support of CIVD as a protective response, humans living in or native to a cold environment seem to have enhanced CIVD, marked by shortened onset times and higher amplitudes, compared to people in tropical environments or nonadapted individuals. For example, Arctic natives, such as Inuit and Lapps, generally have higher mean finger temperatures and CIVD responses than control populations from more temperate regions (Krog et al. 1960). The same enhancement has been seen in individuals working in environments with repeated local cold exposures, such as fishers or fish filleters. One limitation of such studies, however, is that comparing two distinct groups provides little information on the time course or mechanism of any adaptations to cold, or indeed on the ability of CIVD to be trained and improved. A summary of key field and laboratory studies on the trainability of CIVD is presented in table 5.1. Longitudinal acclimatization studies, in which a subject group is exposed to cold and tested over a prolonged period, are logistically more difficult to perform and have to date presented equivocal results. Tropical inhabitants (soldiers from the plains of India) exhibited an improved peripheral blood flow and CIVD response after seven weeks of exposure to the Arctic environment (Purkayastha et al. 1992), but this remained below the level found in Arctic natives and suggests that full adaptation requires much longer exposure periods. Opposing data were obtained in a group of Canadian soldiers with greater vasoconstrictive tone following two weeks of cold exposure (Livingstone 1976), suggesting that the risk for cold injuries may actually increase with repeated cold exposures.

To increase methodological control over what is possible with field studies, another option is to perform laboratory acclimation studies. Yet again, however, laboratory studies present equivocal results. Four daily immersions of the index finger in ice water for a month elicited faster onset of CIVD and a decrease in pain in the index finger compared to nontrained digits (Adams and Smith 1962). However, the majority of studies, including a large range of work in our laboratory, have shown essentially no difference in thermal responses or CIVD improvements in either the fingers or toes with two to three weeks of daily 30 min whole-hand or whole-foot immersion in 8 °C water. For example, the incidence of CIVD in the foot actually tended to decrease over the course of three weeks of foot immersion in 8 °C water with no improvements in mean or minimum skin temperature (Reynolds et al. 2007). Therefore, it appears that systematic CIVD improvement may be possible only after a very rigorous acclimation protocol that may not be logistically practical prior to cold expeditions.

Prediction of CIVD

One of the core goals of thermal modeling, in addition to developing exposure guidelines, is attempting to understand who may be at risk from exposure to environmental stressors. Table 5.2 presents some of the factors known to influence the thermal response of the fingers to cold. As we have seen, cold natives seem to have a stronger CIVD response, and both genetic factors and long-term acclimatization may contribute to this increased protection. But is it possible to develop predictive models and simple screening tests for those who may be more or less susceptible to frostbite? One attempt at a screening tool is the Resistance Index of Frostbite (RIF) developed by Yoshimura and Iida (1950). The RIF is calculated based on minimum and mean finger temperature along with CIVD onset time in response to immersion of one finger in ice water. Each factor is scored 1 to 3 based on whether responses are

Table 5.1   Trainability of Cold-Induced Vasodilation (CIVD)

Study	Experimental design	Main findings
Field-based training (or acclimatization)		
Nelms and Soper (1962)	Investigation of CIVD responses of controls ($n = 9$) and naturally acclimatized fish filleters ($n = 11$) upon immersion of the left hand in 0 °C stirred ice water bath.	The fish filleters had significantly shorter CIVD T_{onset} and a greater magnitude of response than the control group.
Purkayastha et al. (1992)	Comparison of CIVD responses among four groups: controls from a tropical region ($n = 10$); subjects from a tropical region temporarily relocated to an Arctic training camp ($n = 10$); subjects from a temperate climate who had migrated to the Arctic ($n = 6$); and Arctic natives ($n = 6$). CIVD responses were tested during right-hand immersion in a 4 °C stirred water bath.	• The tropical subjects relocated to the Arctic environment had significantly improved CIVD responses (T_{fing} and T_{max}) by the 7th week of Arctic training, similar to those of the migrant and native groups. • While differences were nonsignificant, the responses of the native group tended to be superior to those of the acclimatized tropical and migrant groups.
Lab-based training		
Adams and Smith (1962)	Temperate climate-based subjects ($n = 5$) immersed their right index fingers in a 0 °C stirred water bath for 20 min, four times per day, for 1 month.	The exposed right index fingers exhibited significantly improved CIVD responses (including T_{onset} and rewarming rate) following the 1 month of training compared to pretraining responses and the responses of other digits not chronically exposed to the cold water.
Reynolds et al. (2007)	Subjects ($n = 10$) immersed their left feet in an 8 °C water bath for 30 min, 5 days per week, for 3 weeks.	• The presence of CIVD was uncommon with the protocol. • Minimum and mean toe temperatures, as well as CIVD magnitude, remained similar over the 15 training days. • The prevalence of CIVD decreased over the training period.
Mekjavic et al. (2008)	Subjects ($n = 9$) immersed their right hands in an 8 °C water bath 30 min per day for 13 days. CIVD responses were assessed in the trained hand as well as in the contralateral (left) hand on days 1 and 13 to test the transferability of training across limbs.	• CIVD occurred in 98.5% of the measurements. • Training significantly worsened the CIVD response (number of CIVD waves and T_{fing}) in the exposed hand. • In the contralateral hand, both number of CIVD waves and T_{fing} were also reduced following the training period.

T_{onset} = onset time; T_{fing} = mean finger skin temperature; T_{max} = maximum finger skin temperature.

Table 5.2   Known and Suspected Factors Affecting Cold-Induced Vasodilation (CIVD)

Factor	Effect	Notes
Body temperature	High T_{body} augments CIVD response: • Higher T_{fing} during CIVD • Shorter T_{onset} of CIVD	Seasonal variations in ambient temperature may also affect one's CIVD response. Studies examining CIVD responses throughout the year have demonstrated that the CIVD response is more pronounced during the summer months when the core temperature may experience relative elevation.
Acclimatization to cold	Acclimatization to cold augments CIVD response: • Higher T_{fing} during CIVD • Higher T_{min} during CIVD • Shorter T_{onset} of CIVD	Such effects are also observed in populations habituated to cold environments (i.e., Norwegian Lapps). However, Purkayastha et al. (1993) found that acclimatized tropical residents developed enhanced CIVD responses indistinguishable from the response of Arctic residents, suggesting acclimatization over adaptation.
Altitude	High altitude blunts CIVD response: • Lower T_{fing} during CIVD • Reduced magnitude of CIVD (i.e., lower T_{max} during CIVD)	High-altitude effects are likely related to coexistence of cold environment with systemic hypoxia. During extended stays at altitude, elements of the CIVD response (i.e., T_{max} during CIVD) can return to sea level values (Daanen and van Ruiten 2000).

Diet	Dietary compounds may augment CIVD response:
	High protein (150-200 g per day) and salt intake (>45 g per day)
	  Higher T_{fing} during CIVD
	Vitamin C (2 g per day for 1 month)
	  Shorter T_{onset} of CIVD
		A high-protein, high-salt diet may result in diet-induced thermogenesis, causing an increase in core temperature. This increased temperature leads to an improved CIVD response. Vitamin C supplementation may augment the CIVD response via its antioxidant, metabolic, or thermogenic properties (or some combination of these); collagen synthesis; antistress activity and restoration of intercellular substances; or better maintenance of the rheological status of the blood.

Tobacco smoking	Abstinence from tobacco smoking (in habitual tobacco users) may augment CIVD response: • Higher T_{fing} during CIVD • Shorter T_{onset} of CIVD	The reason for this response is unknown. Tobacco smoking temporarily produces vasoconstriction in the periphery; however, frequent changes between peripheral vasodilation and vasoconstriction in abstinent smokers may lead to desensitization of blood vessels to local vasoactive stimuli, resulting in an improved CIVD response upon cold exposure.

Injury and pathology	*Raynaud's disease* may blunt CIVD response:
	  Longer T_{onset} of CIVD
	Previous cold injury may blunt CIVD response:
	  Little or no CIVD present
		While subjects with Raynaud's disease have exhibited an impaired CIVD response during hand immersion in 10 and 15 °C water, the differences from control subjects were insignificant with 5 °C water immersion (Jobe et al. 1985). For those with Raynaud's disease, training may improve their responses to cold exposure: Finger temperatures of patients with Raynaud's disease increased by 2.2 °C during cold exposure with training (10 min hand immersion in hot water [43°C] with the body exposed to cold [0 °C] air, repeated three times a day, 3 days a week for 3 weeks) compared to those in patients without training.

T_{body} = body temperature; T_{fing} = mean finger skin temperature; T_{min} = minimal finger skin temperature; T_{max} = maximum finger skin temperature; T_{onset} = onset time.

worse (1), average (2), or better (3) compared to population norms, for a total scoring range from 3 (poor resistance) to 9 (excellent resistance).

Daanen and van der Struijs (2005) attempted to validate the resistance index to frostbite (RIF) in 206 Dutch soldiers prior to Arctic deployment and found that RIF was higher in Caucasians than in non-Caucasians, and also that RIF was inversely related to pain sensations. From this subject pool, 11 eventually suffered cold injuries, and the RIF for this subgroup was significantly lower than for the remaining subjects. This suggests that a simple screening test such as the RIF can at least provide a broad prediction of cold injury susceptibility.

Our attempts to investigate CIVD prediction for cold injuries using a case study approach yielded some uncertainty. In one unpublished study, we were fortunate to be able to recruit a cohort of elite mountaineers from Slovenia. One group had a history of frostbite and amputation of fingers or toes, while the other group had no history of cold injuries. We tested the thermal responses of both the feet and hands to cold immersion. Both mountaineering groups demonstrated a stronger CIVD response in the feet and hands than a group of control subjects with no mountaineering experience. However, there were no differences in RIF, thermal responses, or CIVD prevalence between the two mountaineering groups. Based on these results and the case study reports of the actual injuries, the cold injuries appeared to be due more to "accidents" than to any systematic physiological differences.

Torso Versus Hand Heating

With the fingers seemingly disadvantaged by their architecture, protective clothing designers are interested in developing countermeasures to maintain function during cold exposure. While heating is an obvious pathway, one question involves the most efficient location for providing heat, especially if the provision of power is a major limitation. Should heating be directly provided to the fingers, hands, or forearms? Or would providing the same amount of heat to the torso or core be capable of maintaining both core temperature and local blood flow to the hands?

Research within our lab and Defence Research and Development Canada demonstrates that the maintenance of whole-body heat balance appears to be a critical countermeasure in any attempt to preserve finger blood flow and manual function in the cold. In our laboratory, we demonstrated this direct connection by passively exposing individuals to –20 °C for 2 h while they wore protective clothing and mitts, and constantly monitoring finger temperature and periodically testing manual function (Flouris et al. 2006). In one condition, we prewarmed body core temperature by 0.5 °C prior to exposure, while another condition involved no prewarming but added heat to maintain body temperature at thermoneutral values throughout exposure. Both conditions provided higher finger temperatures and manual function than with no heating. This supports the ideas proposed by Brajkovic and Ducharme (2003), who compared direct finger heating to indirect heating via the torso during passive cold exposure to –25 °C. While finger blood flow was significantly lower with direct hand heating, finger temperatures were similar with the two manipulations, and in neither condition did manual dexterity become significantly impaired. Therefore, the authors concluded that indirect heating of the torso is just as effective as direct heating for maintaining manual function in the cold. In addition, torso heating has a number of other potential benefits, including the maintenance of core temperature, a larger surface area and volume for heating, and no loss of manual dexterity from the bulkier gloves required for the heating itself.

COLD INJURIES TO THE EXTREMITIES

If local blood flow and thermal responses such as CIVD are inadequate in preserving tissue integrity, the extremities can incur risk of injury and damage. Local cold injuries can arise from cold–dry or cold–wet environments and can entail both freezing (e.g., frostbite) and nonfreezing (e.g., trenchfoot) conditions (Hamlet 1988). Therefore, temperature by itself is not the sole determining factor for likelihood of injury. Other environmental factors can potentiate the risk of cold injuries, including wind speed and precipitation. In addition, individual characteristics and circumstances, such as race and level of acclimatization, can affect the risk of injury. It remains difficult to predict individual susceptibility, making awareness of predisposing factors and planning for injury prevention critical—because, when they happen, cold injuries are debilitating to the individual and also can deplete mission resources needed to assist the injured party.

Cold–Wet Injuries

The dominant sites for cold–wet injuries are the feet and legs, as these injuries are typically caused by prolonged immersion of the extremities in cold water. Freezing of the extremity does not cause the damage; rather, likely causes are the edema or high rates of sweating when the feet are wet or are in very humid environments (Hamlet 1988). The most problematic condition is popularly known as trenchfoot (see figure 5.2), the term stemming from initial mass diagnoses during World War I, in which troops were forced to live and fight in muddy terrain and flooded trenches. Immersion foot, a similar condition, has also been reported in shipwreck survivors even in fairly warm waters. Therefore, the prime predictors for the onset of cold–wet injuries are the water temperature and duration of exposure; the following list of injuries represents a continuum from initial to most severe (Hamlet 1988).

1. **Chilbain.** The initial manifestation of cold–wet injury is the presence of lesions on the dorsal surfaces of the hands and feet. This superficial injury represents damage to superficial blood vessels, resulting in local edema and inflammation. Features of this condition include redness, swelling, itching, and soreness. In worsening cases, chilbains can progress to more severe blisters or ulcers, and it may be months to years before the symptoms subside.

2. **Pernio.** The continuation from chilbain emerges with further ulceration and the initiation of skin necrosis, again primarily based in the dorsal surface of the hands and feet.

3. **Trenchfoot.** The culminating cold–wet injury, trenchfoot or immersion foot occurs with severe damage to the local vasculature and likely the nerves. An initial prehyperemic phase leaves the extremity numb, swollen, and discolored. Afterward, a hyperemic phase can lead to severe ulceration, pain, and the risk of infection from gangrene. Even upon eventual recovery, edema, loss of sensation, and severe reactions to cold may remain for the rest of the victim's life.

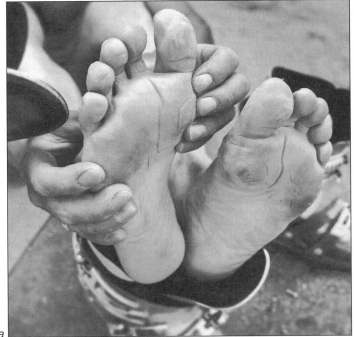

a

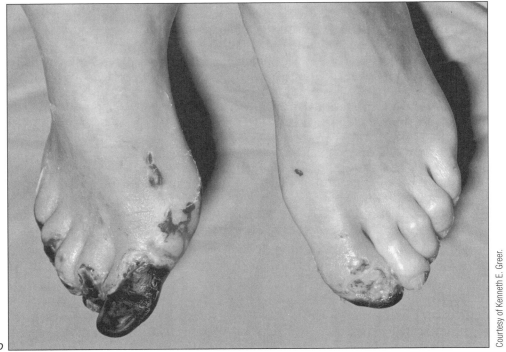

b

Figure 5.2 Severe damage can occur to the peripheries via (*a*) chronic cold–wet (e.g., trenchfoot) and (*b*) cold–dry (e.g., frostbite) exposure.

Cold–Dry Injuries

Unlike cold–wet injuries, cold–dry injuries involve the actual destruction of cells from freezing and crystallization. The rate of local heat loss and cellular damage from convective heat loss are the prime determinants of frostnip and frostbite, and the onset can be much more rapid than for cold–wet injuries (Hamlet 1988).

1. **Frostnip.** The initial freezing of the epidermis and superficial skin tissue is painful but typically does not produce long-term damage. However, depending on the depth of freezing, it is possible for the superficial capillaries and nerves to become damaged. If this is the case, subsequent risk of frostnip and frostbite in the same region may increase due to decreased sensation of cold and also decreased blood supply.

2. **Frostbite.** Continued cooling and freezing of cells can lead to their crystallization, with the damaged regions often becoming waxen and insensitive to touch. Major clinical problems arise with rewarming, with intense pain, inflammation, and the threat of gangrene. Due to the risk of infection, it is critical that rewarming and frostbite treatment occur in medical settings rather than being attempted in the field.

PHYSIOLOGICAL RESPONSES TO EXERCISE IN THE COLD

An important fundamental question in the modeling of thermal response is whether heavy or prolonged exercise negatively affects the body's ability to maintain thermal balance. Consider the grueling effort involved in mountaineering or transpolar missions, and also during intensive military training programs. In such situations, a multitude of nonthermal factors, including exercise-induced fatigue, depleted carbohydrate or lipid stores, altered nutrition, and sleep deprivation can influence the body's ability to sense cold, vasoconstrict, or elevate heat production. Mountaineers also face the additional factors of reduced ambient pressure and hypoxia, which may additively or synergistically impair thermoregulation. Furthermore, these individual factors may impair cognitive capacity and lead to faulty decision making or an increased risk of accidents. For these reasons, individuals performing prolonged exercise during winter or in situations in which cold conditions can suddenly arise need to be aware of these contributing factors and take steps to minimize their risks.

Field Studies of Expeditions in Cold Environments

As is true in much of environmental physiology, scientists can design studies that track participants in "live" field settings, or more closely isolate individual variables and minimize confounding factors in a laboratory setting. One advantage of the former approach, of course, is that it typically involves exotic travel opportunities for the scientists! However, a common limitation in studying expeditions such as polar treks is the typically small sample size of only one or two subjects, making statistical analysis or generalization difficult. Two studies that tracked larger expeditions highlight the intense effect of prolonged exertion on thermoregulatory control in the cold. Savourey and colleagues (1992) were fortunate to be able to systematically test eight (five male, three female) volunteers prior to and following a three-week ski expedition across Greenland, where ambient conditions ranged from –20 to –30 °C. In laboratory tests, metabolic rate increased in thermoneutral environments following the journey. However, with cold exposure, the absolute rectal temperature for shivering onset decreased by approximately 0.5 °C (e.g., shivering was initiated at 36.0 rather than 36.5 °C). Such a hypothermic insulative adaptation, whereby individuals appear able to tolerate lower core temperatures prior to initiating heat production, may serve to minimize metabolic heat production and hence overall caloric requirements. A similar adaptation has been observed in specific groups native to cold environments, such as Lapps in Scandinavia and Aborigines wearing minimal clothing in the cold night in the Australian desert (Scholander et al. 1958).

Another example of the impact of prolonged exercise on cold tolerance was seen in U.S. Army Ranger recruits (Young et al. 1998). Eight candidates were tracked prior to and following 61 days of strenuous field training, over which they had an average daily caloric deficit of 800 kcal and slept around 4 h per day. Upon testing the recruits' cold tolerance with 4 h of passive rest in 10 °C air immediately following the field training, and also 2 and 109 days afterward, the researchers found that wild swings in body weight occurred over the first two days of recovery, with the entire 6.4 kg (14 lb) body mass deficit restored but body fatness remaining low until longer in recovery. Rectal temperature during cold exposure was lower immediately and two days after training, demonstrating that chronic exertion can impair cold tolerance. This appeared again to be due to a depression in the core temperature threshold for shivering onset. At the same time, metabolic rate was lower immediately following training and at 109 days compared with 2 days afterward. Overall, this suggests that a brief period of recovery and heavy feeding can restore heat production capacity, but that tissue insulation and thermoregulatory control may not be restored until much later in recovery.

Laboratory Studies of Prolonged Cold Exposure

Another issue when one is assessing results from expedition studies is the unique nature of the subjects themselves, both physically and psychologically, which makes it potentially difficult to extrapolate the results to a more general population. For example, the Ranger training is extremely arduous, with <50% of already highly fit and self-selected candidates "surviving" the field training. The multiple stressors inherent in field studies, commonly involving sleep and nutritional deficits, also make it difficult to isolate the underlying mechanisms. Therefore, another research approach toward understanding the physiological effects of prolonged exercise in the cold has been laboratory studies isolating exercise, cold stress, or both as the primary variables. However, while this allows the maintenance of better experimental control, it is generally impossible to replicate the extreme or prolonged cold exposures of field studies.

Current laboratory-based studies provide equivocal evidence on the effects of prolonged cold exposure on the ability to maintain thermal balance. Studies investigating multiple cold immersions over the course of a single day demonstrate progressively lower metabolic rates, suggesting an impaired heat production capacity (Castellani et al. 1998). However, it is uncertain whether this is due to factors such as circadian shifts in hormonal balance or core temperature, or actually due to exercise-related fatigue from the prolonged shivering.

In one of the more unpleasant studies that I participated in as a graduate student, the effects of short-term intense fatigue on cold tolerance were tested in a group of young healthy males. Fatigue was induced by 5 h of mixed aerobic and strength exercise with only water provided. Upon subsequent exposure to a cold (10 °C) room and a continuous cold (18 °C water) shower and wind, the mean response of the 16 subjects did not suggest an effect of acute fatigue on cold tolerance compared to that in a control condition in which subjects had not exercised first (Tikuisis et al. 1999). As a caveat, a graph from that study illustrating the individual responses during the control (no exercise) condition (see figure 5.3) shows that some subjects were able to tolerate the entire 4.5 h cooling protocol with only minor decreases in core temperature, while others (including myself) experienced rapid decreases in core temperatures and termination times as low as <90 min. This wide range of interindividual variability in response to environmental stress, common to many studies in environmental physiology but often masked behind group means, makes predicting individual tolerance or setting appropriate exposure limits extremely tricky. For example, while body fatness and shivering intensity were positively correlated to tolerance times and may be used for modeling or predicting response, it is difficult to correlate or predict an individual's shivering capacity based on any easily measurable anthropometric characteristic (Tikuisis et al. 1999).

Shivering Thermogenesis

Shivering, the asynchronous contraction of skeletal muscle resulting in muscle metabolic energy conversion to heat while minimizing mechanical movement, is a highly effective method of endogenous heat production. One of the primary drivers

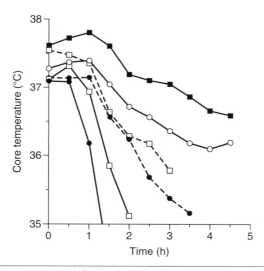

Figure 5.3   Rectal temperature plotted for each individual subject over the course of passive exposure to cold rain and wind. Note the wide range of interindividual variability in thermal response, making for difficulty in predicting individual responses to any environmental stress.

Adapted from P. Tikuisis et al., 1999, "Physiological responses of exercised-fatigued individuals exposed to wet-cold conditions," *Journal of Applied Physiology* 86: 1319-1328. Used with permission.

for shivering is skin temperature, explaining why we typically begin shivering very rapidly upon entry into a cold room. With prolonged cold exposure, shivering recruits progressively greater amounts of musculature, typically beginning in the jaw and trapezius muscles, then proceeding over the trunk muscles and ultimately to the large muscle mass of the limbs (Tikuisis et al. 1991). Different muscles shiver at different intensities, with the central muscles of the trunk such as the pectoralis major, and also the large muscles such as the rectus femoris, contracting at 5% to 16% of their maximal voluntary contraction (MVC) levels. In contrast, smaller and more peripheral muscles, such as the gastrocnemius and brachioradialis, may shiver at only 1% to 4% of MVC (Bell et al. 1992). Bell and colleagues theorize that this pattern and intensity of shivering across the body may reflect a defense strategy to maximize central heat production (recruiting large muscle) while also minimizing heat loss (preferentially recruiting trunk muscles near the vital organs).

At its most intense, shivering can be a highly effective means of elevating endogenous heat production from a basal level of approximately 100 W up to in excess of 500 W. As it is highly difficult to quantify the rate of whole-body muscle contractions in a lab or field setting during exercise, shivering intensity and heat production are

CAN SHIVERING FATIGUE OCCUR?

Given the extreme survival situations and stories of death from hypothermia that we read about in the media, it may appear obvious that shivering will most likely eventually become depressed or will cease altogether. However, understanding when this shivering fatigue may occur becomes very important when one is developing hypothermia survival models because the cessation of shivering, and its attendant high heat production, often precipitates a rapid and potentially fatal decrease in core temperature. This question is typically not addressed in most hypothermia studies because of the design goal of determining time for individuals to reach a state of hypothermia. Therefore, the severe cold stress and high rates of heat loss generally "overwhelm" heat production, and core temperature drops to an ethically mandated termination point based largely on the rate of heat debt.

Tikuisis and colleagues (2002) developed one of the more unpleasant protocols I can imagine, adjusting water temperature throughout immersion to maintain a core temperature of about 35 to 36 °C and therefore prolonged intense shivering. Over the immersion period, ranging from 105 to 388 min, the absolute rate of shivering did not decrease and was maintained at 60% to 70% of maximal shivering intensity. Indeed, these results exceeded previous mathematical models of shivering endurance. However, when normalized for the core temperature decrease observed in the later stages of immersion, the intensity of the central drive to shiver decreased gradually by nearly 20% per hour. The underlying mechanism behind this attenuated drive remains unknown; it may involve glycogen depletion, diminishing cold sensitivity, or conversely an acute habituation to cold stress. Regardless of mechanism, the observation of an impaired shivering drive is important because it demonstrates that the body is clearly unable to respond appropriately to a prolonged cold stimulus, and the time at which shivering drive begins decreasing may be modeled as a "failure" point for thermal balance in survival models.

typically calculated indirectly through calculation of oxygen uptake (see chapter 2 on the heat balance equation), which is in excess of $1 \text{ L} \cdot \text{min}^{-1}$ of oxygen with moderate to intense shivering (Toner and McArdle 1988). Considering that a lean, healthy adult male may have a maximal oxygen uptake of 3 to $4 \text{ L} \cdot \text{min}^{-1}$, it becomes obvious that the simple cost of maintaining thermal balance in the cold represents a major additional energetic and physiological load on the body.

Fuel Utilization During Cold Exposure

Fatigue is a multidimensional concept involving metabolic, muscular, and possibly neural components. During prolonged exercise, one of the dominant factors eliciting fatigue is the depletion of carbohydrate stores, through the increased reliance on carbohydrate metabolism with increasing exercise intensity. In a thermoneutral environment, this can lead to the feeling of "bonking" and inability to sustain high or even moderate exercise workloads. In a cold environment, the increased rate of substrate oxidation to fuel shivering may lead to a more rapid depletion of energy stores, presenting difficulties in keeping adequately fueled during exercise and logistical challenges involving the provision of supplies during extended expeditions. At the same time, from a countermeasures perspective, it would be interesting to investigate whether specific nutrients or compounds, such as caffeine or ephedrine, could increase cold tolerance. This might occur via increase in the rate of shivering or nonshivering thermogenesis or alteration in vasomotor tone to minimize heat loss. If such an effect is possible, then such nutrients or compounds could be targeted for use by individuals performing in cold environments or as emergency supplies for survival situations.

Prolonged exercise is marked by the eventual depletion of carbohydrate stores and inability to maintain blood glucose. In cold environments, this depletion may be accelerated by a greater reliance on carbohydrate oxidation to fuel shivering; estimates are that >50% of the total energy metabolism is derived from carbohydrate during passive cold exposure (Weber and Haman 2005). In contrast, the rate of lipid metabolism increases only marginally in the cold (see figure 5.4 for a general schematic) and appears unaffected by supplementation of additional lipids. While a manipulated increase in circulating nonesterified fatty acids promoted the rate of fat oxidation in temperate (20 °C) conditions, the same stimulation of fat oxidation was not found during exercise at 0 °C, suggesting a cold-induced uncoupling between lipid availability and metabolism (Layden et al. 2004a). A similar protocol also resulted in no difference in intramuscular triglyceride utilization in cold versus temperate conditions (Layden et al. 2004b). Beyond the increased rate of glucose metabolism, stress on the glucoregulatory system may synergistically interact with and further impair the thermoregulatory system. In another of the many experiments I participated in during my graduate studies, Passias and colleagues (1996) used an insulin-clamp technique to manipulate blood glucose concentration and tested the effects of 5 mM (euglycemia) and 2.8 mM (hypoglycemia) on the shivering, sweating, and vasomotor thresholds in healthy males. Cessation of sweating and vasodilation did not alter with hypoglycemia, but the core temperature threshold for the initiation of shivering was significantly decreased by approximately 0.5 °C.

This combination of findings suggests that the accelerated hypothermia with hypoglycemia appears to be driven primarily by a depression in heat production rather than an elevation in the rate of heat loss. Overall, current consensus would suggest that enhanced carbohydrate intake may be required for prolonged exposure or exercise in the cold, both due to a possible shift in substrate utilization and also in order to sustain carbohydrate availability for shivering (Haman 2006).

Nonshivering Thermogenesis

As an alternative to the previously discussed hypothermic insulative adaptation that has been noted in some cold-native populations or following cold expeditions, an elevation in metabolic thermogenesis and basal metabolic rate is another potential symptom of cold acclimatization. Such an increase in nonshivering thermogenesis has been reported in reptilian and mammalian species (Watanabe et al. 2008). In rodent species, a significant proportion of the increase arises not from shivering in the muscles, but rather from an elevated metabolism in the adipose tissue. While typical adipocytes do not have a high rate of metabolism, there appears to be specific brown adipose tissue (BAT) that has a high concentration of uncoupling proteins (UCP) in the inner mitochondrial membrane. These UCP function as a bypass site for hydrogen protons that have been pumped into the lumen as part of aerobic metabolism and the

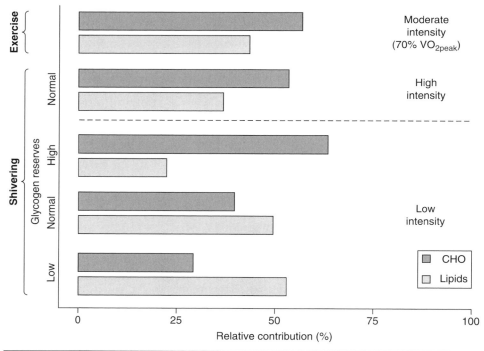

Figure 5.4   Substrate utilization during low- and high-intensity shivering in the cold and during exercise in the cold (0 °C) at moderate intensity (70% of peak oxygen uptake). The horizontal axis represents the relative contribution (%) of carbohydrates (CHO) and lipids to total metabolic heat production (shivering) and energy expenditure (exercise). During shivering, substrate utilization is clearly altered depending upon an individual's glycogen reserves (data from Weber and Haman 2005; as protein utilization was reported in addition to CHO and lipids, the shivering values do not equal 100%). During moderate-intensity exercise in the cold, lipid oxidation is reduced compared to that during exercise at the same intensity in warmer ambient conditions (data from Layden et al. 2004b).

electron transport chain (see figure 5.5). Therefore, aerobic metabolism and thermogenesis can be maintained without the actual production of adenosine triphosphate (ATP), serving as a source of additional heat (Dulloo and Samec 2001).

While there is some evidence of BAT in human neonates, it appears to degenerate during infancy, and the presence of UCP and BAT in adolescent and adult humans is debatable or minimal (van Marken Lichtenbelt and Daanen 2003). However, elevated nonshivering thermogenesis can still occur by similar mechanisms at multiple sites throughout muscle and adipose tissue. For example, the calcium uptake pumps within the sarcoplasmic reticulum are critical to the rapid regulation of calcium within the cytosol in muscles to control actin–myosin binding, and their activity can place significant demands on ATP stores during both rest and exercise (van Marken Lichtenbelt and Daanen 2003). If a cold stimulus decreases the efficiency of these calcium pumps, this may lead to increased metabolism and a potential mechanism in cold adaptation. Similarly, it may be possible for a drug to selectively target such sites to elevate heat production during cold stress. An interesting linkage between thermal physiology and clinical medicine is the connection between the expression and activity of UCP and overall metabolism in humans, as well as the potential role of UCP as a mechanism for obesity in some individuals (Costford et al. 2007).

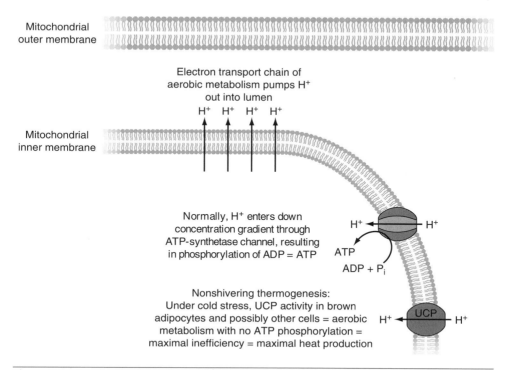

Figure 5.5   Uncoupling proteins within the mitochondria create inefficient aerobic metabolism by providing an alternate pathway back across the inner mitochondrial membrane for hydrogen ions created during the electron transport chain. Such pathways may contribute to the elevated heat production from nonshivering mechanisms during chronic exposure to cold.

AUTONOMIC NERVOUS SYSTEM CHANGES
WITH COLD ADAPTATION

Another pathway of physiological response and adaptation to cold exposure is via hormonal changes. A study by Makinen and colleagues (2008) showed that acute cold exposure elicited an elevated sympathetic response, evidenced by increases in norepinephrine release and increases in total heart rate variability as well as high- and low-frequency domain variability (while consensus remains unclear on the analysis and meaning of heart rate variability, low and high frequencies are thought to represent elevated sympathetic and parasympathetic activity, respectively). However, following 10 days of passive 2 h daily exposure to 10 °C air, norepinephrine response to cold was decreased in these subjects, while high-frequency activity of heart rate variability increased. Therefore, the authors concluded that cold acclimation elicits a blunting of sympathetic drive and a slight stimulation of parasympathetic activity. The effects of these hormonal and neural changes with acute and chronic cold exposure on other physiological systems and long-term health, such as potential shifts in fuel utilization and changes in immunological responses, remain unknown.

Too Cold to Exercise?

Thanks to public health messages and the latest revisions to the windchill index in North America (see chapter 2), awareness of the potential dangers from freezing cold injuries such as frostbite is strong in the general population. For recreational and competitive athletes, another question during winter months is whether it becomes unsafe to exercise outdoors because overly low temperatures increase the risk for clinical symptoms or damage to the body. Concerns that have been raised include whether the breathing of cold air could damage the airways or lung tissue or could trigger exercise-induced dyspnea or asthma attacks. Another potential danger may be direct effects on the cardiac muscle and damage to the heart. Overall, it appears that these possible dangers of exercising in cold air are overstated for asymptomatic individuals and should not be a deterrent to exercise in the cold as long as adequate clothing is worn.

The direct effect of exercise in the cold is the breathing of cold and dry air. As alveolus tissue is thin, moist, and fragile, the inhaled air must be warmed and humidified prior to its entry into the lungs. Therefore, one potential hazard is hyperresponsiveness of the respiratory tract, leading to bronchoconstriction and asthmatic attacks. The incidence of exercise-induced bronchoconstriction may be underreported in healthy individuals, as 37.5% of elite Norwegian Nordic skiers experienced hyperresponsiveness—a 20% reduction in forced expiratory volume—with a laboratory-based methacholine challenge compared to only 10% in the same subjects following an outdoor ski test (Stensrud et al. 2007b). However, not all cases of exercise-induced dyspnea are necessarily due to asthma; bronchoconstriction is exacerbated by cold air in individuals with exercise-induced dyspnea but no clinical diagnosis for asthma (Ternesten-Hasseus et al. 2008). In any case, for asthmatic individuals and those with exercise-induced bronchoconstriction, the breathing of cold air can trigger dyspnea and reduced exercise capacity. When exercising in 20 versus –18 °C environments, individuals with exercise-induced bronchoconstriction experienced a 6.5% decrease in maximum oxygen uptake and had lower running speeds in the cold. These findings were supported by greater impairments in spirometric values following exercise in the cold, with 24% and 31% reductions in forced expiratory volumes in warm versus cold conditions, respectively (Stensrud et al. 2007a). However, while lower humidity levels at 20 °C temperature directly decreased exercise capacity and increased the incidence of exercise-induced bronchoconstriction (Stensrud et al. 2006) in diagnosed individuals, the exact mechanism for the triggering of dyspnea with cold air remains unclear. Dessication of the respiratory tract does not appear to be the primary mechanism, as no changes in the quantity or hypertonicity of the airway surface fluid were observed with bronchoconstriction elicited by hyperventilation of cold air in persons with asthma (Kotaru et al. 2003).

In asymptomatic individuals, exercising in cold temperatures does not appear to pose a significant physiological or clinical risk. Hartung and colleagues (1980) had subjects engage in moderate exercise while breathing ambient or cold (–35 °C) air and concluded that there were minimal differences in physiological responses. Shave and coworkers (2004) investigated the risk to the cardiovascular system from prolonged exercise in the cold by having very fit athletic subjects perform a 100-mile (160 km) cycling test in 0 and 19 °C environments. Extensive echocardiographic imaging of the heart revealed no changes in ventricular filling or contractility before or following exercise in the two environments, along with no differences in either systolic or diastolic blood pressure. Creatine kinase and cardiac troponin T, blood markers for cardiac muscle damage, were also not different across temperatures, suggesting that high-intensity exercise in either temperature did not appear to elevate the risk for exercise-induced cardiac damage. If anything, the authors concluded that the cold environment served a protective function, as the greater sweat rates in the 19 °C condition elicited a greater level of dehydration and higher postexercise heart rates, along with changes in the pattern of ventricular filling. Anecdotally, Nordic skiers train and compete outdoors in temperatures that are extremely cold; but the effects of different inspired temperatures on exercise capacity in healthy individuals remain unknown, along with whether countermeasures such as warming masks may be an effective ergogenic aid.

For all athletes, another issue with inspiring cold and dry air may be that, unlike what occurs in other systems such as the kidneys, which conserve water and electrolytes, there is minimal recovery of water and heat from the respiratory system. Therefore, continued expiration in cold temperatures at high ventilation rates during exercise can result in significant heat loss and potentially dehydration. Cain and colleagues (1990) found that maximal expired air temperature varied only very slightly regardless of ambient temperature ranging from –40 to +20 °C; the expired air was fully saturated and there was little difference due to exercise and ventilation rate. Thus respiratory heat loss can range up to 25% to 30% of resting and 15% to 20% of exercise metabolism, and adequate hydration remains an issue during exercise in the cold. Overall, one potential countermeasure for enhancing exercise tolerance and capacity for both symptomatic and asymptomatic individuals may be the use of a mask to assist in heat and water recovery; at present, some such masks are commercially available. However, the mask must enable adequate ventilation without adding to the resistance and work of breathing, along with moisture and ice buildup.

In summary, for asymptomatic individuals with no history of exercise-induced dyspnea or asthma, exercise even in extremely cold environments can be achieved safely without significant health risks. Even for people with respiratory issues aggravated by cold air inhalation, the health benefits of physical activity likely outweigh any potential health risks as long as appropriate precautions are taken. The following are important points about some of these preventive measures (Castellani et al. 2006):

- Clothing insulation is dependent on the exercise intensity and environment, and too much clothing insulation can lead to heat stress.
- In high-windchill conditions, people should ensure that exposed skin is kept to a minimum (see chapter 2 for a detailed discussion of the 2002 windchill index revisions).
- The extremities, along with the face, are the locations at highest risk for frostbite and other cold injuries because often they have the majority of the exposed skin surface and also because of their high surface area to volume ratio, as well as blood flow changes with cold exposure. Especially given the survival importance of the hands and feet in maintaining manual function and movement, care must be taken to provide additional insulation or supplemental heating to the extremities.
- A higher core temperature promotes blood flow to the extremities, so it is important to ensure that body temperature is preserved.
- Dehydration does not appear to enhance the risk of cold injuries by impairing vasoconstriction or shivering (see chapter 4 for general hydration guidelines).
- Carbohydrate ingestion should be emphasized to ensure adequate substrate availability for increased glucose metabolism and shivering in the cold.

SUMMARY

Working or exercising in the cold brings about unique challenges to the thermoregulatory system. During cold exposure, the accelerated heat loss to the environment places additional requirements on endogenous heat production through shivering. In turn, this raises the rate of substrate utilization and the metabolic cost of exercise, thus increasing nutritional intake requirements and logistical demands during prolonged expeditions. Further research on cellular and biochemical responses to cold may lead to drugs or supplements that can alter the rate of metabolism in the cold, an important concept in both the development of survival rations and the larger issue of energy balance. Manual function is a critical issue in cold environments, and further research is required to increase understanding of the nature of blood flow changes to the extremities with acute and chronic exposure to local cold stress in order to minimize the risk of cold injuries.

Cold Water Immersion

Marine accidents are, unfortunately, not rare or isolated incidents that are somewhat predictable. Rather, they are common events that can occur suddenly with minimal to no warning, resulting in a high level of terror and panic for unprepared and even prepared individuals that greatly elevates the risk of tragedy. In Canada, approximately 500 drowning deaths occur annually, in situations ranging from actual marine scenarios (e.g., swimming, recreational boating) through to unplanned exposure to water (e.g., car or snowmobile accidentally landing in water) (Canadian Red Cross 2006). Marine accidents can be significant threats to specific populations, such as those employed in marine transport, navies, fishing, and gas exploration and extraction, due to the risk inherent in operating over water. For example, over 10 million personnel are transported annually to and from offshore oil rigs using helicopters, and this risk is exacerbated by the nearly 10-fold rate of accidents for helicopters compared to fixed-wing aircraft operating over water (Brooks et al. 2008).

As has been the case with much of environmental physiology research, a major imperative for investigating cold water immersion has been driven by military contexts. In this vein, it has been estimated that nearly a third of overall Royal Navy deaths during World War II were due to hypothermia. However, high-profile accidents such as those involving the *Titanic* in 1912 (~1500 deaths) and the ferry *Estonia* in 1994 (852 deaths among 989 passengers and crew) brought the threat of marine accidents and cold water immersion directly into the public consciousness. In recent years, one of the major paradigm shifts among both scientists and safety officials has been the transition away from categorizing the primary cause of death in many accidents as simply "hypothermia." This generalization is somewhat understandable, as the obvious commonality among many victims is an extremely low core temperature by the time the bodies are recovered. However, awareness is now emerging that the causes of such deaths may be much more diverse, and especially that the initial phases of an accident are critical in determining the odds of survival. Therefore, marine accidents have been broken down into at least four separate but interrelated phases, each with its own unique threats, underlying physiological mechanisms, and potential countermeasures (Golden and Tipton 2002). The four phases involve (1) cold shock and drowning, (2) muscle or swimming failure, (3) hypothermia, and (4) postrescue collapse. Each phase is introduced next and is then detailed in later sections of this chapter.

Humans are air breathers, such that the obvious initial danger with marine accidents is the threat of drowning, as only approximately 150 mL of water entry into the lungs is sufficient to cause fatality (Golden et al. 1997). The first broad phase of marine accidents therefore encompasses the initial entry into water through the first few minutes. During this time, apart from the psychological panic, the primary physiological dangers are the difficulty in controlling breathing and tachycardia leading to drowning. In the second phase of marine accidents, lasting from 10 to 60 min following immersion and prior to the actual dropping of core temperature, the cold can rapidly cause impairment of muscle function, which in turn makes it more difficult to perform necessary survival tasks such as swimming to land or hauling oneself into a life raft, opening flare packages, and so on. The third phase encompasses the actual physiological battle between heat production and heat loss in order to ward off hypothermia, with dangerous body temperatures possibly occurring after 60 min. As should be evident, this third phase is greatly dependent on actions during the second phase, with hugely different potential consequences such as successful deployment and entry into a covered life raft with a group of survivors and signaling for rescue, or drifting alone in the cold water hanging onto flotsam and jetsam. It should also be evident that accurate scientific modeling of survival times in this third phase can be extremely difficult, as these are greatly affected by a multitude of individual (e.g., body mass and fatness, age) and situational (e.g., water temperature, insulation) factors.

The final dangerous phase of marine accidents occurs during the initial period following rescue, as evidenced by the large number of reports of individuals who are conscious and coherent prior to rescue, but who may suddenly collapse during the rescue or rewarming phase. The physiology of this sudden collapse likely revolves around sudden additional strain to a cardiovascular system that is already stressed by hypothermia. One concept that has been debated scientifically is the contribution of the "afterdrop" phenomenon (whereby core temperature seems to rapidly decrease in the early stages of rewarming) to this collapse. Coupled with this have been investigations into the safest and most efficient method of rewarming hypothermic individuals in field situations prior to delivery to a medical setting.

The purpose of this chapter is to examine each of these four phases of a marine incident, focusing on the physiological factors that influence survival in cold water immersion. This will reflect the discussion on modeling human response in the previous chapters on thermoregulation. The issue of marine safety is a multidisciplinary merging of environmental and exercise physiology with the appropriate engineering design of workspace and equipment. While a full examination of water safety, cognitive and physical ergonomics, and industrial design is beyond the scope of this text, this discussion highlights some of the major principles.

SUDDEN IMMERSION AND COLD SHOCK

You are sailing with some friends on a lake. The weather is sunny and the wind is calm. As you're walking along the deck, however, a sudden gust of wind causes the boat to lurch and you lose your footing, tripping over the railing and falling into the 10 °C water. What happens during the first minutes of this accident?

The direct physiological response of immersion into cold water is typically a rapid tachycardia and intense hyperventilatory drive (see figure 6.1). This "cold shock" response can be extreme, going from resting values to more than 30 breaths and 80 L · min⁻¹ in ventilation and 150 beats · min⁻¹ in heart rate in seminude individuals rapidly immersed to the neck in 10 °C water (Tipton et al. 1998b). Obviously, this strong hyperventilatory drive is counterproductive to individuals trying to hold their breath when submerged and trying to regain the surface, or when strong winds are driving high waves against an immersion victim. For individuals who are not physically fit or who have underlying cardiovascular issues, the sudden tachycardia can also bring about the risk of cardiac events. Furthermore, the panic and psychological stress of sudden immersion exert an additional strong sympathetic neural drive, possibly producing a feedforward spiral of increasing hyperventilation and tachycardia (Barwood et al. 2006).

The primary stimulus for the cold shock response appears to be the rapid drop in skin temperature upon immersion, triggering a strong stimulus in the sympathetic nervous system, with plasma norepinephrine rapidly increasing upon cold water

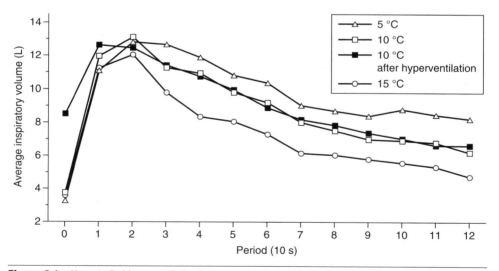

Figure 6.1 Uncontrolled hyperventilation is a common characteristic of the initial phase of cold shock upon cold immersion (Tipton et al. 1991), making it difficult to keep from ingesting water into the stomach or airways. This can lead to drowning before the onset of hypothermia. Victims of marine incidents should prepare themselves psychologically for the cold shock, delay entry into water as long as possible, enter water gradually if possible, and not attempt to self-rescue in the first 1 to 3 min unless immediate escape is possible.

Reprinted from M.J. Tipton, D.A. Stubbs, and D.H. Elliott, 1991, "Human initial responses to immersion in cold water at three temperatures and after hyperventilation," *Journal of Applied Physiology* 70: 317-322. Used with permission.

immersion and rapidly dropping with rewarming of the skin prior to core temperature recovery (Johnson et al. 1977). This is supported by findings that breath-hold times are very closely correlated to the surface area exposed, the rate at which exposure occurs, and also the water temperature. In a classic study, Hayward (Hayward et al. 1984) immersed subjects in water with temperatures ranging from 0 to 35 °C without their prior knowledge of temperature, and developed a regression line relating breath-hold (BH) times with water temperature:

$$BH \text{ duration (seconds)} = 15.01 + 0.92 \, T_{water} \quad (Eqn. \ 6.1)$$

From a safety perspective, the imperative for public education and training appears to be, wherever possible, to progressively "stage" entry into cold water rather than encounter sudden immersion. In the same study, when subjects were able to enter the water gradually and perform a brief voluntary hyperventilation, breath-hold times remained linked to water temperature but increased dramatically according to the equation:

$$BH \text{ duration (seconds)} = 38.90 + 1.70 \, T_{water} \quad (Eqn. \ 6.2)$$

COLD SHOCK TRAINABILITY

Can people be trained to minimize the cold shock response? Given its neural basis, it seems probable that both the physiological and psychological components driving cold shock responses can be attenuated. Mike Tipton's research group at the Institute of Naval Medicine and University of Portsmouth in the United Kingdom has been one of the leading labs involved in investigating the trainability of the cold shock response. One important finding has been that it is not necessary to have the full intensity of exposure to promote habituation. Namely, a brief though still uncomfortable program involving four days of six repeated 3 min immersions up to the neck in 15 °C water was sufficient to significantly decrease, though not eliminate, both the hyperventilatory and tachycardic responses in 10 °C water (Tipton et al. 1998b). This appears to be a centrally driven reduction in neural drive, as training immersions on one side of the body were able to transfer over to reduced cold shock responses when the other side was immersed (Tipton et al. 1998a). Most importantly, the brief training program appeared to have long-lasting protective benefits. After the original whole-body training study (Tipton et al. 1998b), the majority of the subjects were tested again with a 3 min 10 °C immersion 2, 4, 7, and 14 months following the initial training (Tipton et al. 2000). Somewhat surprisingly, despite no experience with cold water in the intervening period, attenuations in hyperventilation and tachycardia were retained at 7 and 14 months, respectively, suggesting a long-term benefit to a brief stimulus. Recently, the same group has extended their work to psychological interventions, finding that a period of visualization and other mental skills training directed toward reducing cold shock responses was able to prolong head-out breath-hold times upon 10 °C water immersion (Barwood et al. 2006). On a methodological note, the beauty of researching cold shock responses is that the experiments themselves are very quick to conduct!

Breath-Holding

The primary physiological stimulus for ventilation is the arterial concentration of carbon dioxide (CO_2), detected by chemoreceptors within the arteries. Therefore, the initial phase of breath-holding is characterized by an increasing level of CO_2 within the blood and alveoli until it reaches a critical threshold for stimulating ventilation (see figure 6.2). Throughout this phase, actual voluntary resistance against breathing and respiratory muscle contractions are minimal. Beyond this point, however, the individual must voluntarily resist the urge to breathe, and respiratory muscle activation is strong. Logically, the determinants of the first phase are primarily starting lung volume and CO_2 concentrations along with buffering capacity. In contrast, the later phase is mainly determined by psychological ability to tolerate the discomfort with apnea (Barwood et al. 2006).

Breath-holding, or voluntary apnea, is a critical component for success in sports such as synchronized swimming and free diving. It has also been important in occupations such as pearl diving in Korea and Japan. With the world record for static apnea (breath-holding under water) at 8:00 for women and 9:08 for men (as of January 2009), it is clear that apnea is a trainable response. Erika Schagatay's research group at the Mid-Sweden University has been a leader in the scientific investigation of apnea, the dive response (described later), their underlying physiological mechanisms, and their trainability. One study dealt with the separate contributions of apneic versus fitness training and showed separate mechanisms of improvement (Schagatay et al. 2000). Two months of aerobic training did not affect any of the physiological factors measured during apnea but extended the "voluntary" second phase beyond the physiological

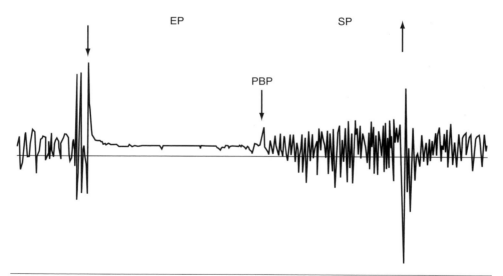

Figure 6.2   Respiratory transducer recordings during an apneic episode. The initiation and termination of apnea are indicated with arrows. The easy-going phase (EP) represents the gradual buildup of PCO_2 levels, and there is minimal urge to breathe. Beyond the physiological breaking point (PBP), the buildup of PCO_2 generates a strong stimulus to breathe in the struggle phase (SP).

With kind permission from Springer Science+Business Media: *European Journal of Applied Physiology*, "Effects of physical and apnea training on apneic time and the diving response in humans," Vol. 82, 2000, pgs. 161-169, E. Schagatay, M. van Kampen, S. Emanuelsson, and B. Holm, figure 1.

threshold of apnea. One of the reasons may have been that physical training enhanced the subjects' awareness and their ability to tolerate discomfort, pain, or both. In contrast, specific apneic training prolonged the physiological threshold for ventilation, likely through an increased respiratory buffering capacity, and also potentiated the bradycardic component of the dive response. It is likely that both mechanisms are important in the large extension of apnea times seen in sporting competitions like free diving and synchronized swimming.

Helicopter Ditching and Supplemental Air

In two helicopter ditching disasters off the coasts of Newfoundland and Scotland in 2009, only 1 of 18 and 0 of 16, respectively, survived the accident, again illustrating the dangers of helicopters flying offshore. Even following a noncatastrophic helicopter ditching (i.e., the aircraft has not broken apart during landing) in water, the survival rate of individuals not mortally injured by the impact is low, ranging from 50% to 85% globally and at 78% over 46 Canadian civilian helicopter ditching incidents involving 124 crew and passengers between 1979 and 2006 (Brooks et al. 2008). One possible cause for the low rate is that the crew and passengers cannot hold their breath underwater long enough to make the often difficult escape from an inverted and submerged helicopter. Advances in survival equipment have been made over the past two decades with the development of various emergency breathing systems (EBS) for use in marine aviation. These systems function either by providing a small supply of compressed air or else by allowing the individual to rebreathe the exhaled lung volume; hybrid systems incorporate both methods. An EBS removes the constraint of a single breath-hold as the limit to underwater survival time and can significantly increase the total underwater time of individuals, enhancing their ability to escape and survive (Tipton et al. 1997). Until recently, EBS were used by a number of military forces but not, with some exceptions, in general marine aviation. A series of studies from our lab and collaborators in Nova Scotia helped to instigate a change in policy within Canada, with EBS now a required piece of survival equipment for all offshore employees flying to Canadian petroleum platforms and vessels.

In our evaluation of the need for EBS for helicopter passengers, the first step in assessing the breath-hold requirements in helicopter escape is to understand the typical time required for escape from the submerged helicopter. While it is impossible to accurately predict the exact time required for escape, simulations provide a good baseline for judgment. Brooks and colleagues (2001) evaluated the breath-holding requirements for instructors evacuating from the Modular Egress Training Simulator (METS) configured as a 15- and 18-passenger Super Puma helicopter. Depending on the ditching scenario, the breath-holding time required for the last person to exit ranged from 28 to 92 s. The subjects were physically fit and healthy divers and wore an EBS for use in the event that they ran into a problem, and the ditching occurred in warm water. Given the "ideal" conditions, these times are likely very optimistic; the times required in an actual emergency may be much longer.

Following up this study, we tested the breath-hold capacity in warm water of 228 trainees undergoing mandatory offshore survival training (Cheung et al. 2001). In this realistic sample, 77 of 228 (34%) subjects could not hold their breath underwater (BHT_w) for 28 s, the absolute minimum time required for complete evacuation of the Super Puma in the pool simulations just discussed. In addition, only 6 of 228 (3%)

could maintain a BHT_w of 92 s, the maximum evacuation time in the same study. Since the study of Brooks and coworkers (2001) and our investigation both involved "best-case" scenarios (e.g., minimum required escape times vs. maximum BHT_w, both in "ideal" circumstances), the disparities between these values can be expected to be greatly magnified in actual emergency situations. Overall, these reports have led Natural Resources Canada to mandate the use of EBS for all marine helicopter passengers. At the same time, it is critical to continue optimizing EBS design and the training for its use, along with other solutions to enhance survivability. Such solutions include increasing the speed of escape by shortening and simplifying the escape route as well as investigating the possibility of deliberately floating the ditched helicopter on its side and providing an air gap.

Diving Response

In contrast to the cold shock response from immersion of the body, submersion of the head into cold water in humans can trigger a somewhat paradoxical prolongation of breath-hold durations and bradycardia. This appears similar to the dive response found in aquatic mammals, such as seals and whales, which enables these air breathers to dramatically reduce oxygen demand underwater. While at a much lower magnitude than that in aquatic mammals, the dive response in humans appears to be a potential factor behind some cold water survival reports. Whereas the large sympathetic drive from an overall drop in skin temperature is the primary stimulus for cold shock, the diving response seems to be triggered by a direct reflex stimulation of the trigeminal nerve, one of the cranial nerves emanating directly from the brain. Similar to the situation with the cold shock response, the magnitude of the skin temperature decrease of the face is a determinant of dive response intensity, with reports of both colder water and warmer air temperatures magnifying the response

THE SPLEEN IN PROLONGED APNEA

Another potential mechanism behind prolonged voluntary apnea that has been explored in diving mammals, and that is now gaining some attention in humans, is the role of the spleen as a reservoir of erythrocytes. During prolonged dives, seals appear to contract their spleen, ejecting extra red blood cells into their circulation and raising both hemoglobin and hematocrit levels. Schagatay's group used ultrasound imaging during repeated apneas in air in 10 male and female subjects with a variety of backgrounds in underwater exercise, ranging from those with no background (four subjects) to underwater rugby and apneic and recreational SCUBA divers (Schagatay et al. 2005). Three maximal apneas with a brief 2 min recovery in between were able to elicit strong splenic contractions of ~50 mL volume, with a concurrent rapid increase in hemoglobin (2.4%) and hematocrit (2.2%). At the same time, breath-hold duration increased by approximately 20 s over the course of the three apneas. The spleen recovered to its normal size 8 to 9 min following the final apnea, as did hemoglobin and hematocrit. Therefore, splenic contractions may be another contributing factor in the prolonged breath-hold durations in underwater athletes.

upon submersion (Schagatay and Holm 1996). The potential interactions between the conflicting cold shock and diving responses in humans are unclear, though it is evident that training and familiarity with cold immersion along with breath-holding and submersion are likely to improve overall odds of survival (Schagatay et al. 2000; Tipton et al. 1998b). In addition, the possibility that direct stimulation of the face alone may prolong breath-hold times is further rationale for the use of immersion suits that leave the face exposed but otherwise provide full thermal coverage and an air barrier with the water for the rest of the skin surface.

MUSCLE FAILURE IN THE COLD

Assuming that you are able to escape from a ditched helicopter or survive the shock of initial immersion, the next approximately 10 min becomes critical to your ultimate long-term survivability. The ability of your muscles to function properly prior to becoming incapacitated by the cold can enable successful entry into the shelter of a life raft, swimming to shore, climbing out of a hole in the ice, or operating flare packages and other signaling equipment. In contrast, the loss of muscle function during this period can result in drowning or eventual hypothermia from continued prolonged immersion. As we shall see, making correct decisions during this time will also play a major role in determining long-term survival odds.

Time Line for Self-Rescue

Due to public awareness about the threats from hypothermia, a perception exists that one must immediately strive to get out of the water at all costs or risk a rapid death from hypothermia. This idea places an additional time pressure on the process of self-rescue, adding to the risk of fatality by increasing stress and sympathetic drive. Specifically, it can cause individuals to try to get out of the water immediately, in the midst of the most intense phase of cold shock response, when the high heart rates and uncontrolled breathing along with panic make coordinated movements difficult. The threat from short-term hypothermia is likely exaggerated, as it is nearly impossible for the human body to lose enough heat for core temperature to drop to critical levels in the first few minutes of immersion.

One primary safety message arising from investigation into cold shock and degradation of muscle function in the cold is the extent of the window of opportunity for successful self-rescue from sudden cold water immersion. While this window is indeed finite, it likely extends well beyond the initial cold shock phase of immersion to approximately 10 min, even in near-freezing water temperatures. In less extreme temperatures of 10 to 15 °C, functional swimming ability may extend to 45 to 60 min or more (Tipton 1995). Rather than hypothermia, the issue may be swimming failure and critical impairment of gross motor function to the point that an individual cannot perform major rescue tasks (swimming or treading water, climbing a ladder onto land). Then, beyond this stage, drowning from inability to keep the airway clear of water or prolonged exposure and hypothermia become dominant dangers (Golden and Tipton 2002). Overall, scientific research suggests that the more appropriate strategy for self-rescue is to avoid struggling excessively to get out immediately. If possible, expect and psychologically brace yourself for the cold shock response; then give yourself 1 to 2 min for the hyperventilation and tachycardia to begin subsiding. During this time, establish inflation (e.g., inflate your life jacket or locate floating debris), and plan your escape and action over the remaining window for functional ability (Golden and Tipton 2002).

Swimming Failure and Self-Rescue

The difficulty posed by cold water to swimming ability should not be underestimated. The strong redistribution of blood to the core results in intense peripheral vasoconstriction and reduced muscle blood flow, exacerbated as the reflex contraction of skeletal muscles further impairs circulation through the limbs. The contraction and muscle stiffness also make proper neural coordination difficult, such that proper flotation (treading water) and swimming quickly become impaired. This is illustrated in videos recorded by Mike Tipton's lab of British Olympic medalist swimmers performing tethered swimming in 10 °C water while wearing typical clothing without a life jacket. Swimming ability rapidly deteriorated, with a slowing of stroke rate, reduced range of motion of the arm and leg strokes, and an increase in vertical angle in the water that added to hydrodynamic drag (see figure 6.3). Overall, despite their strong swimming backgrounds, these subjects could tolerate less than 10 min in these conditions prior to removal. Indeed, fitness and even overall body fatness may have very small roles in determining the distance that can be swum in survival situations. In a study on the rate of swimming decay in cold (14 °C) water with the use of protective flotation devices, Wallingford and colleagues (2000) found that neither maximal

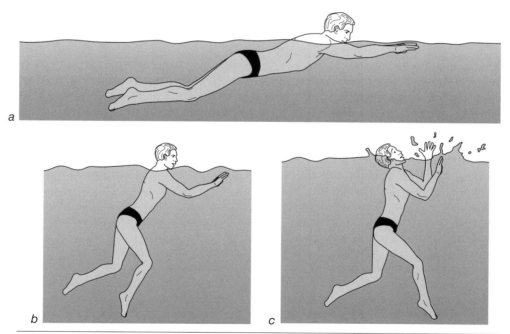

Figure 6.3 Swimming failure, especially the inability of the upper body and arms to contract to maintain adequate swimming speed or buoyancy, is a major determinant of drowning with prolonged cold immersion. A normal swimmer, shown in (a), is relatively horizontal in the water. As the leg muscles weaken (b), swimming speed is slowed, resulting in a more vertical posture and greater drag resistance. Finally, when swimming failure is imminent (c), the individual is almost vertical in the water, struggling and unable to lift the arms above the surface.

Adapted, by permission, from F. Golden and M. Tipton, 2002, *Essentials of sea survival* (Champaign, IL: Human Kinetics), 73.

oxygen uptake nor body fatness correlated with swimming distance when subjects were wearing normal civilian clothing and a life vest. Indeed, the only significant correlate was tricep adiposity, suggesting that the primary determinant of swimming capacity was the ability of the arms to maintain temperature and function.

In cold water situations, one of the most critical decisions made by survivors may be whether to stay with the incident vehicle (i.e., remain relatively stationary and await rescue) or to attempt self-rescue by swimming for shore or safety. The physical difficulty of swimming in cold water is compounded by the increased rate of convective heat loss with increased flow of water over the body. Furthermore, perfusion of the skeletal muscles during swimming decreases the insulative value of subcutaneous fat and muscle. Along with pivotal work in the United Kingdom during the 1950s and 1960s, a famous study at the University of Victoria by John Hayward, a pioneer in studying cold water survival strategies and modeling hypothermia curves, highlighted the seriousness of the "stay or swim" decision. In Hayward's study, subjects were provided with life jackets and immersed off the coast of Vancouver Island in the Strait of Juan de Fuca (water temperature ~12 °C, minimal wind or current), and then proceeded to swim at a slow pace of 15 m · min⁻¹ (49 ft · min⁻¹) (Hayward et al. 1975). In these conditions, core temperature decreased at a rate of 3.44 °C · h⁻¹, a rate up to 50% greater than when subjects were simply floating with the life jacket. At this swimming rate, theoretical incapacitation from hypothermia (assumed at 33 °C) was achieved within 1.4 km (0.87 miles). Largely on the basis of this study, public awareness messages were produced that warned of the dangers of swimming and self-rescue.

The idea of remaining stationary and avoiding swimming in cold water immersion has been challenged by recent work in Canada from Michel DuCharme's group at the Defence Research and Development Canada laboratories in Toronto and Valcartier, and figure 6.4 outlines a revised decision tree for "swim or stay" from their research. In agreement with previous data, DuCharme and Lounsbury (2007) summarized typical swimming distances in cold (10-15 °C) water with a life jacket as ranging between 800 and 1500 m (875 and 1640 yd), with a duration of approximately 45 min. However, DuCharme contends that the perspective has changed with a closer understanding of the different phases of cold water immersion. Specifically, earlier studies had subjects initiate swimming immediately upon immersion, during the peak of the initial cold shock responses. This increased activity on top of the initial hyperventilation and tachycardia may have contributed to unnecessary panic and premature fatigue. Therefore, if immersion victims are permitted sufficient time for the cold shock response to subside, it is possible that swim distance can exceed that found in studies to date.

As mentioned previously, the largest challenge in the initial phases of survival scenarios is appropriate decision making, and the first minutes following cold shock are the optimal time to assess the situation and select appropriate behaviors. Based on the time and distance limitations for cold water swimming already discussed, one important consideration, besides the possibility of external assistance and rescue, is accurately determining the distance from shore. Waves and the location of the head at sea level make depth perception and distance estimation extremely difficult, especially across a monotonic scene of water. Importantly, subjects tended to overestimate distance, thinking it was three times farther than the actual distance, even though they were highly accurate in assessing potential swimming speed and time required to swim a particular distance (Lounsbury 2004). This inability to accurately assess

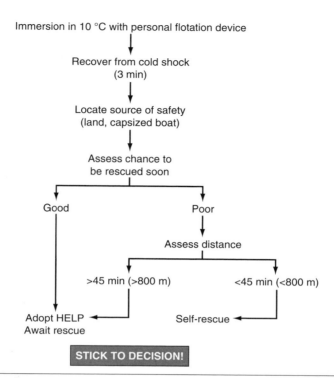

Figure 6.4   Potential decision tree in the aftermath of a marine accident, demonstrating the different considerations involved in the decision between remaining at the incident site and self-rescue. If individuals choose self-rescue, they should avoid leg-only swimming and should swim at the fastest pace possible given their fitness and the distance to safety. HELP = heat escape lessening position (arms and legs curled up, covering armpits and groin).

distance may further contribute to the fear of attempting swimming for self-rescue. Interestingly, the same study showed that most subjects (86%) could predict quite accurately the distance they were capable of swimming in cold water when questioned within 3 min of immersion. However, the proportion dropped dramatically to 32% when they were questioned after 30 min of immersion. This suggests that there is a finite time frame for "functional" mental performance, and also that the strategy initially decided upon should be strictly adhered to and not altered partway through.

Upon the decision to swim for self-rescue, another area in which DuCharme challenged previous consensus was the reason for swimming failure. Similar to the conclusions from Tipton's lab, the dominant mechanism for swimming failure in DuCharme's study was not whole-body hypothermia. Rather, limb cooling and muscle incapacitation, especially of the arms, were the primary causes of swimming failure and inability to keep the airways clear. Using the arms for treading water or propulsion also has the disadvantage of perfusing a region with a large ratio of surface area to volume, enhancing convective heat loss. Therefore, further research is required to investigate the potential strengths and limitations of different swimming styles. For example, using only the legs for propulsion may decrease the rate of heat loss, but the trade-off includes slower speeds and much higher oxygen costs. The optimal swimming stroke may also be dependent on body composition, distance, water temperature, and life jacket design and consequent posture and buoyancy.

HYPOTHERMIA

Unfortunately, the circumstances may dictate that you are stuck at the incident site and immersed in cold water. Obviously, this is an untenable situation long-term, and your survival is ultimately predicated on whether help has been contacted. Assuming this, the actual behavior adopted by a victim can greatly influence the rate of heat loss and therefore hypothermia onset and survival time. Investigation of these individual and situational characteristics highlights both the different physiological responses to cold exposure and the difficulty in scientifically modeling survival time and informing search and rescue policy.

Hayward and associates (1975) investigated the efficacy of different survival strategies in prolonged immersion. The behaviors ranged from wearing a life jacket and adopting a stationary "open" position (e.g., limbs straight and with maximal surface exposure to the water) to individual and group huddling (limbs flexed and curled in a fetal position, see figure 6.5). Furthermore, where no life jackets are worn, flotation can be maintained by either the active mode of treading water or by "drown-proofing," which refers to passively floating while periodically submerging the head. Analysis of the results in the life jacket conditions made it clear that minimizing the surface area for convective heat exchange was critical to minimizing heat loss. This was especially true for areas such as at the neck, armpits, and groin, where large blood supply runs relatively close to the skin surface or where minimal vasoconstriction occurs. Therefore, maintaining a near-fetal position, termed by Hayward a heat escape lessening position or HELP, slowed the rate of rectal temperature cooling by 66% compared to maintaining an open posture. In reality, this position may be difficult to perform in choppy waters or because of the life jacket's design and buoyancy, which may cause

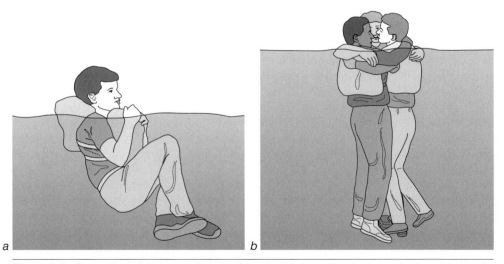

a   *b*

Figure 6.5   One method of decreasing the rate of heat loss during cold water immersion is to *(a)* adopt the heat escape lessening position, or HELP. Proposed by John Hayward, this involves curling up into a fetal position with the limbs protecting the high heat loss areas. Another possibility is *(b)* huddling in small groups, which has the additional advantages of providing psychological support and a larger surface area for visibility to rescuers.

Reprinted from J.S. Haywood, J.D. Eckerson, and M.L. Collis, 1975, "Effect of behavioral variables on cooling rate of man in cold water," *Journal of Applied Physiology* 38: 1073-1077. Used with permission.

MODELING LIFE RAFT SURVIVAL

Current collaborative work to model the thermal characteristics of life rafts in North Atlantic conditions illustrates the complexity of accurately predicting responses in survival situations. Our large multidisciplinary team, based in Atlantic Canada and funded by the Search and Rescue Secretariat of Transport Canada, includes engineers, offshore survival trainers and researchers, ergonomists, and mathematical modelers in addition to exercise and thermal physiologists. Currently, despite the reliance on life rafts as the primary emergency vessel in the majority of marine transport, no accurate model exists to predict the survival times for individuals inside these life rafts (see figure 6.6). At the same time, though engineering standards exist for life raft approval, no equivalent thermal testing standard is in place. One goal of the project is developing thermal standards that can be tested using only manikins. This has the potential to minimize risk and speed up certification. Therefore, testing has included both human testing and the use of thermal manikins to enable cross-validation of data.

In targeting our objectives, we needed to test a large array of potential thermal and nonthermal variables, including the following:

- The effects of floor insulation (uninflated, partly, and fully inflated). We found that an uninflated floor greatly potentiated heat loss compared to one partly or fully inflated, with about 50% inflation needed to reach a plateau in insulation.
- The effects of wave height or speed through the water. We found that floor insulation was a more important factor than wave height or tow speed, within reason given the extreme conditions possible.
- The effects of heat production from a sparsely or fully loaded raft. We found that the heat production from life raft passengers did not significantly increase the ambient temperatures inside the raft even when it was fully loaded. However, one unanticipated finding was that the carbon dioxide levels inside the raft very rapidly exceeded toxic levels when it was fully loaded, even with passengers not moving about. In real use, this would require the opening of ventilation ports within the life raft, possibly greatly decreasing internal temperature and accelerating heat loss.
- The effects of clothing wetness. In real situations, it is almost impossible to keep the floor of a life raft from becoming soaked, which results in wetting of the passengers' clothing. Not surprisingly, wet clothing resulted in much higher rates of heat loss, and also shivering and discomfort, compared to dry conditions.
- The effects of water and air temperatures. Our initial phases of testing were performed in relatively mild conditions of 20 °C water and 15 °C air. Subsequent phases of the study decreased temperatures down to 5 °C for both water and air to simulate typical Atlantic conditions and showed greatly accelerated rates of heat loss.
- The effects of nutrition. While nutrition is an important component of heat production capacity, we were not able to incorporate either nutrition or hydration status into the testing design.

(continued)

Modeling Life Raft Survival *(continued)*

- The effects of different clothing. On the basis of their normal use in marine vessels, we assumed the use of life jackets over underwear and cotton coveralls. We were not able to test the effects of different clothing or survival suits.
- The size of the life raft. We tested a 16-person life raft that is currently in use in the North Atlantic. We are assuming that our results can be extrapolated to both smaller and larger (up to 150-person capacity) life rafts, though this may not be the case.

The preceding list should make evident the large number of variables that may be involved in developing thermal models or survival curves. In addition, the listing illustrates the multidisciplinary approach required to address even apparently straightforward questions in environmental physiology.

Figure 6.6   The first challenge for survivors of marine incidents may be getting themselves out of the water or off of a sinking vessel and into life rafts. Research is ongoing to model the environmental and thermal characteristics of both small and large life rafts used in cold or polar environments. Such work is being done in conjunction with ergonomic research on designing life rafts that are easier to deploy.

the individual to tip and roll. In group situations, huddling closely slows the rate of core cooling to a similar extent. Even though the individual posture in group huddling is relatively open, the linking together of bodies minimizes water circulation through the high heat loss areas. Huddling also has the added benefits of providing psychological support for survivors and forming a larger target for search and rescue.

The importance of wearing life jackets in all marine situations is readily evident when one compares the effects of treading water and drown-proofing. Beyond any thermal comparisons, both behaviors require additional energy and make keeping the airway clear extremely difficult compared to floating in an open position with a life jacket. With treading water, convective heat exchange is greatly enhanced due to the movement of the limbs; in Hayward's study, this resulted in a 1.34 times faster core cooling rate than an open position. Inevitably, however, muscular fatigue will determine inability to maintain flotation. The worst condition in Hayward's investigations was drown-proofing, in which the high heat loss regions of the head and neck were submerged, with a cooling rate of 1.82 compared to control. Therefore, drown-proofing is recommended as effective only where water temperatures are extremely warm, as in the Caribbean. Overall, such a wide disparity in cooling rates and potential survival times strongly demonstrates the importance of developing correct public awareness messages and training programs to ensure appropriate behaviors in survival situations.

POSTRESCUE COLLAPSE

The fourth danger phase of cold water immersion occurs not in the water itself but upon rescue (Golden et al. 1991). In a number of anecdotal cases, victims seen waving to signal their rescuers, and many even able to assist in their own rescue or act coherently in the first minutes following rescue, collapse into unconsciousness or death shortly upon being brought to safety in a helicopter or ship. Scientific investigation of the causes behind such collapses highlights the multifactorial responses to environmental stress and may lead to better search and rescue practices to enhance the odds of survival.

Afterdrop Phenomenon

One phenomenon that has been repeatedly observed in studies on hypothermia is that, in the first minutes of rewarming, the rectal temperatures of subjects drop very rapidly. This rate of drop often greatly exceeds the rate that occurred during the actual cooling, and the drop can continue a further 1 to 2 °C within the first 5 to 10 min of rewarming. This is the case even if the subjects enter a hot tub where their skin is quickly rewarmed and they feel comfortable and cease shivering (Giesbrecht 2000).

Such observations raise concerns that if the body is already in a state of hypothermia and the cardiovascular system is sluggish and stressed, further core cooling could precipitate a cardiovascular collapse. Therefore, much has been made in the media and public awareness campaigns about the issue of core temperature afterdrop upon rescue and rewarming. Scientists have also striven to understand the mechanisms underlying afterdrop, while the rate of afterdrop has often been used as a key determinant of the efficacy of different rewarming techniques. Two primary events, one physiological and one physical, have been advanced as the main mechanism eliciting an afterdrop during rewarming:

1. **Hemodynamic.** Upon exposure to cold, the immediate cardiovascular response is vasoconstriction in the peripheries and limbs, resulting in a pooling of blood and heat in the core regions of the body. Therefore, skin and limb muscle temperatures can be much lower than the "core" temperature measured by rectal or esophageal temperature probes. Upon rewarming by external sources of heat (e.g., warm bath,

forced air), skin temperature rapidly rewarms and peripheral vasodilation occurs. This leads to the opening up of limb blood flow, such that "warm" blood from the core is circulated out to the colder limbs and peripheries. This blood becomes cooled and then circulates back to the core, along with the initial surge of cold blood, thus producing an afterdrop in core temperature. The same mechanism of increased circulation to the limbs is proposed to occur during active (e.g., exercise) rewarming, again resulting in the cooling of blood and an afterdrop.

2. **Biophysical.** This mechanism, argued in studies by Golden and Hervey (1977), relies primarily on basic biophysical thermodynamics in its explanation of afterdrop. Similar to the process just outlined, vasoconstriction and pooling of blood and heat in the core occur with cold exposure and descent into hypothermia. Upon the initiation of rewarming, the warming of the skin results in the removal of the drive for sustained vasoconstriction and core pooling. When this occurs, the simple temperature differences between the core and peripheries would necessitate establishment of a thermal gradient. According to thermodynamics, heat energy must travel down the gradient; therefore the heat from the warmer core must flow outward to the limbs and peripheries, resulting in an afterdrop that will be sustained until the thermal gradient is eliminated and then ultimately reversed.

A logical extension of the biophysical mechanism of afterdrop is that, because of the immutable movement of heat energy along the thermal gradients established from sustained cooling, afterdrop is not a phenomenon that is open to a high degree of manipulation by rewarming techniques. In addition, it can be argued that the large degree of core temperature drop required for a core temperature of >30° C (which is likely in a conscious or semiconscious survivor) to decrease to deep body temperatures of <25 °C (required for cardiac events) cannot be explained by afterdrop. Finally, the phenomenon of afterdrop appears to be highest when measurement is at the rectum, where conductive heat flow predominates, and is much smaller at the esophagus with its higher rate of blood flow and circulatory heat transfer. Therefore, it is argued that afterdrop may not be a major source of concern in rescue situations, though it remains important to consider methods of rescue and rewarming that minimize shock to the victim (Golden and Tipton 2002).

Cardiovascular Redistribution

Even if it remains a minor factor in actual postrescue collapse, the physiological model of afterdrop highlights the possible threat from sudden redistribution of blood flow or volume during rescue and rewarming. Indeed, as this section will outline, cardiovascular stress may be the predominant catalyst of postrescue collapse rather than any further thermal damage to the body.

In prolonged immersion situations, the victim would ideally be dressed in a survival suit that provides thermal insulation and also flotation. Due to the supine or semisupine posture brought about by the buoyancy and flotation of the clothing, its bulkiness, and the extended exposure period, activity may be minimal. Combined with vasoconstriction of the large muscle mass in the limbs, even if the core and heart are being gradually cooled, the overall cardiovascular demand for blood flow may be quite low.

This situation changes dramatically upon rescue. After a prolonged period of inactivity, the survivor may begin waving to rescuers or engaging in activity (swimming toward a boat, climbing up a ladder), elevating heart rate and stimulating circulation to the limbs. The survivor may be hoisted up into a helicopter using a winch and a brace underneath the arms and around the torso, resulting in a sudden transition from a horizontal to a vertical posture. The postural change may cause a sudden drop in blood pressure as seen with cases of orthostatic intolerance. Especially if the victim is not actually secured in a rescue vessel (e.g., still in the process of being winched into a helicopter, climbing a rope ladder), even a brief episode of orthostatic intolerance will cause the body to go limp, sending the victim back into the water and possibly causing drowning. Therefore, whenever possible, rescuers eliminate or minimize the participation of the victim in the rescue process, for example by using rope stretchers that wrap around and "roll" the victim onto the rescue craft. If winches are used, postural changes could be minimized by stretchers or the use of a second brace under the knees.

Even with successful movement into the rescue vessel, the process of rewarming, as already noted, may bring about sudden circulatory redistribution. Irrespective of thermal afterdrop, this increased demand on a cold and sluggish heart may be sufficient to cause a cardiac episode. Finally, it is hypothesized that relief at the prospect of rescue may bring about a psychological "letdown" that precipitates a drop in sympathetic nervous system response, which in turn predisposes individuals to physiological collapse. The linkages between psychological, hormonal, and physiological responses to cold water immersion and its aftermath, especially in actual field situations, remain unknown. However, from a practical perspective, it is essential that survivors be monitored continuously throughout the postrescue period in case of collapse, and that they are not placed in situations of sudden shock or conditions requiring extensive and sudden activity.

SUMMARY

The potential for death from cold water immersions is very real, and the threat of incidents and entry into water is always present in any marine activity. Unfortunately, human nature can be very good at rationalizing why survival equipment is not necessary and at minimizing the risk for accidents. In addition, interpretation of the early scientific data on hypothermia and cold water immersion may have promoted some misconceptions about the dangers and appropriate behaviors. Therefore, continued scientific research into hypothermia and cold water immersion and its translation to public health messages are a critical application of environmental physiology. Chief among the recent advances in this field has been the categorization of cold water survival situations into distinct phases, each with its own particular physiological and behavioral responses. Further research on these individual phases has the potential for developing better safety equipment, training programs, public messages, and first aid and rescue responses.

CHAPTER

7

Diving and Hyperbaric Physiology

The popularization of the Aqua-Lung, which regulates the pressure emerging from compressed air tanks to ambient pressures, by Jacques Cousteau and other underwater explorers in the middle of the 20th century made the oceans and seas accessible to the general population. Before then, the underwater environment was largely the realm of commercial divers, with limited research using diving bells and bathyspheres. In stark contrast to the situation with human achievements in space, the underwater environment is relatively unexplored and logistically more difficult to attain. This was dramatically illustrated by the 2000 disaster of the Russian submarine *Kursk*. With the vessel damaged from an explosion and grounding in ~100 m (328 ft) depth on the seabed, the trapped submariners were not able to escape, and all hands had perished by the time external assistance was coordinated and reached the vessel about a week later.

The difficulties of working in a hyperbaric environment are numerous. Humans are able to survive and thrive in permanent settlements from below sea level to in excess of 4000 m (13,120 ft) elevation, and the summit of Everest (8848 m or 29,030 ft) has been reached without assistance from supplemental breathing apparatus (see chapter 9). In contrast, humans are not able to breathe underwater except with supplemental breathing systems. A snorkel is one underwater option, but the maximum depth at which a snorkel can be utilized is only 1 m (3 ft). The reason is that the hydrostatic pressure from the surrounding water quickly exceeds the capacity of the inspiratory muscles to expand the thoracic cavity. At the same time, extending a snorkel's length or diameter to increase diving depth or decrease breathing resistance, respectively, serves only to increase the anatomical dead space—the ventilated volume that does not directly perform gas exchange with the blood, such as the nasal passages and trachea in normal breathing. Below this depth, breathing systems must be used at the cost of additional weight from gas tanks, increasing the metabolic cost of exercise, though this is diminished somewhat by the buoyancy of water.

In view of the numerous challenges to continued underwater activity and exploration, the purpose of this chapter is to explore the physiology of working in the underwater environment. Beginning with an examination of the physics of pressure influences on air volumes, we examine the physiological basis for many of the clinical issues associated with diving, from decompression sickness and barotraumas to inert gas narcosis and high-pressure neurological syndrome. The respiratory and cardiovascular challenges from underwater activity, along with advances in specialized equipment such as closed-circuit rebreathers and diving gases, are also outlined. The reader is referred to *Bennett and Elliott's Physiology and Medicine of Diving* (Brubakk and Neuman 2002) for a classic and comprehensive text in this field.

THE PHYSICS OF DIVING

Sport and commercial undersea diving is governed by the physical differences between an air and a water environment, and also by several physical relationships between pressure and gases. The underwater environment is also a difficult ergonomic environment to work in, with visibility impediments from both low lighting and turbulence that make it very challenging to see and to navigate. Other factors affecting function include buoyancy factors and difficulty in communication. As we have seen in the chapter on cold water immersion, the higher heat conductivity of water and the low temperatures of a large part of the oceans and seas pose challenges in maintaining thermal homeostasis for divers, possibly exacerbated by the effects of breathing gases compressed at high pressures.

The Water Environment

Humans live in a pressurized air environment, with the pressure at sea level standardized to 760 mmHg or 1 atmosphere (ATA) and reflecting the total weight of the column of air in the atmosphere above the individual. Upon diving, the total pressure becomes a combination of the weight of the air column in the atmosphere and the weight of the water column above the diver. The first obvious change in a water environment is the much greater density of water compared to air. Because of this density, the pressure from 10 m (33 ft) of seawater (msw) is equivalent to another 1 ATA of pressure. Water is relatively noncompressible, such that diving pressures correlate directly with increasing depth. Therefore, each additional 10 msw depth results in an additional 1 ATA of pressure.

The human body is largely composed of water, and depth and pressure do not exert a direct effect on this aqueous component. Therefore, problems such as decompression sickness (DCS) and barotraumas are not due to squeezing or compression of the cells or organs. However, major problems arise from the fact that the human body has large air cavities within many tissues and organ systems, along with the human requirement to breathe air. Examples of relatively large air cavities include the respiratory airways, lungs, and gastrointestinal spaces. Within the head, air cavities are also present in the sinuses and the middle ear; and it is important to note that small air cavities exist in most tissue, including bone, as well. The volumes of space within these cavities are finite and constrained (e.g., lung volume, skull), so only minimal change or expansion is possible without severe damage. It is this disparity in compressibility between air and water, and the inability to equalize pressure and volume within the air spaces, that are the bases for many of the problems associated with hyperbaria.

Boyle's Law

In the 17th century, the scientist Robert Boyle first outlined the relationship between pressure and gas volume. What he proposed was that, *in a closed system where temperature remains constant, the volume is directly and inversely proportional to the pressure.* Simply put, as pressure doubles, the volume of the gas decreases by half. In addition to its impact on diving physiology, this law has major logistical implications in that it affects the absolute amount of gas consumed during diving and therefore the duration of diving possible with air tanks. With self-contained underwater breathing apparatus (SCUBA), air is breathed at ambient pressure. Therefore, at a pressure of 2 ATA, assuming a similar tidal volume and minute ventilation, the rate of air usage is twice that at surface (1 ATA) pressure, in turn halving the time a diver may operate (see figure 7.1). As we shall see, Boyle's law and the imbalances between pressures in different air cavities are also the prime determining factors underlying a multitude of diving medical issues and pathologies, including air embolisms, pneumothorax, and other barotraumas.

Henry's Law

William Henry was responsible for proposing another gas law with high relevance to hyperbaric physiology. Henry's law can be summarized as follows: *At a given temperature, the amount of gas dissolved into a fluid is directly proportional to the pressure.* In simple terms, the higher the ambient pressure, the greater the amount of gas dissolved into a liquid. Additionally, the amount of gas dissolved is also dependent on the solubility of the gas into the liquid. In basic respiratory physiology, carbon dioxide has a much higher rate of solubility in water than oxygen, which explains why the arteriovenous pressure differential is much smaller for carbon dioxide than for oxygen. In combination with Boyle's law, Henry's law and the dissolution of nitrogen at high pressures into the body tissues become the primary mechanism behind DCS (see later section on DCS) when there are overly abrupt changes in pressure.

RESPIRATORY RESPONSES TO A HYPERBARIC ENVIRONMENT

The respiratory consequences of diving revolve around several important changes: (a) the requirement to breathe through a regulator, (b) the higher density of gases and changes in hydrostatic forces with depth and pressure, and (c) the lower temperature of inhaled air at pressure and in the cold.

One of the requirements of recreational diving and most commercial diving is the use of a regulator mouthpiece and SCUBA to provide a regulated flow of air to the lungs. However, an inevitable consequence of a supplemental air source is an increase in the work of breathing. This arises from the increased anatomical dead space in the air system and also the higher inspiratory pressure required to overcome the pressure within the regulator. As a result, the energy required to exercise at the surface with SCUBA is elevated compared to that for swimming without supplemental air at the

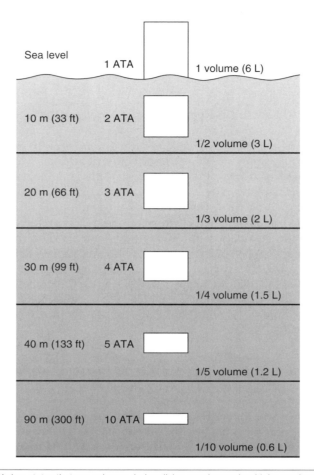

Figure 7.1   Boyle's law states that gas volumes during diving vary inversely with increasing depth, with each 10 m (33 ft) of seawater (msw) equaling an additional atmosphere (ATA) of pressure. A free diver taking a full breath (e.g., 6 L) at the surface would find the lung volume decreased to 3 L at 10 m depth (2 ATA) and 2 L at 20 m (66 ft) depth (3 ATA). Conversely, a SCUBA diver taking a full breath (e.g., 6 L) at 10 m depth and then holding that breath upon ascent would find the volume expanding dangerously to a theoretical 12 L at the surface, resulting in dangerous or even fatal barotrauma.

Data from McArdle, Katch, and Katch 2007.

surface. In addition to any further factors, this increase in the work of breathing and metabolic rate also causes a greater rate of air usage, thus decreasing dive duration.

As evidenced by research on firefighters exercising in air with self-contained breathing apparatus (SCBA), the magnitude of the additional metabolic cost in exercising with a regulator system can be substantial. This can be true irrespective of any additional effect from added weight or changes in ambient pressures or temperature. Figure 7.2 illustrates the ventilation rates in a simulated firefighting task, in which subjects exercised continuously at a steadily increasing intensity. In all conditions, they wore the full firefighter turnout kit and SCBA equipment (Butcher et al. 2006). The difference in ventilation rates at a given exercise intensity between breathing ambient air through a low-resistance valve and breathing compressed air through a

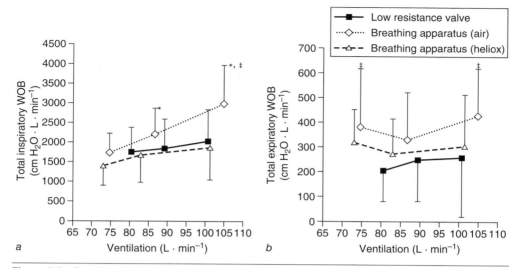

Figure 7.2 Breathing through regulators can add significantly to both the *(a)* inspiratory and *(b)* expiratory work of breathing (WOB) and consequently the metabolic demands of exercise, as demonstrated in this experimental study on firefighters. In contrast, the lower density of heliox reduces the work of breathing back to baseline (low-resistance valve) levels and may be a potential ergogenic aid for occupational work requiring the breathing of supplemental air.

With kind permission from Springer Science+Business Media: *European Journal of Applied Physiology,* Impaired exercise ventilatory mechanics with the self-contained breathing apparatus are improved with heliox, Vol. 101, 2007, pgs. 659-669, S.J. Butcher, R.L. Jones, J.R. Mayne, T.C. Hartley, S.R. Petersen, figure 3.

CLOSED-CIRCUIT SCUBA

In military special forces diving, one key goal is to avoid detection by the enemy during missions. Therefore, another parallel goal in diving technology has been the development of closed-circuit SCUBA breathing systems, whereby exhaled air is not bled through a regulator and into the water (see figure 7.3). Instead, the exhaled air is recycled and rebreathed, and the carbon dioxide produced by metabolism is removed by a scrubbing system. Besides eliminating the threat of detection from bubbles at the surface, another major advantage of closed-circuit systems is that a much smaller amount of gas—only the oxygen that is actually consumed—needs to be carried. Also, because typically only oxygen is inhaled, the risks of nitrogen narcosis and DCS are removed.

However, closed-circuit SCUBA is not without its physiological problems, as oxygen at high pressures can be toxic to the body. One reason is that the supersaturation of oxygen in the blood and hemoglobin reduces the uptake of carbon dioxide from cellular metabolism by the formation of carboxyhemoglobin, producing high CO_2 buildup within the cells. Oxygen toxicity can also be marked by central nervous system seizures. Therefore, specific dive tables for oxygen diving have been created, and dive depths are typically limited to 6 m (20 ft) or less. To minimize the risk of oxygen toxicity, closed-circuit systems may also employ helium or nitrogen within the gas mixture to decrease the partial pressure of inspired oxygen.

(continued)

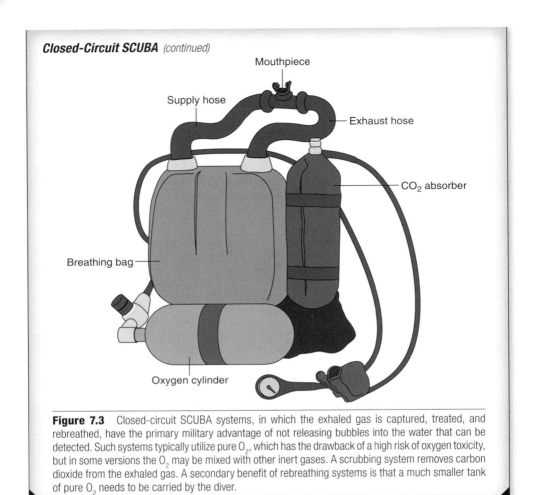

Closed-Circuit SCUBA *(continued)*

Mouthpiece

Supply hose

Exhaust hose

CO_2 absorber

Breathing bag

Oxygen cylinder

Figure 7.3   Closed-circuit SCUBA systems, in which the exhaled gas is captured, treated, and rebreathed, have the primary military advantage of not releasing bubbles into the water that can be detected. Such systems typically utilize pure O_2, which has the drawback of a high risk of oxygen toxicity, but in some versions the O_2 may be mixed with other inert gases. A scrubbing system removes carbon dioxide from the exhaled gas. A secondary benefit of rebreathing systems is that a much smaller tank of pure O_2 needs to be carried by the diver.

regulator was substantial, especially at higher exercise intensities. However, there is an additional benefit in technical diving with regulators using special gas mixtures like helium-oxygen (heliox), as the lower density can decrease the ventilatory rate and the work of breathing. In the same test on firefighters, breathing heliox through a regulator reduced ventilation to levels similar to those from breathing ambient air without a regulator (Butcher et al. 2007).

BAROTRAUMA

The problems relating to Boyle's law and rapid changes in pressure are the most immediately dangerous effects from diving (DeGorordo et al. 2003). Therefore, the absolutely primary safety rule when diving with SCUBA is this: *Never hold your breath!* With breath-hold diving, the absolute amount of air in your lungs remains constant because you are not taking further breaths underwater. Therefore, according to Boyle's law, the increased hydrostatic pressure compresses your lung volume, but the volume will increase as you ascend and will return to normal at the surface. However, as SCUBA systems are designed to regulate inspired air to the ambient pressure, your lung volume at 2 ATA is full at normal surface levels. If you held your breath as you ascended, Boyle's law dictates that the pressure would be halved and the relative volume of gas in the lungs would double. The resulting rupture and collapse of the lung tissue (pneumothorax), in addition to being problematic by itself, causes an air embolism from the release of large volumes of air into the pulmonary circulation. The bubbles rapidly expand throughout the body and are especially dangerous if they reach the cerebrospinal fluid and the brain cavity inside the hard enclosure of the skull.

Apart from air embolisms from improper ascent, barotraumas can occur with rapid compression and inability to equalize pressures during descent. Clinical barotraumas associated with breath-hold or SCUBA diving include the following (see figure 7.4):

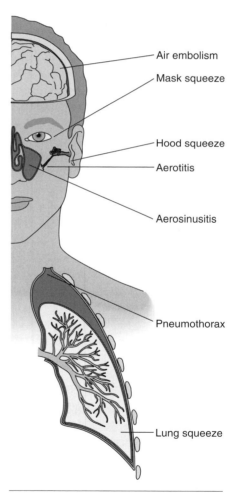

- **Lung squeeze.** Rather than problems with expansion, lung squeeze is associated with compression of respiratory spaces. With breath-hold diving, the hydrostatic pressure with depth compresses the lung tissues, and the potential exists for the total lung volume to be compressed to the level of residual lung volume or below. If this occurs, the lung tissue and alveoli can become damaged and the thin pulmonary capillaries may burst, causing blood and fluid leakage into the lungs and drowning.

- **Mask squeeze.** The volume and pressure within the diving mask must be equalized during changes in pressure. The eyes are highly sensitive tissue with a rich blood supply, and capillary rupture can occur during descent. Equalization of pressure can be readily achieved via exhaling through the nose, thereby adding air volume into the mask space. Therefore, breath-hold and SCUBA diving must not be performed with ordinary swimming goggles.

- **Aerotitis.** The tympanic membrane (eardrum) is a thin and sensitive membrane in the aural canal separating the outer and middle ear, and pressure differences between these two spaces must be equalized during diving. The common means of doing this is to force air through the Eustachian tubes, as during descents in commercial airplanes. In severe cases, the tympanic membrane may rupture.

Figure 7.4 The failure to equalize pressure within the various gas spaces inside the body during ascent or descent can result in barotraumas.
Adapted from McArdle, Katch, and Katch 2007.

SUBMARINE UNDERWATER ESCAPE

Unlike what is depicted graphically about wartime submarine life in films like *Das Boot*, modern submarines are relatively safe vessels during peacetime. However, accidents from mechanical failures and impact with other ships or the seafloor are still possible and frequent. In this century, the most infamous accident was the death of all 118 crew aboard the Russian submarine *Kursk* in August 2000 following an accidental explosion. Even though the *Kursk* was aground in only ~100 m (328 ft) of water, international rescue efforts were ultimately ineffective. Since then, accidents have continued to occur throughout the world's navies, with 13 reported accidents involving submarines from 2000 through 2007 and the complete loss of several crews.

In addition to potential flooding and fire, submarine accident survivors face several immediate problems that compound and intensify with time. Some of the inevitable environmental issues are the depletion of limited oxygen supply, the continued buildup of carbon dioxide from respiration, and the cold environment once the battery and heating systems fail. In addition, depending on the situation, there may be smoke, carbon monoxide, and other toxic fumes, along with the risk of atmospheric changes from flooding or hull damage. The latter are especially problematic because they raise the concentration of gases, potentiating the effects of hypercapnia and also increasing oxygen use and exacerbating climate control issues, as well as producing the risk of DCS with subsequent rescue or escape.

With short-duration exposure, research in my MSc lab and elsewhere on the effects of hypercapnia and hypoxia suggests that thermoregulation is not significantly impaired by these additional environmental stressors (Lun et al. 1993). However, the situation obviously remains ultimately fatal. Faced with these threats, survivors have two primary options for rescue. The first is to await external assistance, which would include either raising the submarine or using deep-submergence rescue vehicles to dock with the vessel and evacuate survivors back to the surface. The second option is to self-rescue by performing emergency ascents from the submarine up to the surface. Using special escape hatches built into the submarine, escapees exit into the water and then ascend to the surface. Ascent can be performed freely with no breathing assistance, relying on the escapee's consistently exhaling a single breath throughout to avoid barotrauma. Alternately, emergency breathing hoods permit inhaling and exhaling for a brief period of time. Such emergency ascents are practiced in special training facilities, and epidemiological reports conclude that such training is relatively benign and presents low risk for barotraumas and other clinical problems for trainees with no existing respiratory conditions (Yildiz et al. 2004). Nevertheless, the reality of donning survival suits and attempting to ascend 100 m (328 ft) or more through frigid waters and into a hostile surface environment remains a daunting prospect.

The most recent development in submarine escape has been the integration of a complete individual survival system, culminating in the Submarine Escape Immersion Equipment (SEIE) MK-10 that has become standard issue in both the Royal Navy and the U.S. Navy. The SEIE completely encloses the submariner, providing buoyancy to enable reaching the surface at a controlled rate and breathing support during escape. Upon reaching the surface, the system inflates to become a single-person life raft–dry suit, providing thermal insulation and flotation. Such systems are rated for escapes from 183 m (600 ft) depth, greatly extending the possibility of escape from submarine disasters.

- **Aerosinusitis.** The sinus cavities within the head are another potential source of air expansion and capillary rupture. Both aerotitis and aerosinusitis can be especially prevalent when the membranes are already inflamed, and diving with a common cold or sinus infection is generally not recommended.

- **Hood squeeze.** Although similar to a mask squeeze, a hood squeeze is caused by a diving hood that creates a tight seal around the flexible portion of the ear (pinna). With the seal present, a volume of air is trapped in the external auditory canal. As the external pressure increases during descent, this volume of trapped air can be squeezed to the point that the tympanic membrane ruptures. Regardless of whether the tympanic membrane ruptures as a result of aerotitis or a hood squeeze, the result is an in-flush of water from the surrounding environment. As this inrushing water envelopes the inner structures of the ear and cools the portion of the Eustachian tube next to the semicircular canals, impairment of the vestibular system (vertigo) may occur, causing the diver to become disoriented and lose spatial awareness.

DECOMPRESSION SICKNESS

Fittingly, given his work in developing the eponymous gas law, Robert Boyle was the first to report evidence of decompression sickness (DCS) in his laboratory. Using an air pump to compress and decompress animals, Boyle reported the development of bubbles in the eye of a viper. With the continued evolution and use of diving bells and underwater breathing apparatus along with caissons (dry chambers for underwater construction of bridges and tunnels), hyperbaric workers began developing symptoms of lethargy and fatigue, muscle and joint pain, and difficulty in standing up. The latter symptom, with victims limping and appearing bent over upon return to the surface in early reports, led to the popularization of the term "the bends" to refer to DCS. In more serious cases, victims may experience severe pain throughout the body, neurological disorders, cardiac arrest, and death.

As discussed in connection with Boyle's law, the major risk inherent in diving is the overly rapid changes in pressures within the body. The method I use to teach this to my physiology class is dramatic and simple. I take a couple of bottles of soda pop (champagne in the classroom is frowned upon by administration!) and discuss the fact that the "fizz" consists of carbon dioxide dissolved into the water at high pressures, with the bottle then capped. Then with one bottle, I unscrew the cap slowly, which typically results in the formation of a few small bubbles. With the other bottle, I give it a little shake to accentuate the bubble formation, then quickly remove the cap and send the bubbles rushing out. As I'm making a sticky mess all over the classroom floor, I tell the students that this is their blood and tissues if they decompress too rapidly and suffer an extreme case of the bends.

In normal dives, the unopened bottle is analogous to the diver's body in which nitrogen is compressed at high pressures and dissolves into blood and body tissues. When the diver ascends within the relevant time guidelines and at a moderate rate, bubbles are gradually produced as the nitrogen comes out of solution and need to be eliminated. This elimination generally occurs via filtering through the pulmonary circulation into the lungs and eventually out of the body through the exhaled breath, both during ascent and once the diver is at the surface. Imagine now the shaken and quickly opened soda or champagne. The sudden drop in ambient pressure when a diver rapidly ascends to the surface after a prolonged period at depth causes the dissolved nitrogen to come out of solution and bubble—in the skin, the muscles, the synovial fluid in the joints, the blood, and the nervous system. This forms the basis for DCS, and the symptoms and severity depend on the location, along with the size and frequency, of the bubble formation (see table 7.1).

Dive Tables

With compression and nitrogen as the prime determinants in DCS, the major predictors for bubble formation and DCS logically become the depth (pressure) and the duration of the dive, along with the rate of ascent. Using these parameters, the rate of gas dissolution can be derived mathematically, leading to the development of dive tables for compressed air diving. The first attempt at calculating nitrogen retention rates and dive tables was made by the father–son scientific duo of J.S. and J.B.S. Haldane in the early part of the 20th century. The basic premise of diving medicine for the general public is to maximize safety by erring on the side of being as conservative

Table 7.1   Symptoms and Gradations of Decompression Sickness (DCS)

	Symptoms	Notes
Mild (type I) DCS		
Skin	Itchiness, mottled patterns	Generally minor but can be prelude to more severe symptoms.
Joints and muscles	"The bends," fatigue, joint pain, difficulty maintaining upright posture	Generally appears 4 to 6 h postdive; can be mild and intermittent, ranging to more severe and constant. Generally not life threatening but can be prelude to more severe type II DCS.
Severe (type II) DCS: Major organs and nervous system		
Ear	Vertigo, "the staggers"	Bubbles form within the neurovestibular system.
Lungs and heart	• "The chokes": coughing, choking, dyspnea, chest pain • Circulatory collapse	Bubbles travel through bloodstream and enter into the lungs.
Central and peripheral nervous system	• Tingling sensation to loss of sensation or motor function in limbs • Urinary incontinence • Temporary to permanent paralysis • Double vision, headache, confusion, difficulty speaking • Seizures	Severe cases can result in permanent neurological problems and paralysis.
Bone	• Chronic arthritis • Bone wasting (dysbaric osteonecrosis, avascular bone necrosis)	Long-term destruction of bone tissue, especially around the hips and shoulders, produces chronic pain and disability.

as reasonable. Therefore, the majority of dive tables are for a "square" dive profile; that is, the assumed depth for the entire dive duration is the maximum dive depth achieved (e.g., if the dive is for 10 min at surface, 10 min at 20 m [66 ft], and 10 min at 30 m [98 ft], the depth is assumed to be 30 m for the entire 30 min). Recreational diving tables are also generally developed as "no-decompression" profiles; that is, the maximum time at depth is designed to permit direct ascent back to the surface without any decompression stops. As we will see in the section on saturation diving, the pressure within the body and the external environment will eventually equalize if a diver stays at a depth for a sufficient length of time, and the amount of dissolved nitrogen will max out. When this occurs, the time required for ascent will not increase any further. It is this process that is employed by commercial divers at great depths.

Despite the safety designed into dive tables, the profiles are guidelines only, and individual and situational factors can affect the risk of DCS. The following are some of the primary variables:

- **Body fat content.** Solubility of gases depends not only on ambient pressure and the gas itself, but also on body content. Fat tissue has a slower rate of equilibrium for nitrogen than water. Hence, nitrogen is also retained in fat tissue longer upon decompression; thus individuals with higher body fat content may eliminate nitrogen more slowly from the body and therefore have a greater propensity for DCS development. However, integrated analysis of a number of individual characteristics revealed that increasing age, weight, and aerobic capacity had a greater relationship with DCS risk than did body fatness (Carturan et al. 1999).

- **Multiple dives.** Simpler dive tables are designed for a single dive each day. Performance of multiple dives over the course of a day typically means that residual nitrogen remains dissolved in the tissues, such that subsequent dives begin with excess nitrogen already in the tissues and incur a greater risk for DCS. The amount depends not just on the previous dive but also on the recovery period at surface, and needs to be accounted for in repetitive-diving tables.

- **Diving at altitude.** Due to the lower barometric pressure at surface when one is at altitude, the absolute pressure differential is greater and the risk for DCS is elevated. Similarly, air travel shortly after diving can increase the risk for DCS development due to the lower cabin pressure maintained in private and commercial airplane cabins. Therefore, people are generally advised to avoid flying within 24 h of a dive.

- **Patent foramen ovale (PFO).** Up to 25% of all adults may possess a PFO, a small channel between the right and left atria. Functionally, the PFO generally raises little concern in most activities. However, with diving, it is possible that this shunt can cause the passage of some nitrogen bubbles from the venous blood returning into the heart directly into the arterial circulation, rather than through the pulmonary circulation, thus exacerbating the risk of DCS (Bove 1998; Gempp et al. 2009).

If DCS is detected and diagnosed, the only effective treatment involves recompression of the victim and subsequent gradual decompression with close physiological and medical monitoring. For this reason, many major commercial diving sites, such as the North Atlantic off the east coast of Canada, have hyperbaric chambers based at hospitals or universities in a nearby city (e.g., St. John's in Newfoundland) on emergency standby in case of accidents.

Bubble Detection Methods

With the danger from DCS, an important tool in prevention and proper medical care is the accurate detection, monitoring, and quantification of bubble formation and DCS severity. As DCS symptoms can be extremely well localized to specific tissues, it is difficult to develop a standard generalized detection method. Besides a qualitative analysis based on symptom location and magnitude, the primary method of quantifying bubble formation has been through the use of ultrasound recordings of the heart and major blood vessels. In this technique, a Doppler audio recording is made of the veins returning to the heart. A trained technician then analyzes the recording, listening for the "popping" sound signature of bubbles. The frequency and loudness of these sounds can then provide some quantification of denitrogenation. A drawback with this technique remains that advanced training is required to perform the analysis, analogous to training of the ears of sonar operators onboard submarines, yet the results remain highly subjective. Another drawback is that there is often a time lag between recording and analysis unless the technician is on-site. Newer ultrasound imaging techniques permit visual analysis and image recording and can provide more direct and rapid analysis both on- and off-site.

The eyes are a promising site for nitrogen bubble analysis, as the frequent production of fluid from the tear glands provides a rapid time course for quantifying bubble formation. Indeed, as already mentioned, Boyle's original report of DCS in 1670 noted bubble formation in the eyes of a viper upon rapid decompression. However, despite this history, eye measurement has been largely neglected until recently. Mekjavic and Mekjavic (2007) systematically performed simulated chamber dives on subjects to 2 ATA for 60 min using compressed air and also pure oxygen. In the compressed air condition (i.e., nitrogen present), microscopic examination revealed bubbles and increased ultrasound reflectivity of the vitreous humor, whereas no changes were observed with pure oxygen. In addition, microscope and ultrasound analyses detected progressive changes in the eye up to two days following both a no-stop dive and a dive requiring decompression stops. Overall, these data are promising in their specificity to nitrogen and their sensitivity to the time course of denitrogenation. Therefore, further investigations fine-tuning the use of ocular analysis may provide for simpler and faster monitoring of decompression safety.

INERT GAS NARCOSIS

The existence of impaired mental acuity and performance in humans exposed to a compressed air environment for prolonged periods, such as divers and tunnel workers, has been generally recognized since the initial documentation by Junod in 1835 (referenced in Brubaak and Neuman 2002). Jacques Cousteau coined the term "the rapture of the deep" to refer to this problem. With compressed air pressures exceeding 4 ATA, symptoms of intoxication, euphoria, and narcosis can be observed (Fowler et al. 1985). As pressure further increases to 7 to 10 ATA, mental processes and decision-making ability slow, and loss of memory and neuromuscular coordination occurs; at these pressures, the diver may suffer loss of consciousness.

Many causes for this impairment at depth were initially proposed, including the effects of pressure itself, latent suppressed claustrophobia, altered circulation leading to blood stagnation, and impurities in the breathing mixture. Behnke and colleagues

(1935) were the first to deduce that the impairment resulted as the increased nitrogen pressure in the compressed air acted as an anesthetic agent. This compressed air intoxication has since been termed inert gas narcosis and has been compared to the symptoms present during the early stages of anesthesia. With increasing awareness of the cause of this intoxication at high pressures, diving regulations have been adopted that recommend against the use of compressed air for diving below 6 ATA except in extreme circumstances.

The chemical structure of an inert gas is unaltered in the body, and no covalent or hydrogen bonds are formed with the body tissues under biological conditions. Inert gases include the clinical anesthetics nitrous oxide (N_2O), ethylene, and cyclopropane, along with the noble gases nitrogen, xenon, argon, krypton, neon, and helium. Of the noble gases, some (e.g., xenon, argon, and krypton) are narcotic at atmospheric pressures; nitrogen is narcotic at supra-atmospheric pressures, while other of these gases (e.g., helium and neon) are much less narcotic than nitrogen. From a performance perspective, the primary danger of inert gas narcosis comes from its effect on mental acuity and function, making it extremely difficult for divers to follow work plans or to adapt to new situations (Fowler et al. 1980). Therefore, it becomes critical to ensure that a detailed work plan be developed and that extensive training in planned and contingency maneuvers be performed prior to missions. Furthermore, communication between the diver and the base must be maintained throughout missions unless silence is absolutely necessary. Therefore, as outlined next, helium has been used extensively in deep and saturation diving situations as a replacement for nitrogen.

The exact cause and mechanism of inert gas narcosis remain contentious (Sessler 1993). Due to the nonreactivity of inert gases, biophysical rather than biochemical causes of inert gas narcosis have been stressed. As early workers observed a strong correlation between the solubility of an anesthetic agent or noble gas in lipid and its narcotic potency, one theory suggests that the site of anesthetic action is within the lipid membrane. Thus, narcotic action may derive from dissolution into the lipid membrane region, resulting in the expansion of the membrane and disruption of neural functioning. The expansion of neural membranes would alter the ability to conduct ions across the membrane, resulting in alterations to the propagation of action potentials. In contrast, hydrostatic pressure may offset the expansion and narcotic action somewhat via compression of the lipid membrane. It therefore appears attractive to suggest direct opposing sites of action between these two factors, such that pressure and narcosis would be directly antagonistic and one may be used to counteract the other. However, this may be a gross oversimplification, as isolation of the site of action for each remains elusive. Other possible mechanisms for the anesthetic action of inert gases may be through the alteration of transmitter release from the presynaptic terminal or a decrease in postsynaptic electrogenesis.

Another safety issue arises from the possibility of undetected hypothermia during diving, as both N_2O and compressed nitrogen appear to inhibit shivering responses. In cold water immersion either with 30% N_2O or at 6 ATA pressure in a dive chamber, the threshold for the initiation of shivering decreased to a lower core temperature (Mekjavic et al. 1995). However, once shivering was initiated, the sensitivity or the gain of the shivering response to further core temperature decreases was not affected. Beyond the impaired physiological response to hypothermia, another issue is the altered subjective perception of thermal status. Due to the anesthetic action, subjects experiencing narcosis have also exhibited an attenuated sensation of cold, such that they rated themselves as comfortable even as core temperature was decreasing. Interestingly, the decrease in thermal comfort was progressive with increasing N_2O concentrations from 10% to 25%, whereas shivering impairment plateaued and did not change further from 15% to 25% N_2O (Cheung and Mekjavic 1995); this suggested differing physiological and behavioral impairment patterns with compressed N_2. Operationally, the behavioral inability to appropriately sense cold makes it difficult for divers to self-regulate their thermal status via methods such as manual heating valves. Therefore, the development of automated thermal countermeasures for commercial divers, and education about the potential for undetected hypothermia from nitrogen narcosis for commercial and recreational divers, should be avenues for investigation.

AN ANALOG FOR HYPERBARIC NITROGEN?

Due to the logistical and safety issues involved in performing experiments underwater or at pressure inside a dive chamber, research on inert gas narcosis could be greatly enhanced if an analog could be found that elicits behavioral and physiological effects similar to those of compressed nitrogen at normal pressures. Experiments at normal pressures would also enable the isolation of narcotic effects from confounding factors such as water immersion, hydrostatic pressure, and hyperbaria. One readily accessible candidate appears to be nitrous oxide (N_2O), the "laughing gas" used as an anesthetic in many dental procedures. Early behavioral studies suggest that 30% N_2O at sea level pressures is roughly equivalent in terms of mental and task performance impairment to compressed air at 6 ATA pressures (Fowler et al. 1985). My MSc research lab had a strong research focus on diving physiology. In this vein, N_2O was used to investigate inert gas narcosis effects on thermoregulation and also muscle function during compressed air exercise. Dive chamber research with compressed air at 6 ATA demonstrated significantly greater cooling rates and lower perception of cold compared to 1 ATA (Mekjavic et al. 1995), and these changes in autonomic and subjective cold response were similar to that from 30% N_2O at sea level pressures (Passias et al. 1992).

It has been generally assumed that the various inert gases induce qualitatively similar effects on humans, though little work has directly compared and correlated the effects of two inert gases such as N_2 and N_2O (Fowler et al. 1980). For example, while my own MSc project addressed the dose–response nature of thermoregulatory impairment from N_2O (Cheung and Mekjavic 1995), the exact physiological and behavioral correlation between N_2O concentrations and hyperbaric N_2 remains unclear. Overall, an important but not necessarily valid assumption of experiments on inert gas narcosis is that the mechanism of action and the physiological effects of these gases are similar to those of other general anesthetics and that results can be transferred. In addition, much of the research on the effects of inert gases has concentrated on qualitative behavior or mental performance, and has deliberately employed high concentrations to elicit an unambiguous level of narcosis. Therefore, the degree of narcosis in such studies is typically much higher than that faced by hyperbaric workers, who are legally limited to 6 to 7 ATA compressed air exposure, and the effects of subanesthetic levels of inert gas narcosis remain unclear.

SATURATION DIVING

In recreational and in much of commercial diving using SCUBA, the overall dive duration is relatively short, requiring either no or only short decompression stops to eliminate dissolved gases from the body. Commercial diving, however, is not restricted to depths of only 6 to 7 ATA. Construction and inspection of structures such as oil rigs and pipelines can be done with machines to a large extent but still place a heavy emphasis on human intervention. However, great depths require exponentially longer decompression stops, making a rapid dive to depth and return to surface prohibitive logistically. Saturation diving (see figure 7.5 for a sample dive profile) refers specifically to a protocol in which the diver stays at a particular depth long enough that the body reaches equilibrium with the ambient pressure. At this point, the maximal quantity of gases is dissolved into the tissues, and further duration at that pressure does not add to the decompression times required. Therefore, logistical rather than physiological determinants dominate diving schedules beyond the point of saturation.

A typical oil platform dive profile may consist of several days to compress down to 30+ ATA inside an integrated diving bell and hyperbaric chamber docked onto a specialized tender boat or platform. Once compression is achieved, the diving bell and chamber is lowered into the appropriate depth under the water, whereupon divers may exit the chamber to perform diving work. At the end of each work bout, the chamber is brought back up to the tender boat while the pressure inside is maintained. This compressed phase may last for days to several weeks, during which the equilibrium point between the body and the environment may be reached or exceeded. Eventually, the divers inside the chamber begin a prolonged and very gradual decompression phase that may last one to two weeks depending on depth, resulting in a three- to four-week overall dive phase for each group of divers. During and beyond this time, divers are carefully monitored for the symptoms associated with DCS, and a recompression chamber is made available in case of emergencies.

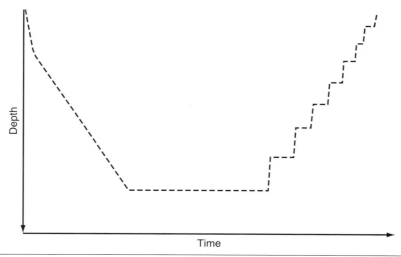

Figure 7.5   Generalized profile of compression and decompression schedules for a saturation dive. To minimize the risk of high-pressure neurological syndrome, the rate of compression is initially rapid and then progressively decreases. A series of stops during ascent to avoid decompression sickness can take nearly two weeks depending on depth and duration.

As already noted, the threat of mental and physiological impairment that occurs when the high partial pressure of nitrogen causes inert gas narcosis prevents the use of compressed air beyond 5 to 6 ATA pressure for commercial diving. Beyond these pressures for both "bounce" (rapid descent and ascent) and saturation diving, gases must be employed that replace the nitrogen with a less narcotic alternative (see table 7.2 for a summary of the major diving gases). Of these, the most popular gas mixture is heliox, a mixture of 98% or more helium and a small percentage of oxygen. Helium has the advantage of being an inert gas that is light and much less narcotic than nitrogen. However, one disadvantage of the low density is the effect on vocal cords and therefore communications. Other gases employed in saturation diving include

Table 7.2   Common Gas Compositions Used in Diving

Gas	Composition	Uses and benefits	Side effects
Compressed air	Standard air	• Recreational and commercial diving • <5 to 6 ATA • Cheap and readily available	Nitrogen narcosis
Nitrox	Nitrogen-O_2 in varying mixtures dependent on depth and duration	Commercial diving <5 to 6 ATA	Percentage of gases a compromise between O_2 toxicity versus nitrogen narcosis and decompression times
Heliox	Helium and oxygen: % O_2 adjusted to maintain PO_2 ~1.3 ATA for saturation diving	• >5 to 6 ATA with technical and saturation diving • Helium: an inert gas with much lower (~4.3 times less) narcotic effect than nitrogen • Low density of helium ↓ breathing resistance	• Heat loss from high heat conductivity of helium • Speech impairment hampering voice communication • High-pressure neurological syndrome (HPNS), especially with overly rapid compression
Trimix	Heliox mixture with small amount of nitrogen	• >5 to 6 ATA with technical and saturation diving • Low density of helium ↓ breathing resistance • Seems to lessen risk of HPNS	• Heat loss from high heat conductivity of helium • Speech impairment hampering voice communication
Oxygen	Pure O_2, nitrogen-O_2, or helium-O_2	• Closed-circuit SCUBA, primarily for military diving (no bubbles) • Pure O_2 <6 m • Gas mixture adjusted to maintain PO_2 <2 ATA • Rebreathing gas mixture ↓ tank size and weight	• ↑ O_2 toxicity risk when PO_2 >2 ATA, especially with exercise and high ventilation rates • CO_2 scrubbing system required

Many exotic gases (e.g., hydrogen, argon, neon) not listed here may also be used for specific applications.

hydrox (hydrogen and oxygen), which carries with it an elevated risk of combustion, and "trimix" (helium with small amounts of nitrogen and oxygen). As outlined in the following sections, heliox and other gas mixtures are not without physiological and clinical risks.

Thermal Conductivity and Heat Loss With Inert Gases

The use of helium as the primary component of heliox and trimix has been largely successful in ameliorating the dangers of inert gas narcosis and high-pressure neurological syndrome (HPNS) during saturation diving. However, one side effect of a compressed helium environment is a higher risk of hypothermia. A depressed thermoregulatory ability, as evidenced by an increase in the range of preferred thermoneutral temperatures, has been observed and calculated for animals in a laboratory setup for heliox compared to compressed air (Clarkson et al. 1972). Thermal tracking of humans during saturation diving or prolonged hyperbaric chamber dives also demonstrates a lower core temperature and higher rates of heat production via shivering (Brubakk et al. 1982; Nakayama 1978). The mechanism underlying this impairment remains unclear.

Biophysically, helium has much lower density than nitrogen and also has higher thermal conductivity and heat capacity. Exposure to a compressed helium environment, irrespective of ambient temperature, can also contribute to altered thermoregulatory status during saturation diving. For example, while helium does not have the strong narcotic effect of compressed nitrogen, its dissolution in the body may alter neural function or sensitivity to a thermal stimulus, depressing vasoconstrictory tone and shivering onset. The higher thermal conductivity of helium may also contribute to a higher conductive and convective heat loss from the body. Additionally, the higher conductivity can lead to greater rates of respiratory heat loss, especially if the gas mixture is at a low temperature when inspired and ventilation rates are high with exercise (Brubakk et al. 1982). In one of the only studies on the effects of breathing heliox at surface pressure on thermoregulatory variables during exercise, Spitler and colleagues (1980) reported no changes in core temperature compared to breathing air. However, the exercise was brief (10 min), and no details were provided about how respiratory heat exchange was calculated. Cain and associates (1990) carefully measured ventilation rates and inspired and expired air temperatures during breathing cold air ranging from –30 to 0 °C at surface pressure, and reported that respiratory heat loss can constitute 25% to 30% of metabolic heat production at rest and 15% to 20% during a prolonged submaximal treadmill walk.

High-Pressure Neurological Syndrome

While saturation diving has eased the ability to work at great depths, one of the major and relatively unknown risks with high levels of compression is the development of HPNS. The initial symptoms of HPNS include lethargy and fatigue, further progressing in severity to motor dysfunction, tremors, nausea, and seizures. Cognitive ability also can be impaired. The dangers with HPNS come not just from the symptoms themselves, but also from the fact that the victim is in a hyperbaric environment, making rescue via decompression or reliance on prompt medical support extremely difficult. The onset and pathophysiology of HPNS are completely different from those for DCS. The latter, as we have seen, is due to the breathing of high pressures of nitrogen, the dissolution of the gas into body tissue over time, and then an overly quick reduction in the ambient pressure (i.e., ascending too rapidly to the surface) that produces air bubbles. Therefore, DCS occurs only upon ascent to the surface and acts on all body tissues. In contrast, HPNS occurs during the descent (compression) phase and appears to act directly on the nervous system.

The mechanism behind HPNS remains unclear, though it is evident that diving mammals and other aquatic creatures are largely unaffected by repeatedly descending to depth (Talpalar and Grossman 2006). The prevailing consensus is that HPNS may be caused by an overly rapid rate of descent (compression), possibly affecting the structural integrity of the neural tissue or specific ionic channels and transporters within the membrane and resulting in hyperexcitability of the neurons (Daniels 2008). Consequently, saturation diving typically follows an exponential rate of compression, with the rate of descent progressively slowing at greater depths. This manipulation of the compression schedule has greatly reduced the incidence of HPNS, although

ATMOSPHERIC DIVING SUITS

The costs associated with saturation diving are immense, as a crew needs to be compressed, maintained at pressure, and then decompressed over each work cycle. The cost of maintaining the hyperbaric facility and supporting systems, in addition to the high costs for heliox or trimix gases, can be tremendous. Furthermore, in addition to the safety of the divers in a hyperbaric environment, the compression and decompression phases both require days to weeks to achieve safely, with this time serving as nonfunctional time for operation. With the challenges from saturation and other forms of commercial diving, alternatives to compression diving have been an important parallel thrust for engineering invention and design throughout the past century.

The ideal diving system would be an autonomous one-person suit (see figure 7.6), analogous to the spacesuits used by astronauts during extravehicular activities (Carter 1976). Each of these units would have life support requirements similar to those of spacesuits: oxygen supply and carbon dioxide clearance, temperature control, communication, and a continuous power supply lasting 8 to 10 h in extremely harsh and hazardous environments. The fundamental goal would be a 1 ATA environment maintained inside the suit, whereby a diver could quickly descend to great depths, perform a prolonged period of work, and then ascend to the surface without any decompression phase. The other primary requirement would be for a high level of mobility and flexibility within the suit, permitting rapid movement at depth and also a high degree of manual function. The primary engineering challenge is designing joints that are able to withstand the enormous external pressures while permitting maximal range of motion for the upper body.

Important milestones in the development of an atmospheric diving suit (ADS) include use of the Tritonia suit in exploration of the wreck of the RMS *Lusitania* in 1935. This was followed by the JIM suit from the United Kingdom in the 1970s and the Newtsuit from Canada in the 1980s. The Newtsuit has been further developed by the U.S. Navy as the ADS2000, forming the basis for a small specialized diving unit based out of San Diego.

(continued)

Atmospheric Diving Suits (continued)

AP Photo/The News Tribune, Jill DiPasquale

Figure 7.6   An atmospheric diving suit has the major physiological advantage of maintaining 1 ATA pressure inside regardless of ambient pressure, thus removing the need for compression or decompression. One engineering challenge of such suits is designing joints that maintain easy movement at pressure. As with a spacesuit, a full life support system needs to be incorporated into such suits.

These ADS systems have leg joints to permit walking on the seafloor or an underwater platform, but also are able to self-propel. An alternative is the WASP unit, which lacks traditional legs and relies on self-propulsion.

refinement and optimization of the schedule would assist in making saturation diving more efficient. The type of gas breathed in saturation diving also seems to influence the risk of HPNS. Compression to high pressures is typically performed in a heliox environment, and some work has addressed the efficacy of using alternate gases. Of these, a trimix gas of helium and oxygen with a small amount of nitrogen appears to provide some protective effect, though the mechanism of this attenuation remains unclear. Overall, due partly to the lack of understanding of the mechanisms underlying HPNS, little to no information is available on whether any individual variability in susceptibility exists or on potential countermeasures.

SUMMARY

The basis of underwater diving is determined by relatively simple physical laws of gas pressure, volume, temperature, and solubility. However, the realities of these laws impose extreme physiological stresses and clinical problems on the body, in turn requiring immense engineering and logistical effort to overcome. Much remains unknown about the human body's responses to hyperbaria and high gas pressures, including individual variability in gas retention and the mechanisms of inert gas narcosis and HPNS. Field studies are difficult due to the technical challenges of collecting data underwater and at pressure, often in remote locations. Laboratory-based research includes the use of hyperbaric chambers, which impose their own logistical challenges, or else surface simulations or analogs of diving (e.g., nitrous oxide for compressed nitrogen) that may or may not directly translate to actual underwater physiology. Research on the ergonomic challenges imposed by the underwater environment (buoyancy, visibility, communications) also needs to be merged with physiological and clinical research and engineering advances to ensure the health and safety of recreational and commercial divers.

Training and Performing at Moderate Altitude

The year 1968 was a pivotal year in the 20th century, marked by civil unrest and the assassinations of Martin Luther King and Robert Kennedy. And much as the launch of *Sputnik* in 1957 ushered in the space race and spawned the field of microgravity physiology, the 1968 Summer Olympics in Mexico City, at an elevation of 2240 m (7350 ft) above sea level and featuring Bob Beamon breaking the long jump world record by a phenomenal 55 cm (21 in.), helped to launch the field of altitude physiology research. While air at altitude has the exact same composition (20.93% oxygen) as at sea level, the lower total atmospheric pressure results in fewer molecules of oxygen in each liter of inspired air. This can severely reduce the oxygen saturation levels within the body and oxygen's availability within the muscles, resulting in impaired oxidative metabolism and lower performance. Therefore, the ability of athletes to perform optimally at altitude took on significant interest for scientists, coaches, and athletes.

At the same time, spurred on by the emerging dominance of middle- to long-distance running events in Mexico City by altitude natives from Kenya and Ethiopia, scientific research began focusing on the concept of altitude training and its potential ergogenic benefits on exercise performance at lower altitudes. The theoretical basis is simple: With lower oxygen availability, the body responds by stimulating greater production of red blood cells, resulting in greater oxygen-carrying capacity. This should translate to an improved aerobic capacity upon return to sea level. The results of such research are somewhat equivocal, however, with studies split between improved performance and no benefits from altitude training.

IDEAL ALTITUDE FOR COMPETITION

Assuming ideal acclimatization and preparation, another important question for athletes, especially those planning world record attempts, is, *At what altitude is there an optimum trade-off between aerodynamic gains and physiological impairment?* The lower total atmospheric pressure at altitude results in lower air resistance—the holy grail of aerodynamics. This allure of a significant reduction in wind resistance is the reason why many world records, notably the 1 h cycling time trial records by Ole Ritter (1968), Eddy Merckx (1974), and Francesco Moser (1984), have been based at altitude, though decreased air resistance is not an all-encompassing panacea for performance. The trade-off for the lowered air resistance is increasing difficulty in getting sufficient oxygen at altitude. Notably, no world records were set at Mexico in 1968 in any events of >4 min duration, though this may also reflect the lack of knowledge and understanding at the time concerning how to prepare for altitude. Mathematical models for running propose that the 400 m event is the threshold distance for improved performance at a range of altitudes, and that 800 m and longer events are impaired at all altitudes above sea level (Peronnet et al. 1991). Other models calculate the ideal altitude for a 1 h cycling time trial as ranging from 2000 to 2500 m (2187 to 2734 yd) for unacclimatized and acclimatized athletes, respectively (Bassett et al. 1999). Finally, one must keep in mind that attempting a record at altitude is not a basic requirement for success. Notably, the past nine world records for the 1 h cycling time trial have all been accomplished at altitudes <200 m (656 ft) in elevation.

Most of the studies in the initial era of altitude research focused on individuals adapting by both training and living at altitude. The second wave of research has branched away from this, exploring such concepts as "live high, train low." This trend has been aided by technological advances in the development of relatively portable and inexpensive altitude tents or rooms, permitting athletes to control oxygen levels during both training and recovery without the logistical issues and financial expense of protracted field-based training camps. At the same time, scientists have also explored the possibility of hyperoxic training to maximize training stimulus. Many questions remain about altitude training, including individual variability in response, the rate and physiological mechanisms underlying adaptation, and the optimal implementation of such training for different sports. This chapter surveys the physiological bases for altitude training, explores the latest research and sport science trends in the area, and outlines our current understanding of best practices in implementation. The reader is also referred to a comprehensive text on the use of altitude training in sports (Wilber 2004) for further information on the ergogenic benefits and implementation of hypoxic training.

MODELS OF PHYSIOLOGICAL RESPONSES TO HYPOXIC STIMULUS

As with any of the environments discussed in this text, altitude exposure has systemic effects throughout every organ and tissue in the body. Since this chapter focuses on the use of altitude to enhance athletic performance at sea level (0-500 m [0-1640 ft]) and at low (500-2000 m [1640-6560 ft]) or moderate (2000-3000 m [6560-9840 ft]) altitude levels, this section deals primarily with the hematological model proposed for improved subsequent exercise at sea level, along with a discussion of several potential nonhematological factors that may also contribute to enhanced exercise capacity (Levine et al. 2005). Chapter 9 contains a more detailed discussion of the acute physiological responses to high altitude, especially in relation to performance at extreme altitudes (>4000 m [13,120 ft]) and the mechanisms underlying acute mountain sicknesses.

The Hematological Model of Hypoxic Stimulus

One of the primary physiological adaptations underlying the concept of altitude training is an elevation in hemoglobin and red blood cell mass (see figure 8.1). At a theoretical level, the decreased oxygen availability, due to lower partial pressure of oxygen (PO_2) at altitude, results in local arterial hypoxia within the kidneys and stimulates an elevated secretion of erythropoietin (EPO). In turn, EPO acts upon the marrow within the long bones to increase erythrocyte production (Mackenzie et al. 2008). This hematopoiesis or polycythemia (i.e., increase in hemoglobin or red blood cell mass, respectively) should therefore contribute to an increase in blood volume

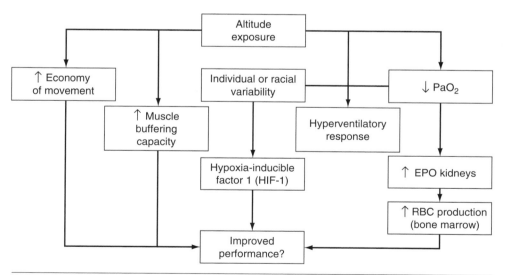

Figure 8.1   Summary schematic of major potential hematological pathway for aerobic improvement from hypoxic exposure, along with other potential pathways for improvement and individual variability in response.

and oxygen-carrying capacity, ultimately producing an increase in aerobic capacity.

Independent of the use of altitude or hypoxia, the linkage between red blood cell levels and aerobic capacity appears unequivocal. In a classic study manipulating erythrocyte levels, the benefits of autologous blood transfusions (the reinfusion of an individual's own blood) were demonstrated in a group of national- and international-caliber runners (Buick et al. 1980). Subjects were tested for $\dot{V}O_{2max}$ and running time to exhaustion (1) before blood collection, (2) following seven weeks of recovery to regain normocythemia, (3) with sham infusion of saline, (4) with erythrocythemia upon reinfusion of the two units (~900 mL) of stored erythrocytes, and (5) upon return to normocythemia 16 weeks after erythrocythemia. The post-reinfusion tests were specifically performed at the 24 h time point to elicit a state of normovolemic erythrocythemia (i.e., no major readjustments by the body in total blood volume to compensate for the additional red blood cells). Compared to baseline tests before blood collection, $\dot{V}O_{2max}$ and exercise tolerance times were significantly increased with the erythrocyte reinfusion condition, with improvements found at both one and seven days post-reinfusion and no benefits following sham infusion. Interestingly, after 16 weeks and the natural reinstatement of baseline hematological levels, $\dot{V}O_{2max}$ remained significantly elevated. The mechanism for the maintained increase in aerobic capacity is unclear, but may stem from the athletes' ramping up their training regimen due to the initial beneficial results. Overall, this study confirms the importance of erythrocyte levels in oxygen-carrying capacity and endurance performance. Unfortunately, such clear evidence may have also confirmed the determination of some athletes to employ this technique, which was not explicitly banned until the creation of the World Anti-Doping Agency (WADA) in the late 1990s, to gain a competitive advantage.

Other methods besides autologous transfusion can be employed to raise erythrocyte levels. Transfusion of homologous blood from donors with similar blood typing can also be performed, and this is common practice for a variety of clinical conditions in medical settings. However, the lack of available donated blood, along with infection risks from transfusions, led to the development in the mid-1980s of artificial, recombinant EPO (rhEPO), primarily for cancer and kidney patients suffering from reduced hemopoietic capacity and subsequent anemia. In an athletic setting, the administration of rhEPO definitively produced hematological and aerobic improvements in a group of fit but nonelite individuals (Berglund and Ekblom 1991). With no specific training but three injections of rhEPO weekly for six weeks, hemoglobin and hematocrit values both increased by >10%, while $\dot{V}O_{2max}$ rose from 4.52 to 4.88 $L \cdot min^{-1}$. Importantly, tolerance time in a progressive run to exhaustion was dramatically extended by nearly 90 s, or 17%. At the same time, submaximal cycling elicited lower heart rates, indicative of a reduction in physiological strain. Another important finding from this study was that, unlike the variability in response to altitude exposure discussed later in the chapter, improvement was consistent and observed in all 15 subjects without any specific training manipulation. Similar hematological benefits have been consistent in other studies investigating the efficacy of EPO in relation to physiological responses in healthy individuals. Overall, therefore, it is clear that rhEPO and blood transfusions can boost red blood cell volume, oxygen-carrying capacity, and aerobic performance.

In summary, the hematopoietic model of altitude training for athletic performance proposes that the primary effect of hypoxic exposure during training or recovery is to induce an elevation in EPO production and release. This EPO secretion stimulates greater reticulocyte and red blood cell production within the bone marrow, resulting in a higher oxygen-carrying capacity and improved aerobic performance.

Nonhematological Mechanisms of Hypoxic Benefit

While the hematopoietic pathway for altitude training is compelling, and the individual links have been clearly demonstrated in a number of studies, other mechanisms may also be of equal or greater prominence (Gore et al. 2007). This contention is supported by cross-sectional population studies on high-altitude natives, which do not uniformly show a hematological adaptation (Rupert and Hochachka 2001). For example, Andean natives generally exhibit a higher hemoglobin concentration than lowland individuals. However, Tibetan natives generally do not demonstrate this response; their hemoglobin concentrations are lower than those of Andeans and not significantly different from typical lowland-native values. Such disparate findings in two populations, both of whom have had the opportunity of natural acclimatization to altitude exposure over millennia, clearly promote the notion that a multiplicity of potential pathways of altitude adaptation are possible (Moore 2001).

In relation to athletic applications, evidence for the erythropoietic pathway is also incomplete. This is especially notable in a number of studies that demonstrated an improvement in aerobic capacity or exercise performance without any changes in hematological variables (Gore et al. 2007; Hahn et al. 2001). Indeed, some analyses of the correlation between changes in red blood cell volume and $\dot{V}O_{2max}$ following altitude manipulations are calculated with a correlation coefficient (r^2, a measure of closeness of fit between two variables) value as low as 0.14 (Gore et al. 2007), suggesting that the large majority of the changes in aerobic capacity must be due to nonhematological factors. Another sign of incomplete linkage is that many studies quantify postaltitude improvement using $\dot{V}O_{2max}$, which in itself may not correlate well with performance results in actual race situations such as 5 or 10 km run times. If altitude training benefits are not achieved solely via hematological mechanisms, the obvious question becomes, what other factors may be responsible? Two potential pathways are explored here; but many others may be involved, including changes in the hypoxic ventilatory response, the production of 2,3-DPG, and changes in the Na^+/K^+-ATPase (adenosine triphosphatase) channels, and have been surveyed in a recent review (Gore et al. 2007).

Changes in Economy of Movement

Endurance performance is not solely a function of maximal aerobic capacity as embodied by $\dot{V}O_{2max}$. Rather, two other primary determinants exist. The first is the fraction of $\dot{V}O_{2max}$ that can be sustained for prolonged periods of time. This intensity, pace, or power output has been variously quantified as lactate threshold, the work intensity at which blood lactate begins to accumulate and rise, or maximal lactate at steady state, the highest workload that can be sustained without a continued rise in blood lactate. Secondly, an often overlooked factor is the economy of movement, typically quantified as the amount of oxygen required to perform a set workload. In turn, economy can be improved by two paths: (1) technical improvements, such as perfecting a swimming stroke, resulting in overall reduction in net cost of movement; and (2) improved cellular efficiency, resulting in less metabolic cost in adenosine triphosphate (ATP) production (Saunders et al. 2004a).

Intriguingly, hypoxic exposure appears to elicit a marked improvement in economy of movement, with reductions in submaximal oxygen requirements of 3% to 10% having been reported in a number of independent research laboratories (Neya et al. 2007; Saunders et al. 2004b), though contradictory conclusions were drawn in a review of studies from four different research groups (Lundby et al. 2007). At the elite level, the potential performance benefit from such improvements in efficiency is enormous. In a nonhypoxic example, the top professional cyclist Lance Armstrong recorded an 8% improvement in economy through his prime years of physical maturation from ages 21 to 27 (Coyle 2005), a period in his career in which he transformed from an elite cyclist to the overwhelmingly dominant athlete of his sport. Breaking down cellular efficiency further, improvements in economy can be achieved via a greater production of ATP per oxygen utilization or else reduced ATP costs of muscular contraction. One potential mechanism for enhanced cellular efficiency is via improvements in the electron transport chain, for example minimizing the amount of H^+ leakage through the mitochondrial membranes during oxidative phosphorylation. A number of uncoupling proteins (UCP) have been identified as increasing the rate of membrane H^+ leakage (thereby reducing mitochondrial efficiency); and UCP3, present in human skeletal muscle, is reported to decrease with physical training (see figure 5.5 for a schematic of how UCP function within the mitochondria). As the process of cellular metabolism is complex, it appears highly possible that other molecular factors within the skeletal muscles themselves may also serve to modulate metabolic efficiency and that hypoxic stress may alter the relative activity or sensitivity of these factors.

Muscle Buffering Capacity

The ability of the body and the skeletal muscles to tolerate and buffer against lactate may be another nonhematological mechanism for altitude training benefits. One of the acute and chronic responses to hypoxia is hyperventilation, which occurs in an attempt to maintain alveolar PO_2 levels. At the same time, hyperventilation potentiates the removal of alveolar CO_2, resulting in a state of respiratory alkalosis. Ultimately, this can lead to enhanced muscle buffering capacity from a lowered pH and additional renal excretion of bicarbonate. Studies on acclimatization to moderate altitude levels in athletes appear to support the enhanced buffering capacity model, with two to three weeks of altitude exposure at above 2000 m (6560 ft) resulting in enhanced muscle buffering in trained individuals (Gore et al. 2001).

GENETIC BASIS OF ALTITUDE PERFORMANCE

As with any study aimed at finding differences across sexes or racial groups, the understanding of racial or population-specific responses to altitude is fraught with methodological and also political difficulties. An overt emphasis on genetic differences can appear to belittle the effects of athletic effort and affront cultural sentiment, such as the extreme national pride of Kenyans and Ethiopians in their running history and accomplishments. Such athletic success also likely contributes to a positive spiral of strong levels of participation and support in that sport (Onywera et al. 2006; Pitsiladis and Scott 2005). To date, while the prevalence of certain genetic polymorphisms has been noted in individual elite running athletes from different regions of Africa, any broad racial generalization is impossible due to the vast genetic diversity within even local geographic regions like East Africa, intermixed with other environmental and sociocultural factors affecting athletic development (Onywera et al. 2006; Scott and Pitsiladis 2007).

Nevertheless, population studies are valuable tools that yield important insight into physiological mechanisms of responses to extreme environments. One such study, presented in table 8.1, explored the separate and combined effects of racial groups and long-term acclimatization versus training by comparing hematological parameters in sedentary individuals and professional cyclists from both a lowland area (Germany) and altitude (Colombia) (Schmidt et al. 2002). Plasma volume was increased by training but not by altitude. However, total hemoglobin and red cell volume were separately and synergistically elevated by both training and altitude. Therefore, a net higher blood volume was observed in trained lowlanders and in sedentary highlanders compared to sedentary lowlanders, with a further increase seen in trained highlanders. However, no differences in $\dot{V}O_{2max}$ were found between the lowland and highland cyclists, while the sedentary lowlanders actually had a higher $\dot{V}O_{2max}$ than sedentary highlanders. Unfortunately, the logical extension of this project, determining the effects on aerobic capacity with the highland groups at sea level and lowland groups at altitude, was not performed.

Even within populations or racial groups, a high level of individual variability in responses to an altitude stimulus appears to be present. This variability was investigated retrospectively in 32 elite male and female middle- to long-distance collegiate runners who performed different training protocols, including live high, train low; live high, train high; and live high + train high base + train low interval, based around four weeks of living at 2500 m (8200 ft) altitude (see table 8.2) (Chapman et al. 1998). A distinct categorization into "responders" and "nonresponders" was observed regardless of sex or training program at altitude. The responder group ($n = 17$), who decreased their 5000 m (5470 yd) run times by 4% compared to prealtitude times, demonstrated a significant elevation of serum EPO concentrations within 30 h of initial arrival at altitude, and this increase was maintained after 14 days of exposure. At the conclusion of the four weeks of exposure, responders exhibited an increased red blood cell volume and $\dot{V}O_{2max}$. In sharp contrast, nonresponders ($n = 15$), who actually

Table 8.1   Hematology and Aerobic Capacity in Sedentary and Trained Lowland and Altitude Natives

Subject group	$\dot{V}O_{2max}$ (mL · min^{-1} · kg^{-1})	Total hemoglobin (g · kg^{-1})	Plasma volume (mL · kg^{-1})	Hematocrit (%)	Blood volume (mL · kg^{-1})
Untrained, lowland	45.3 ± 3.2	11.0 ± 1.1	45.7 ± 5.3	45.8 ± 2.0	78.3 ± 7.9
Untrained, altitude	39.6 ± 4.0	13.4 ± 0.9	48.1 ± 3.1	48.5 ± 1.7	88.2 ± 4.8
Cyclists, lowland	68.2 ± 2.7	15.4 ± 0.9	60.8 ± 3.0	47.4 ± 1.9	107.0 ± 6.2
Cyclists, altitude	69.9 ± 4.4	17.1 ± 1.4	64.9 ± 7.3	47.8 ± 1.5	116.5 ± 11.4

$n = 12$ in each group.

Data reported as means ± standard deviation. Total hemoglobin demonstrated differences across both training and altitude status, while plasma volume demonstrated only differences based on training. In total, this generally resulted in higher blood volumes across both training and altitude status, with an apparent synergistic effect. No differences were reported in the aerobic capacity across altitude, however.

Data from Schmidt et al. 2002.

Table 8.2   Individual Variability in Performance Following Variations of a 28-Day Live High, Train Low Altitude Training Protocol

	Nonresponders ($n = 15$)		Responders ($n = 17$)	
	Prealtitude	**Postaltitude**	**Prealtitude**	**Postaltitude**
5000 m run time (min:s)	17:24 (:91)	17:38 (:98)*	17:11 (:76)	16:34 (:75)*†
$\dot{V}O_{2max}$ (mL · min^{-1} · kg^{-1})	64.1 (4.4)	64.4 (4.7)	65.0 (5.8)	69.2 (6.8)*†
Maximal steady state $\dot{V}O_2$ (mL · min^{-1} · kg^{-1})	50.2 (8.5)	52.4 (4.9)	54.2 (5.3)	59.0 (9.1)*†

Means (standard deviation).

*Significantly different from prealtitude.

†Significantly different from nonresponders.

Data from Chapman et al. 1998.

averaged a 1% slower 5000 m time postaltitude, demonstrated a slight EPO increase at 30 h into exposure, but the magnitude was less than that seen in the responders. Beyond this initial elevation, no hematological or aerobic changes occurred at any further point of the altitude exposure in this group; the EPO concentrations decreased back to near sea level values by 14 days following altitude exposure. The difference in performance and training capacity between the responders and nonresponders was also evident in the nonresponders' significantly lower average velocity during interval training at altitude after two weeks. Following up this retrospective analysis, a prospective examination of 22 runners undergoing the same four-week exposure at altitude categorized nine responders and five nonresponders, again on the basis of postaltitude performance, and showed similar hematological and aerobic response patterns in the two groups (Chapman et al. 1998).

The design of Chapman and colleagues' study provides a few interesting ideas for sport scientists. First, it becomes clear that hypoxic exposure cannot be generalized as having a consistent ergogenic benefit, as neither the 2500 m (8200 ft) elevation nor four weeks of exposure appears to be unreasonable or deviates from common practice, with the elevation matching or exceeding the thresholds estimated to be required for significant benefit (Ge et al. 2002). However, it should be noted that a disproportionate percentage of the nonresponders were from the live high, train high group (82% of the responders were from the two variants of live high, train low), such that the nonresponse may be more a reflection of the lack of efficacy of training at high altitudes. Second, no sex-specific response pattern can be generalized; distribution of men and women was roughly equal between the responder and nonresponder groups. The physiological or genetic mechanisms underlying such wide variability in hypoxic adaptation are unclear, however, with minimal association between EPO response to hypoxia and markers linked to the EPO gene or its regulators (Ameln et al. 2005; Jedlickova et al. 2003). Therefore, predicting individual suitability and customization of altitude training remain extremely difficult.

Advances in cellular and genetic approaches to studying exercise physiology have prompted investigations into the genetic variability of responses to hypoxia. This has yielded the identification of the *hypoxia-inducible factor 1* (HIF-1), a transcription factor that is present throughout the body. This factor appears to play a critical

---

HOW TO TRAIN HIGH-ALTITUDE NATIVES?

The dominant cultural slant of applied hypoxia research has involved understanding the effects of altitude as an ergogenic aid for lowland natives preparing for competition at sea level or prior to mountaineering expeditions at extreme altitudes. A flipping of this context can lead to the question of whether training of high-altitude natives is optimized in their natural hypoxic environment or whether they would benefit from training in a normoxic environment with greater oxygen availability. This question was posed by a study on a group of sedentary Bolivian subjects, who possessed the typical Andean hematological adaptation to altitude of elevated hemoglobin and hematocrit (hematocrit >50% and hemoglobin >17.3 g · dL^{-1}) (Favier et al. 1995). Subjects were placed into a hypobaric hypoxic group or else into one of two hypobaric normoxic groups training at either a similar relative or absolute intensity as the hypoxic group. Following six weeks of training, $\dot{V}O_{2max}$ improved equally in all three groups, but both of the normoxic groups demonstrated greater lactate levels during maximal exercise. Such a study leads to the conclusion that the "hyperoxia" training that was provided to the two normoxic groups was ineffective in eliciting a greater aerobic benefit. This was true despite training at an identical relative intensity (and therefore a greater absolute training stimulus) in one of the normoxic groups. In turn, this suggests a finite ceiling in physiological altitude adaptation that is lower than that from training itself. Finally, the equal aerobic capacity but higher lactate levels at maximal exercise in the normoxic compared to the hypoxic group lend support to the proposal that muscle buffering capacity is increased with adaptation to altitude.

---

role in the regulation of oxygen homeostasis within the body and individual tissues, including the stimulation of EPO response to hypoxia (Semenza 2006). Analogous to the expression of heat shock protein by heat stress and its multimodal actions throughout the body (see chapter 3), HIF-1 is minimally present and rapidly degrades in normoxic conditions, but appears to respond very rapidly and sensitively to the presence of hypoxia, with gene transcriptional effects on vascular endothelial growth, cellular glucose transport, nitric oxide synthase, and many other metabolic pathways and hormones (Semenza 2000). Studies on knockout mice with HIF-1 deficiency demonstrate multiple physiological impairments to continuous or intermittent hypoxia, including a possible role in the induction and proliferation of cancer (Semenza 2008). Overall, HIF-1 activation and its effectiveness in stimulating gene expression across a variety of physiological systems may be a key factor in determining individual and racial variability to hypoxic exposure by both hematological and nonhematological mechanisms.

LIVE LOW, TRAIN HIGH

As we have seen, acute exposure to hypoxia or altitude results in a series of physiological changes and impairment in exercise capacity. On initial review, it would appear counterintuitive that training in a hypoxic environment would elicit positive training benefits, given the acute reduced exercise capacity and subsequent training stimulus possible at altitude. However, due to the high plasticity and adaptability of the human body, hypoxic stress during exercise can serve as a supplemental training stimulus for eliciting a compensatory adaptation in the body. Therefore, the proper first approach to determining the effects of altitude as an ergogenic aid is to investigate the effects of chronic hypoxic training on subsequent altitude and sea level performance. This mode of training is also the easiest to implement, making it highly desirable from a practical perspective; hypoxia can be relatively easily achieved in a normobaric laboratory without the need to establish a training base at altitude. Methods to induce hypoxic stress during exercise (intermittent hypoxic training, or IHT) include (a) breathing prepared hypoxic gases and (b) using oxygen filtration or nitrogen dilution methods to lower the inspired O_2 content.

The existing live low, train high (LLTH) literature, in which hypoxia was employed during exercise but subjects otherwise lived in a normoxic environment, is predominantly weighted toward the idea that LLTH has minimal ergogenic benefit. In a recent survey of different altitude training modalities, Wilber (2007) reported only a small number of controlled IHT studies on trained or elite athletes that demonstrated significant improvements in blood parameters, aerobic capacity, or work performance. Specifically, while some studies have shown an increase in hemoglobin concentration, none have measured improvement with more robust hemopoietic markers such as erythrocyte volume or hemoglobin mass. No study employing IHT elicited an increase in $\dot{V}O_{2max}$, and only a few have demonstrated minor improvements in actual exercise performance. Two such studies showed an improved exercise capacity in cycling peak power output, including one using only four professional cyclists (Terrados et al. 1988), while anaerobic capacity during a Wingate sprint was also improved in the other study on trained triathletes (Hendriksen and Meeuwsen 2003). Interestingly, neither study showed improvements in either hemoglobin concentrations or $\dot{V}O_{2max}$ post-IHT to support these improvements in performance tests, implying that the mechanism for performance enhancement may not be the traditionally assumed hematopoietic–oxygen-carrying capacity pathway. Furthermore, neither of these studies utilized a "realistic" field performance test such as a time trial. However, the fact that power outputs improved with a small sample size of only four cyclists remains suggestive that IHT can be of significant physiological benefit, especially given the high level of training and presumed test reproducibility obtainable with professional cyclists.

From a practical perspective, one of the major limitations of altitude and hypoxia as ergogenic tools is that athletes are relatively constrained to a laboratory environment or a training base at altitude during the training sessions. Both of the studies that were just discussed employed 1 to 2 h of IHT in the laboratory daily for two to four weeks to achieve their improved results. While such a level of control is readily accomplished with animal models, it ultimately becomes impractical for human athletes to train exclusively in an artificial laboratory environment. For example, few athletes would have the motivation to log large amounts of training hours on a stationary ergometer or treadmill rather than exercising in a natural setting. In this paradigm, one major criticism of these types of studies is their impracticality for both solo and team athletes. Therefore, the concept of replacing only part of an athlete's training with a hypoxic stimulus has been investigated.

Ventura and colleagues (2003) had trained cyclists perform three 30 min laboratory bouts a week of either hypoxic or normoxic workouts at the estimated anaerobic threshold for six weeks, with the remainder of the training week self-determined by the subjects. Before and following the training program, tests were performed in both normoxia and hypoxia. However, no significant changes were observed in $\dot{V}O_{2max}$ or maximal power output in either the hypoxic or the normoxic training groups in either hypoxia or normoxia. Furthermore, no elevation of hypoxic work capacity during a 10 min self-paced time trial was evident. Overall, one extrapolation may be that partial integration of hypoxic training may be an insufficient stimulus for significant physiological adaptation, though this conclusion is clouded by the lack of improvements in the normoxic training group. In contrast, elite runners training twice per week in a hypoxic (14.5% inspired oxygen) environment experienced a significant improvement in $\dot{V}O_{2max}$ and time to exhaustion at a high-intensity workload, with no improvements observed in a normoxic group (Dufour et al. 2006; Ponsot et al. 2006). Further analysis revealed no correlation between improvements in time to exhaustion and $\dot{V}O_{2max}$, concluding that the observed performance benefits were not due to changes in oxygen-carrying capacity.

Despite its equivocal ergogenic effectiveness to date, training in hypoxic environments can remain a highly practical and relevant training modality for specific conditions and populations, such as lowland athletes preparing for competitions at altitude, soldiers assigned to rapid deployment in mountainous terrain, and mountaineers prior to high-altitude expeditions. Exercising in normobaric hypoxic environments for brief periods prior to competition or deployment may attenuate some of the initial acute and deleterious responses to altitude. In turn, this form of preacclimation, especially if used in conjunction with hypoxic exposure at rest, may shorten the time required to fully acclimatize to altitude, thus enabling a faster return to full work capacity upon arrival. Given the high logistical demands and costs involved in mountaineering expeditions, especially large guided trips (e.g., Everest), preacclimation through hypoxic training by individual climbers may help decrease overall expedition duration and also possibly prescreen for individuals with high susceptibility or risk for acute mountain sickness.

LIVE HIGH, TRAIN HIGH

The traditional modality of altitude training involves the athlete's moving to a training base at altitude and then living and training there for weeks or more prior to returning to competition at sea level. Such live high, train high (LHTH) protocols at hypobaric hypoxia were the earliest ones employed, as this type of protocol was the main practical option prior to the advent of portable hypoxic facilities (see figure 8.2). When evaluating the LHTH literature, one must also consider the nonaltitude-related factors that may have enhanced performance. Namely, even for national-caliber athletes, such altitude camps may be a unique experience in that they have full access to technical support (massage, meals, coaching, equipment mechanics) and a peer group of training partners that they may not have at home. Many of the LHTH studies have not incorporated a control group training at sea level, making it difficult to clearly isolate the effects of altitude. Despite their long history of use and these supplemental benefits, however, the results of LHTH protocols for sea level performance remain equivocal in both the scientific literature and in actual practice.

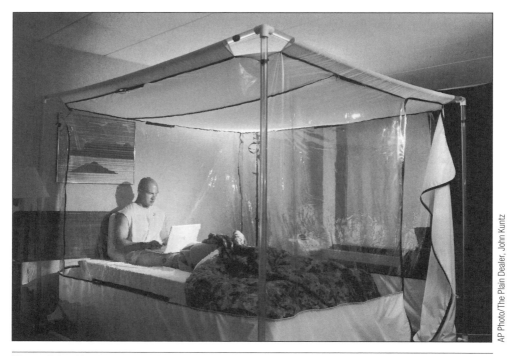

AP Photo/The Plain Dealer, John Kuntz

Figure 8.2 Altitude simulation facilities have greatly increased in availability and popularity in the past decade, from portable tents for sleeping (pictured) through to entire sport complexes designed for hypoxic control.

The classic study concluding no effect of LHTH was conducted in 1975 on trained male distance runners (Adams et al. 1975) (see figure 8.3). The study featured a nice repeated-measures crossover design, in which one group underwent three weeks of sea level training followed by three weeks of training at 2300 m (7545 ft) and the other matched group of subjects had the reverse order of conditions (i.e., three weeks of altitude followed by three weeks of sea level training). The critical finding with aerobic capacity was that, compared to preexperimental baseline values at sea level, neither group demonstrated any increase in $\dot{V}O_{2max}$ when tested at sea level upon return from altitude. Indeed, the group that trained at altitude first had a significantly lower (–4%) $\dot{V}O_{2max}$ upon return to sea level when compared to baseline $\dot{V}O_{2max}$ measured at sea level prior to altitude exposure, and $\dot{V}O_{2max}$ remained lower or at preexperimental values even over the three additional weeks of sea level training following altitude exposure. Performance times over a 2-mile (3.2 km) run test did not significantly differ after altitude training upon return to sea level in either group. These data strongly support the contention that altitude training did not elicit a direct short-term benefit by itself. Furthermore, the lack of $\dot{V}O_{2max}$ or performance time improvement in the altitude–sea level group implies that the altitude training did not enable a greater postaltitude response to normal training at sea level.

Well-controlled studies do exist demonstrating an enhancement of aerobic performance from LHTH protocols (Daniels and Oldridge 1970). U.S. National team distance runners completed two 14-day LHTH blocks at 2300 m (7545 ft); the normal

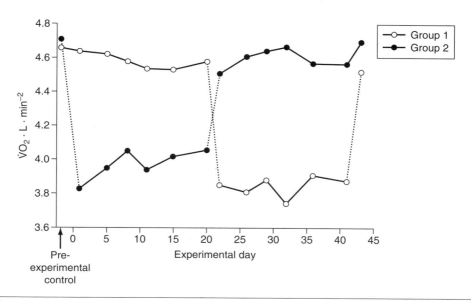

Figure 8.3 The results from a crossover design (each group performed both sea level and altitude training) investigating the effects of altitude training on maximal oxygen uptake. Acute and chronic altitude exposure of three weeks impaired exercise capacity, which remained impaired upon return to sea level in both groups.

Reprinted from W.C. Adams et al., 1975, "Effects of equivalent sea-level and altitude training on VO2max and running performance," *Journal of Applied Physiology* 39: 262-266. Used with permission.

sea level training loads were employed from the beginning of camp. These blocks were separated by a five-day period during which the athletes competed in elite events. Competitions also occurred in the five days following the second 14-day LHTH block. Aerobic capacity testing following the initial and final LHTH blocks elicited a 4% and 5% increase, respectively, in $\dot{V}O_{2max}$ compared to preexperimental values. Exercise performance also benefited, with a significant 3% improvement in 3-mile (4.8 km) race pace at the end of all LHTH blocks. From an applied perspective, the most important finding was that all but one athlete set personal records in competition over the course of the study, with all of these five setting and subsequently resetting their records. Caveats do need to be raised with this study, though, primarily the lack of a control group of elite athletes training at a sea level camp.

Overall, clear consensus on the efficacy of LHTH as an ergogenic tool for athletes remains clouded by widely disparate results across hematological, aerobic testing, and performance measures. Studies present data ranging from a slight impairment through to no effect and a slight improvement in all these measures, with no consistent pattern. Certainly, no controlled study has clearly delineated a hematopoietic effect, linking elevated EPO levels to erythropoiesis, increased $\dot{V}O_{2max}$, and ultimately improved performance. Therefore, according to current knowledge, any efficacy from a LHTH regimen would appear to occur, at least partly, through nonhematological mechanisms. However, research into the other potential mechanisms, such as muscle oxidative capacity and capillarization, also yields equivocal conclusions.

LIVE HIGH, TRAIN LOW

The late 1990s saw a new paradigm emerge in altitude training. In order to maximize the physiological adaptation from exposure to hypoxia yet minimize the reduction in exercise capacity typically observed at altitude, Ben Levine and Jim Stray-Gundersen (1997) proposed a live high, train low (LHTL) altitude training model. With LHTL, athletes experience passive hypoxia during rest or sleep in order to elicit erythropoiesis and nonhematological adaptation, typically 8 to 10 h daily at altitudes ranging from 2000 to 2700 m (6560 to 8860 ft). Regular training is performed at altitudes as close to "normal" or sea level as practical, typically <1000 m (3820 ft). Initially, studies on LHTL were conducted in natural settings, with athletes transported between training and altitude bases. This still required a significant investment in time and logistics. However, with the growing prevalence and popularity of hypoxic facilities, LHTL has come within reach of most elite and even recreational athletes, making it the current dominant mode of altitude training and research.

The classic initial LHTL study investigated 13 male and female collegiate runners over a 28-day period, alternating between residing at Park City, Utah (2500 m [8200 ft]), and training at Salt Lake City (1250 m [4100 ft]) (Levine and Stray-Gundersen 1997). A control group of 13 male and female runners trained and resided at the lower altitude. Unlike the studies on LHTH in the 1970s, this study included the innovation of incorporating both base training and interval sessions into the training regimen. The LHTL group demonstrated a strong linkage between elevated hematopoietic response and ultimately increased $\dot{V}O_{2max}$ and exercise performance. Specifically, red blood cell mass, hemoglobin, and treadmill $\dot{V}O_{2max}$ all increased following the altitude training period, whereas no changes were observed in the control group in these variables or in 5000 m (5470 yd) run times. In the LHTL group, 5000 m run times improved by an average of 13.4 s at three days postaltitude, and this improvement was maintained in continued testing at weekly intervals up to three weeks postaltitude. Such clean demarcation of responses in a well-controlled study certainly boded well for the potential for LHTL training, along with popularizing the hematological model for hypoxic adaptations. As with other modes of altitude training, it is important to highlight that the efficacy of LHTL is not universal in the scientific literature. A number of studies on elite athletes have not been able to demonstrate significant hematopoietic response following a LHTL regimen, showing no changes in variables such as serum EPO, hemoglobin, reticulocyte measures, or red blood cell mass (Gore et al. 2006).

Following the introduction of the LHTL model, much research has gone into refining the application of this research to elite athletes. One LHTL variation by the same research group has involved further partitioning the training component, with base training performed at altitude (2500 m [8200 ft]) and only the interval training sessions at the lower altitude (Stray-Gundersen et al. 2001). The major findings of the original LHTL study were replicated, with elevated EPO, hemoglobin, and $\dot{V}O_{2max}$ coupled with improved 3000 m (3280 yd) run times. Notably, these improvements were found in elite rather than collegiate-level athletes, but one limitation was the lack of a control group for comparison. Other variations of this model have included refining and individualizing the altitude employed for both rest and training, along with the optimal progression of altitude throughout a block of training (Wilber 2007).

— ALTITUDE TRAINING AND ANAEROBIC PERFORMANCE —

Because the hematopoietic model is so prevalent in thinking about altitude training, much hypoxic research has focused on effects on aerobic capacity and performance. In contrast, minimal work has dealt with the potential efficacy of altitude training for maximal or supra-maximal exercise in which anaerobic metabolism dominates. However, such exercise tests may prove an interesting model for isolating hypoxic effects on local muscle (e.g., buffering capacity) rather than cardiovascular factors. To date, the body of literature remains small, but no clear benefit on anaerobic performance has been demonstrated (Roberts et al. 2003). One potential limitation appears to be that much of the work has been designed and performed within the context of endurance events, leading to the utilization of endurance athletes rather than specifically anaerobic athletes. Ultimately, anaerobic efforts such as field events (e.g., discus, high jump) and aerodynamic-dependent (e.g., running, cycling) events lasting short durations may be optimal candidates for benefiting from training at altitude to supplement the reduced air resistance at higher elevations.

HYPEROXIA

As we have seen, the original design of altitude training involved exercising and living at altitude. However, the common complaint of most athletes is a greatly reduced ability to both train hard and recover from hard workouts while at altitude. There-fore, while cardiovascular benefits may occur, this may be negated by the decreased exercise stimulus. As discussed earlier, the advent of hypoxic tents and facilities has made the LHTL regimen logistically feasible.

From the "If a little is good, more must be better" philosophy of training, the rela-tively new concept of hyperoxic training seems quite logical. Because the purpose of "train low" is to maintain the training stimulus by not affecting oxygen availability, it appears reasonable that training in a higher than normal (hyperoxic) environment might permit greater than normal workloads and therefore a greater than normal training stimulus. Some of the earliest dedicated work on this topic occurred prior to the Atlanta 1996 Olympics, as the U.S. Olympic Committee test-piloted the use of hyperoxic training in cycling track pursuiters. Subjects were indeed able to do inter-vals of much higher than normal wattages when in hyperoxia, but as a consequence of initial problems with incorporating such protocols into an overall training plan, many of the athletes seemed to suffer symptoms of overtraining.

Acute exposure to hyperoxia seems capable of providing an immediate boost to performance (see table 8.3). Breathing 40% O_2 resulted in higher self-selected power outputs during a 20 km (12.4-mile) cycling time trial, along with higher electromyo-graphic activity within the quadriceps, while ratings of perceived exertion remained similar to those during normoxia (Tucker et al. 2007). In addition, the rate of muscle glycogenolysis seem to be decreased during 60% O_2 hyperoxia, along with lowered blood and muscle lactate, although the mechanisms for these findings remain unclear

Table 8.3   Summary of Selected Studies on Hyperoxic Training and Performance

Study	Conditions	Main findings
Tucker et al. (2007)	Normobaria, acute exposure to 40% O_2 during 20 km cycling time trial	• ↑ 5% power output and EMG activity with hyperoxia • Non-significant lactate, HR, RPE
Perry et al. (2005)	• Normobaria, 60% O_2 training for 6 weeks • Three times a week, 10 × 4 min intervals at 80% $\dot{V}O_{2max}$ • Ride to exhaustion at 90% $\dot{V}O_{2max}$	• ↑ 8.1% power at same HR to maintain 80% $\dot{V}O_{2max}$ HR with hyperoxia • ↑ tolerance time at 90% $\dot{V}O_{2max}$ with hyperoxia • Non-significant trend to ↑ $\dot{V}O_{2max}$ with hyperoxia
Perry et al. (2007)	• Normobaria, 60% O_2 training for 6 weeks • Three times a week, 10 × 4 min intervals at 90% $\dot{V}O_{2max}$ • Ride to exhaustion at 90% $\dot{V}O_{2max}$	• ↑ 8.1% power at same HR to maintain 80% $\dot{V}O_{2max}$ HR with hyperoxia • Similar ↑ tolerance times at 90% $\dot{V}O_{2max}$ with hyperoxia and normoxia • Similar ↑ aerobic enzyme activity with hyperoxia and normoxia
Stellingwerff et al. (2005)	Normobaria, acute exposure to 60% O_2 during 15 min 70% $\dot{V}O_{2max}$ cycling	• ↓ muscle glycogenolysis • ↓ lactate accumulation • Non-significant pyruvate dehydrogenase activity
Calbet et al. (2002)	Hypobaria: Lowlanders acclimatized for 9 weeks to 5260 m, then tested at normoxia and hyperoxia (55% O_2) submaximal and maximal cycling tests	• ↑ max power output with hyperoxia • ↑ leg $\dot{V}O_{2max}$ with hyperoxia

(Stellingwerff et al. 2005). Even in lowlanders acclimatized to 5260 m (17,260 ft) altitude, acute hyperoxia was capable of increasing maximal power output and leg $\dot{V}O_{2max}$ (Calbet et al. 2002), suggesting that muscle capacity is limited at altitude by central delivery and O_2 availability rather than peripheral muscular oxidative capacity.

With acute exposure appearing beneficial, the next phase becomes testing and implementing hyperoxia as a prolonged training aid. In a difficult-to-achieve repeated-measures design, nine subjects performed six weeks of interval training in both normoxic (21% O_2) and hyperoxic (60% O_2) environments, with 12 weeks of inactivity in between to permit detraining (Perry et al. 2005). The training involved three workouts per week of ten 4 min intervals, with the intensity adjusted to maintain a consistent 80% $\dot{V}O_{2max}$ heart rate throughout. Significantly, an average of 8.1% greater power output was required during hyperoxia training to maintain the same heart rates in the two conditions, and hyperoxia produced a greater tolerance to sustained 90% $\dot{V}O_{2max}$ after six weeks of training. Furthermore, a slight but nonsignificant trend toward a higher $\dot{V}O_{2max}$ posttraining was observed with hyperoxia than with normoxia. Such results in a well-controlled study clearly point toward an ergogenic potential for hyperoxic training.

Overall, this vein of research appears to be full of promise along with unanswered questions. For example, it is somewhat unfortunate that no blood parameters, even relatively basic ones such as hemoglobin and hematocrit, were taken, along with blood markers for immune function, muscle damage, and so on. Therefore, the specific cardiovascular adaptations to hyperoxic training remain unknown. However, a subsequent study by the same research group, using a nearly identical training and testing protocol (Perry et al. 2007), was unable to replicate the improved tolerance to 90% $\dot{V}O_{2max}$ with hyperoxia compared to normoxia previously reported, and also did not detect any differences across hyperoxia or normoxia in the aerobic metabolic adaptations within skeletal muscles when muscle biopsies were analyzed.

CONSIDERATIONS IN APPLYING ALTITUDE TRAINING

For athletes, coaches, and exercise scientists, the vast body of literature available on altitude training, along with the multiple modalities and tools available, makes it very difficult to answer the simple question, *How high and for how long should I use altitude to achieve a basic or optimal adaptation?* The answer is obviously modulated by individual factors, such as individual responsiveness, competitive demands, competition altitude, and training status. While the efficacy of altitude training does not appear to be universal for all athletic endeavors and across all individuals, altitude training is likely to remain popular among elite athletes due to its relative accessibility, generally perceived efficacy (or the idea that it is benign and without major side effects at worst), and legality. The dominant mode of altitude training, as currently employed, is a LHTL approach, though an emerging frontier may be the integration of hyperoxia into the training and altitude regimen. This section briefly summarizes some of the common practices currently in use among scientists and athletes in applications of altitude training (Bartsch et al. 2008; Muza 2007; Wilber 2007).

Altitude Threshold

For situations in which the protocol involves passive exposure to hypoxia (i.e., live high), and assuming hematopoiesis as the dominant desired adaptation, a dose response based on the elevation of exposure appears to exist. Short-term (6-24 h) passive exposure at a simulated or natural minimum of 2100 to 2500 m (6890-8200 ft) was required to stimulate a significant EPO increase in the majority of individuals (Ge et al. 2002).

Duration of Altitude Training

For lowland natives, there is unlikely to be a physiological upper limit to duration of exposure; so the question becomes, assuming that living long-term at the altitude is not feasible, What constitutes the minimal or optimal duration required over a short-term acclimation to achieve enhanced performance? While rapid elevations in EPO concentrations are often observed in the initial hours following altitude exposure, the actual time course for erythropoiesis is much longer, and adaptations may not be fully realized until four or more weeks following arrival at altitude. If a LHTL protocol is being used, the majority, if not all, of the nontraining time should be spent recovering in a hypoxic facility or at altitude to enhance physiological adaptation, with a minimum of 12 to 16 h daily proposed as a threshold and a preferred duration of 22 h. At shorter durations, a compensatory increase in altitude may be required (Muza 2007; Wilber 2007).

Decay of Altitude Adaptations

Upon attainment of an adaptive response, another important consideration is how long the improved response may be maintained upon return to low altitude. There is minimal agreement or consensus on this question, given the high degree of individual variability in response and lack of a single defined mechanism of action. Some studies indicate a very rapid decay of any performance improvements within a week of ending hypoxic exposures, while others show sustained improvements more than three weeks posthypoxia (Levine and Stray-Gundersen 1997). Other questions are whether periodic hypoxic exposures at less frequent intervals can maintain adaptation and what the time course is for reinduction of adaptation. These questions become especially relevant with the emerging prevalence of portable hypoxic facilities, such that athletes may be able to receive periodic doses to "top up" their hypoxic adaptation even while traveling throughout the training year or during prolonged competitions.

Transition From Sea Level to Altitude

For competitions at altitude, it would appear that adaptation should take place at levels as close as is reasonable or practical to the competition altitude. But when this is not possible, when is the best time to arrive at the competition venue? Should athletes arrive at altitude as long as possible beforehand to acclimatize, and possibly suffer the consequences of lower training capacity and fitness, or should they attempt a "commando mission" approach and arrive at altitude as close to competition time as possible? A 2001 South African–Australian study (Weston et al. 2001) addressed this question using 15 rugby players by testing these "native" lowlanders at sea level, then transporting them rapidly up to a test center at 1700 m (5580 ft) altitude, where they underwent testing again at 6 h, 18 h, and 47 h after ascent. In this way, the timing of ascent simulated typical arrival times for a 3 P.M. competition. For example, the 6 h trial simulated arriving at altitude at 9 A.M. for a 3 P.M. event, and the 47 h trial simulated arrival at 10 P.M. two nights before the event. The study therefore also provided information on the time course of performance upon the first few days of altitude exposure.

No change to hematocrit or hemoglobin occurred with arrival to altitude. Heart rates were significantly higher during a submaximal cycle test at 6 h compared with sea level (182 vs. 177 beats · min^{-1}), but returned to "normal" at 18 and 47 h (177 and 176, respectively). Only five subjects could complete the full 5 min at the 6 h trial compared to nine at sea level. The same pattern in heart rates was found for the four trials with the shuttle run aerobic capacity test. The tolerance time for the shuttle test was 37% shorter at 6 h compared to sea level. While it improved with the 18 h and 47 h tests, tolerance times for both remained lower than at sea level. A similar pattern of initial impairment and some level of recovery with 45 h of short-term altitude acclimatization was reported with exposure to 3200 m (10,500 ft) in 5 and 50 min cycling tests (Burtscher et al. 2006). Interestingly, no decrement was reported with the 30 s cycling test upon initial and short-term exposure, suggesting that anaerobic-based exercise is not affected by moderate altitude.

THE ETHICS OF ALTITUDE TRAINING

The relatively clear advantages of elevated red blood cell volume have unfortunately led to abuse by unscrupulous athletes and other individuals involved in sport. This has been seen predominantly in endurance sports such as cycling and Nordic skiing, with major scandals involving the use of blood doping or artificial EPO.

Altitude training and blood doping converge in their similar physiological mechanisms of manipulating EPO and hematopoiesis. This has made for tricky ethical considerations, leading some individuals and agencies to consider regulating or banning the use of hypoxic tents. The finding of nonhematological mechanisms behind performance improvements with hypoxic training has helped to minimize the negative comparisons to blood doping. However, as altitude and hypoxia is ultimately a natural environment and therefore impossible to regulate, it is much more likely that altitude training will be governed by a series of regulations grounded in the safety of athletes and also fairness and accessibility (Loland and Caplan 2008). In the course of this debate, one argument is that, if used solely for performance enhancement independent of altitude or training, hypoxic devices may challenge the notion of athletic autonomy and contravene the spirit of sport. Another argument is that hypoxic training devices help to level the playing field by increasing the accessibility of altitude training for lowland-native athletes or those who cannot perform natural altitude training, though the cost of such devices poses other questions about accessibility. At present, WADA expresses concerns about the potential health risks from hypoxia but does not specifically ban its use. At the same time, the International Olympic Committee has banned the use of simulated altitude devices within the boundaries of the Olympic Village and other sporting venues since the Sydney 2000 Games.

Overall, the results of these studies lead to the conclusion that athletes should try to arrive as early as possible at the competition altitude to give their bodies time to acclimatize to the hypoxia. Fortunately, even if logistics preclude early travel, the popularity of hypoxic tents among elite athletes is making adaptation easier to achieve. Specifically, athletes can set their hypoxic exposure to match what would be expected at the competition venue, performing a "live high, train low" program to maintain maximal fitness. In addition, testing can be performed at the competition altitude to determine when the athlete is truly "ready." However, this "remote" acclimatization still carries with it some problems, namely the potential problems from jet lag if people adapt at home and then have to cross many time zones to compete.

SUMMARY

As oxygen is critical to human survival, the alterations in oxygen partial pressures with altitude changes pose a challenge to the maintenance of homeostasis. In turn, a host of physiological responses occur across a variety of systems in an attempt to adapt to hypoxic conditions. These adaptations have been used by athletes as a means of increasing their exercise capacity at sea level. One primary adaptation may be in the stimulus for greater oxygen-carrying capacity in the cardiovascular system through an increase in the production of red blood cells. Other potential mechanisms of adaptation may include alterations in muscle function, lactate buffering capacity, and improvements in metabolic efficiency and economy of movement. The practical implementation of hypoxic training can be through natural exposure to altitude or, increasingly, through the use of artificial hypoxia to replicate altitude. Various training modalities manipulate the hypoxic exposure during both exercise itself and recovery periods; one popular modality is that of "live high, train low"—exercising at lower altitudes to maximize training stimulus, then recovering and living at higher altitudes to enhance hypoxic stress and physiological response. Much fundamental and applied research is still required to fully elucidate the effects of hypoxic stress on human physiology and ultimately to refine and tailor altitude training to athletes and to occupations such as the military.

Mountaineering and High-Altitude Physiology

When we think of high altitude, most of our images have to do with the giants of mountaineering, such as Sir Edmund Hillary and Tenzing Norgay, who in 1953 became the first to reach the summit of Everest. Other famous individuals include Reinhold Messner and Peter Habeler, who shocked both scientists and the Alpine community by summiting Everest in 1978 without the use of supplemental oxygen; Messner eventually repeated the feat for all 14 mountains above 8000 m (26,246 ft) in elevation. With the establishment and popularity of guided expeditions since the mid-1980s, high altitudes have come within realistic reach of both mountaineers and adventure seekers, sometimes with tragic results. The high mountains annually exact a human toll, both from accidents and from clinical problems such as various high-altitude illnesses. Yet high altitude is not simply the domain of a small group of adventurers, as an estimated 140 million people permanently reside at elevations above 2500 m (8200 ft). While Everest lies at 8848 m (29,030 ft), the highest continuously inhabited human settlements, small mining camps in the Andes, are at or just above an altitude of 5000 m (16,400 ft) (West 2002). At the same time, the military may be tasked with performing missions at altitude on very short notice. Finally, the increasing accessibility of many previously remote areas of the earth has opened up more recreational opportunities (e.g., trekking, skiing) in mountainous regions.

High altitude forms a unique stress on the human organism, making it a superb research environment for investigating the physiological limits of the body. The respiratory and oxygen transport systems within the body are designed around a cascade of oxygen levels from the environment to the lungs, arterial blood, and eventually muscles and tissues. However, the greatly reduced oxygen partial pressures (PO_2) in the ambient environment at altitude disrupt this gradient, resulting in a comprehensive

WHAT'S THE BAROMETRIC PRESSURE ON EVEREST?

The barometric pressure at sea level is standardized as 760 mmHg. However, while it is clear that barometric pressure and partial pressure of oxygen decrease with increasing elevation (see figure 9.1), the very simple question, *What is the barometric pressure at the summit of Everest?* is not so simple to answer. That is because of the extreme difficulty of designing and transporting suitable measurement devices to the summit. In addition, as with any site on Earth, barometric pressure may change day to day and seasonally. Accurately measuring barometric pressure is of great interest to altitude physiologists, as most current models and anecdotal reports suggest that the Everest summit is at or very close to the physiological limits for human functioning. Therefore, such information becomes critical for both research design and subsequent modeling of human physiology (e.g., oxygen transport and $\dot{V}O_{2max}$). For example, a 10 mmHg variation in pressure at such altitudes would result in a calculated reduction in $\dot{V}O_{2max}$ of approximately 12% (West et al. 1983b)!

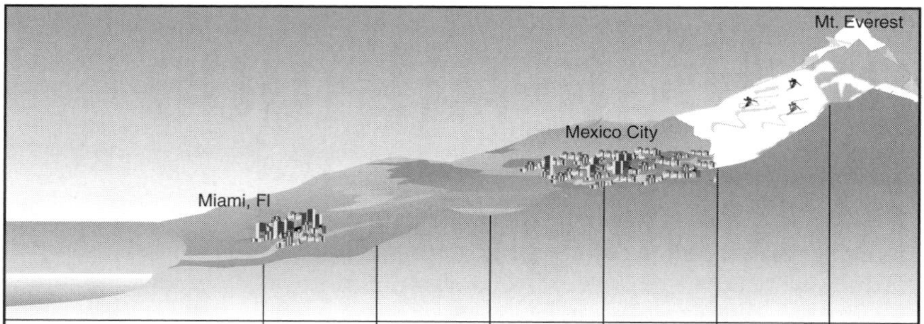

Altitude (ft) (m)	0 (sea level) 0	3280 1000	6560 2000	9840 4000	13,120 4000	29,520 9000
Barometric pressure P_b (mmHg)	760	674	596	526	462	231
% O_2 in the air	20.93	20.93	20.93	20.93	20.93	20.93
Partial pressure of oxygen PO_2 (mmHg) in the air	159	141	125	110	97	48
Typical temperature (°C) (°F)	15 59	9 47	2 36	-5 24	-11 12	-43 -46

Figure 9.1 Barometric pressures and PO_2 values at different altitude and major cities throughout the world.

Reprinted, by permission, from J. Wilmore and D. Costill, 2008, *Physiology of sport and exercise,* 4th ed. (Champaign, IL: Human Kinetics), 281.

The first direct measurement of barometric pressure on Everest yielded a pressure of 253 mmHg and did not occur until October 1981 (West et al. 1983b), 28 years after Hillary's climb and three years after Messner and Habeler had achieved the summit without supplemental oxygen. Surprisingly, this remained the only direct measurement for many years, and was used for a number of large-scale laboratory-based physiological studies and for modeling purposes. During the late 1990s, another direct measurement, weather balloons, and barometers at lower elevations were used to predict a general barometric range of 251 to 253 mmHg from May through October (West 1999). In turn, the inspired PO_2 is modeled at 43 mmHg. Overall, scientists can breathe a sigh of relief for the close matching of the results, indicating that previous studies were indeed representative of actual conditions.

series of adaptations in every physiological system. As with the thermal environment, humans have been exposed to a wide range of altitudes throughout evolutionary history and have therefore developed a high degree of adaptation to hypoxic stress. One method of investigating these adaptations is to study populations who are high-altitude natives and compare their responses and capacity with those of lowland natives. Another pathway is to study the nature, time course, and magnitude of adaptation in lowland natives exposed acutely and chronically to rest and exercise at altitude. The purpose of this chapter is to use such studies to discuss the effects of high (>3000 m [9840 ft]) elevation on exercise capacity and to explore the limits of oxygen transport and dynamics within the human body.

PHYSIOLOGICAL RESPONSES TO EXTREMELY HIGH ALTITUDES

Images of high-altitude mountaineering commonly involve supremely fit and healthy Alpinists who are reduced to repeatedly stopping after every couple of steps to regain their breath. With the reduction in oxygen partial pressures at altitude, the underlying cause for this dramatic reduction in exercise capacity is the hypoxemia and ultimately reduced oxygen delivery to the muscles during exercise (West 2006). This section explores the physiological responses of unadapted and adapted humans to extreme altitudes exceeding 5000 m (16,400 ft), primarily with reference to a series of major field and hypobaric chamber experiments featuring acute and prolonged exposure to high altitudes (see table 9.1).

Cardiorespiratory Capacity

The first series of classic studies of exercise limitations at high altitude in 1960 and 1961 involved lowland native scientists performing tests over eight months of exposure at a base camp of 5800 m (19,030 ft), at a barometric pressure of 388 mmHg or about half that at sea level. This Himalayan project was quickly nicknamed the "Silver Hut expedition" in honor of the prefab building where the camp was established. This extensive expedition generated an enormous amount of new information about cardiorespiratory dynamics at altitude. While many of the findings appear logical and self-evident in the current era, a number of the ideas were only theorized at the time, with no firm data to base models or conclusions on. The following were some of the major findings:

• Oxygen requirements of exercise remain fixed regardless of altitude (Pugh et al. 1964). For example, at the mildest work rate of 300 kiloponds (the only one sustained for a full 5 min at all altitudes), oxygen consumption remained steady at approximately $0.90 \text{ L} \cdot \text{min}^{-1}$ at sea level and at 4650, 5800, 6400, and 7440 m (15,255, 19,030, 21,000, and 24,400 ft). Certainly, as discussed in chapter 8 on altitude training, this is definitely a case in which any reduced air resistance does not compensate for the impaired physiology!

• Amazingly, during this expedition, exercise tests on a cycle ergometer were conducted at an elevation of 7440 m (Pugh et al. 1964). The highest observed value for $\dot{V}O_{2max}$ in the fittest subject was $1.48 \text{ L} \cdot \text{min}^{-1}$ or about five times that of basal metabolism. This obviously is a huge decrease from the typical values of 4 to $6 \text{ L} \cdot \text{min}^{-1}$

Table 9.1 Summary of Selected Field and Chamber Simulation Studies on Physiological Responses to Exercise at Extreme Altitudes

Study	Conditions	Highlights
Everest I, 1946	Simulated altitude exposure inside hypobaric chamber with four subjects.	Heart size was measured as slightly decreased at altitude.
"Silver Hut," 1960-1961	Himalayan base camp (~5800 m) with periodic sojourns to higher altitudes. Lowland native scientists and mountaineers served as experimenters and subjects.	• First set of comprehensive field studies on human physiology at high altitudes. • $\dot{V}O_{2max}$ tests were performed at 5800 and 7440 m, the highest elevation ever for such testing. $\dot{V}O_{2max}$ of 1.48 L · min^{-1} in one subject led to prediction of $\dot{V}O_{2max}$ ~1 L · min^{-1} at summit of Everest.
American Medical Research Expedition to Everest (AMREE), 1981 (see figure 9.2)	• Dedicated mission with physiological research as the primary objective. • Labs were set up at 5400, 6300, and 8050 m and measurements were taken at the summit (8848 m).	• Five team members reached summit; some tests were performed on the summit itself. • First direct measurement of barometric pressure on summit. • Continuous electrocardiograms recorded on two medical summiters.
Everest II, 1985	Simulated acclimatization, summit ascent, and descent inside hypobaric chamber over 40 days in eight healthy male subjects.	• Six subjects reached summit, performed maximal exercise at a $\dot{V}O_2 = 1.18$ L · min^{-1}. • Differential organ susceptibility to hypoxia, with brain most affected, followed by lungs. Heart and skeletal muscle were less affected.
Everest III (COMEX '97), 1997	Preacclimatization period of 7 days on Mont Blanc (4500 m), followed by simulated acclimatization and summit ascent inside hypobaric chamber over 31 days in eight healthy male subjects.	Seven subjects reached summit.

Each expedition performed a vast range of physiological studies.

in recreational to elite athletes, and helps to explain the dramatic reduction in exercise capacity at altitude. Extrapolating from these data, the authors predicted a $\dot{V}O_{2max}$ of approximately 1 L · min^{-1} at the summit of Everest at 8848 m (29,030 ft).

• Reduced oxygen partial pressures at altitude indeed impaired physiological oxygen transport, with oxygen saturation levels of only 67% at rest at 7440 m compared to the near-100% saturation in healthy individuals at sea level (West et al. 1962). Saturation dropped to between 63% and 56% with mild exercise and even below 50% with moderate to heavy exercise. One major proposed mechanism for the maintained desaturation was diffusion limitations within the lung, as the larger oxygen pressure gradients across the alveolar–arterial space would suggest a greater diffusion potential into the bloodstream. No blood sampling has ever occurred at the summit of Everest to directly measure blood gas levels, although a method for an arterial catheter with automated sampling has been proposed (Catron et al. 2006).

• Cardiac output at a given set submaximal effort remained similar at altitude to that at sea level, but the maximum cardiac output decreased from approximately 23 to 16 L · min⁻¹ at 5800 m (19,030 ft) (Pugh et al. 1964). Cardiovascular drift—a decrease in stroke volume at each work rate at altitude with a compensatory increase in heart rate—was evident. This was coupled with a dramatically lower maximal heart rate of 130 to 150 beats · min⁻¹ compared to sea level values of 180 to 196 beats · min⁻¹.

• Blood volume changes were highly variable across subjects and longitudinally within subjects (Pugh et al. 1964). This was likely due to changes in hydration status and plasma volume. In contrast, red blood cell volume and hemoglobin, along with hematocrit, progressively and dramatically increased by 30% to 50% with prolonged stay at altitude. However, these values and those from other studies of lowland natives at altitude remain below those of Andean natives.

Muscle Function

Another series of altitude experiments in the late 1980s reproduced the hypobaric hypoxia of an Everest expedition by having eight subjects simulate the pressure profiles of a summit attempt inside a pressure chamber facility over a course of 40 days. We will use this mission, dubbed "Everest II" (a similar Everest III series of studies was performed in the 1990s), to explore muscle changes at altitude.

Beyond cardiorespiratory factors, changes in muscle morphology, oxidative capacity, or force characteristics may also contribute to the altered exercise capacity at altitude. For example, one early hypothesis was that oxidative metabolism is maximized within the muscle in response to high levels of hypoxic stress, thus forming an adaptive response to the limited oxygen availability. Green and colleagues (1989) measured muscle biopsies at sea level, after adaptation to 380 and 282 mmHg ambient pressure, and again upon return to sea level. In direct opposition to a model proposing increased oxidative capacity, maximal activity levels of a range of metabolic enzymes (including those involved in the citric acid cycle, β-oxidation, glycogenolysis, and glycolysis) remained unchanged at 380 and 282 mmHg compared to initial sea level values. Importantly, some enzymes were significantly reduced after seven days at 282 mmHg and upon return to sea level, suggesting instead a maladaptive response in the muscles to prolonged extreme hypoxia.

Morphological changes within the muscle tissue can also enhance oxygen delivery in hypoxia. One potential avenue is a higher degree of muscle capillarization, decreasing the distance for oxygen exchange between the capillaries and the muscle fibers. The number of capillaries per surface area does appear to increase with prolonged hypoxia in both type I and type II fibers (Green et al. 1989), so oxygen transport may indeed be potentiated. However, rather than the formation of new blood vessels, this change is likely primarily due to a reduction in the cross-sectional area of the individual muscle fibers themselves. No shift in the relative distribution of type I and II fibers was noted. However, the 40-day time line of the expedition may have been insufficient to elicit major structural changes in muscle morphology.

Central and peripheral neuromuscular function was also investigated during Everest II (Garner et al. 1990). This was studied via measurement of the twitch characteristics of the tibialis anterior muscle—responsible for ankle dorsiflexion—in response to isolated electrical stimulation. In addition, maximal voluntary contractions for voluntary force output were tested in conjunction with superimposed electrical stimulation to test for central activation. Torque measured by individual twitches or with tetanic stimulation was maintained at 760, 335, and 282 mmHg pressures. This was also true of torque produced with maximal voluntary contractions. However, data demonstrated that central fatigability appeared to be greater with a decrease in pressure, implying a decreased neural capacity to voluntarily recruit muscles to fire at altitude and a potential pathway for increased muscle fatigue. Also, local muscle fatigue was greater with low-frequency electrical stimulation. No electrical changes in the capacity for neuromuscular transmission were reported, such that the authors concluded that any decreased neuromuscular capacity was likely due to central drive or changes within the muscle fiber itself.

Metabolism and Body Composition

In 1981, an American-led expedition to summit Everest coordinated a series of physiological investigations and was named the American Medical Research Expedition to Everest (AMREE) (see figure 9.2). We will use this mission to highlight some of the metabolic and body composition changes at altitude.

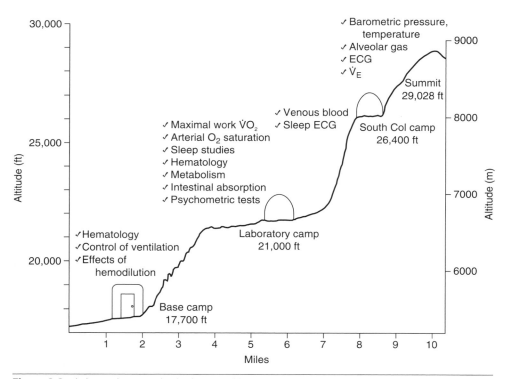

Figure 9.2 Laboratories set up by the American Medical Research Expedition to Everest in 1981, with experiments performed at each laboratory. Despite the huge effort and logistics required to transport a bicycle ergometer and other equipment to the South Col camp at 8050 m, the experiment on maximal aerobic capacity had to be canceled due to weather.

Reprinted from J.B. West, 1981, "American medical research expedition to Everest," *The Physiologist* 25(1): 36. Used with permission.

One of the very characteristic responses to prolonged expeditions at altitude is a persistent weight loss; some mountaineers experience weight reductions of 5 to 10 kg (11-22 lb) over the course of an expedition. This appears especially prevalent with prolonged duration at 6000 m (19,685 ft) or higher. At the most basic level, this weight loss occurs through an imbalance between energy expenditure and intake. As the metabolic cost of exercise remains the same at sea level and at altitude, it is often a marked anorexia and an inability to take in sufficient calories that produce this weight reduction. In turn, the anorexia may be due to a combination of appetite suppression and changes in taste sensations, along with the logistical and practical difficulties in cooking sufficient amounts of food. Further factors regarding weight change are the potential for dehydration from inadequate fluid intake and gastrointestinal distress with travel and altitude. Acute exposure to hypoxia can alter the activity of fluid-regulating hormones involved in an attempt to maintain fluid balance, including an increase in aldosterone, vasopressin (antidiuretic hormone), and plasma renin activity (Okazaki et al. 1984). However, dehydration can still occur, and progressive increases in dehydration were strongly associated with increasing severity of acute mountain sickness (AMS) symptoms in a field study conducted on high-altitude pilgrims (Cumbo et al. 2002).

Another dietary change at altitude may be within the gastrointestinal system, potentially impairing the intestinal absorption of various nutrients. Studies on absorption and excretion of nutrients suggest that the absorption of both fat and xylose is reduced in lowlanders exposed to hypoxia, as evidenced by their increased presence in stool and urine samples (Boyer and Blume 1984). The difficulty in fat absorption poses an additional obstacle to long-term survival at altitude, as the high caloric density of fat also makes it a preferred fuel source nutritionally and logistically. Therefore, one avenue of applied nutritional research may be toward countermeasures that would enhance intestinal absorption of fat and other essential nutrients under hypoxia.

Interestingly, the proportion of mass loss from various tissue stores appears to change over the course of an expedition (Boyer and Blume 1984). During the initial approach march to base camp, fat represented 70% of the total 1.9 kg (4.2 lb) mass loss. However, fat loss accounted for only 27% of the 4.0 kg (8.8 lb) mass loss after arrival above 5400 m (17,715 ft), suggesting that a major metabolic shift occurred toward muscle and protein catabolism at extreme altitudes. This was supported by the observation of significant reductions in both arm and leg circumference. It would thus appear that both muscle catabolism and intestinal malabsorption are additional factors in weight loss at altitude. These changes are not necessarily accounted for by insufficient fat stores, as the Caucasian members had average body fat levels of 18%, in line with population norms. In contrast to these observations in Caucasians, the Sherpa porters, all high-altitude natives, began the expedition with a lower starting body fatness of 9% yet were successful in maintaining weight, body fatness, and limb circumference throughout. Overall, contrary to popular ideas, it appears that gaining weight or body fat prior to expeditions does not protect against muscle wasting at altitude.

Elite Alpinists

As high altitude places humans so close to their physiological limits, it should come as little surprise that some individuals perform better than others in the high mountains. But, besides physical training and mountaineering skill and experience, are there inherent physiological differences determining success at high-altitude mountaineering?

THE LACTATE PARADOX

The increased muscle catabolism is especially problematic during exercise at altitude. As noted earlier, the metabolic cost of exercise remains constant at altitude. However, despite the reduction in overall body mass, the remaining musculature is placed under a greater relative load at both rest and exercise. This produces higher energy demands on the muscles, entailing a shift toward type II fiber recruitment, anaerobic metabolism, and ultimately more rapid fatigue. However, this model is not supported by lactate measurements, which repeatedly show that blood lactate values after maximal exercise generally decrease upon short-term exposure to hypoxia. For example, lactate values after maximal exercise at a normobaric hypoxic level of 6300 m (20,670 ft) (PO_2 = 64 mmHg) in three subjects reached only 3.0 mM (West et al. 1983a), a level far below the 12 mM typical in trained subjects after maximal exercise at sea level, or even the definition of OBLA (onset of blood lactate accumulation) of 4 mM for sustained aerobic exercise. This phenomenon appears to be transient, as peak blood lactate levels dropped after one week at Everest base camp but gradually increased and returned to sea level values over the course of a six-week expedition (Lundby et al. 2000). Peak lactate level in the blood is a dynamic function of production in the muscles, release into the blood, and lactate uptake and metabolism by other tissues. The mechanisms for this "lactate paradox" remain unknown, but likely reside in alterations in the metabolic and physiological control of adenosine triphosphate supply and demand and metabolite homeostasis within the cell (Hochachka et al. 2002).

In one notable study, six mountaineers who had each achieved summits >8500 m (27,890 ft) without supplemental oxygen were characterized for a variety of physiological factors, including aerobic and anaerobic power, muscle morphology, and ventilatory control (Oelz et al. 1986). While these subjects varied in history of AMS and frostbite, all had been born and grown up at low altitudes (<1150 m [3370 ft]) and were tested in a deacclimatized state (2-12 months postexpedition). The results were then compared with those for sedentary individuals and also for long-distance runners and power athletes. Overall respiratory capacities and cardiac dimensions for the mountaineers as well as the control and athletic groups were within normal clinical ranges. However, the mountaineers were marked by a resting hyperventilatory response at both normoxia and mild hypoxia, resulting in a higher oxyhemoglobin saturation. Muscle typing leaned toward type I fiber distribution (70% type I, 22% type IIa, and 7% type IIb), typical of many endurance athletes but not at an extreme range. Aerobic fitness was high at 60 mL · kg^{-1} · min^{-1}, a level higher than sedentary levels but significantly below those typical for elite long-distance runners. Also, anaerobic power was similar to that of sedentary individuals and significantly lower than that of competitive high jumpers.

From these data, it appears that no particular physiological parameter, measured at a baseline or deacclimatized state, distinguishes elite mountaineers from a normal population (Oelz et al. 1986). Therefore, as in the case of sport institutes that attempt to select elite athletes from an early age, it is difficult to predict which young individual possesses the potential to develop into an elite performer at high altitudes based solely on test results in the lab. One potentially significant difference may be in the hyperventilatory response with mild hypoxia. This slight protection of oxyhemoglobin saturation levels may become a critical determinant for success with the small margins for error at these extreme limits of human performance. Specifically, it may enable a slightly higher aerobic capacity at extreme altitude and also minimize the risk for debilitating AMS and impaired decision making. Beyond baseline responses, however, the critical source of differentiation for elite mountaineers may lie in the time course and magnitude of adaptation upon exposure and exercise at altitude, which were beyond the scope of this study.

HIGH-ALTITUDE ILLNESSES

This section outlines a range of increasingly debilitating and dangerous illnesses that may befall individuals trekking or climbing at high altitudes, summarized in the consensus statement published by the International Society for Mountain Medicine (Leon-Velarde et al. 2005). The common mechanism underlying the three most dangerous conditions—AMS, high-altitude pulmonary edema (HAPE), and high-altitude cerebral edema (HACE)—is the lower barometric pressure and partial pressure of oxygen resulting in a reduced arterial PO_2 (Bartsch and Saltin 2008). It should also be kept in mind that, rather than being distinct, the three exist along a continuum, requiring proper diagnosis and careful management upon initial onset in order to prevent deterioration.

Acute Mountain Sickness

The broadest range of mild, nonfatal high-altitude illnesses are grouped together into the umbrella term acute mountain sickness or AMS, which may be present for approximately 50% of people trekking at altitudes over 4000 m (13,120 ft) (Basnyat and Murdoch 2003). Depending on how strictly AMS is defined, 2500 m (8200 ft) is commonly employed as a minimum threshold for occurrence. However, due to the high individual variability in response to hypoxia and also situational variables (see later section on prediction of altitude illnesses), AMS can occur at much lower altitudes of even 1500 m (4920 ft) (Basnyat and Murdoch 2003). Symptoms are also very general and can range widely in both prevalence and magnitude. These symptoms include headache, fatigue, shortness of breath (dyspnea) and hyperventilation, insomnia, loss of appetite or gastrointestinal distress, and decreased thirst (Coote 1995). As should be evident, the commonality of such symptoms can make it very difficult to correctly diagnose AMS. This is especially true with the prolonged travel that is typical for trips to high altitude, coupled with changes in diet and available foods along with the sudden exercise involved with trekking to base camps at altitude.

Due to the broad range of general symptoms, the proper definition and diagnosis of AMS are somewhat unclear and contentious. The most commonly used tool for diagnosis is the Lake Louise subjective questionnaire, first developed and adopted in 1991 at a conference in Canada (Roach et al. 1993) (see table 9.2). The Lake Louise scale is grouped into specific symptoms of AMS, including headache, gastrointestinal, fatigue, sleep, and overall activity. Each of these five categories is subjectively rated from 0 (no symptoms) to 3 (severe symptoms), and mild or severe AMS can be diagnosed with a total score of 3 or 6, respectively, along with the presence of headaches as part of the score. Further symptoms and scoring can be added with clinical assessment, including

Table 9.2 Summary of Lake Louise Subjective Questionnaire for Acute Mountain Sickness

Standard scale	0	1	2	3
Headache	No headache	Mild	Moderate	Severe
Gastrointestinal	No symptoms	Poor appetite or nausea	Moderate nausea or vomiting	Severe to incapacitating nausea and vomiting
Fatigue and weakness	Not tired or weak	Mild	Moderate	Severe to incapacitating
Dizziness and light-headedness	Not dizzy	Mild	Moderate	Severe to incapacitating
Difficulty sleeping	Slept as well as usual	Did not sleep as well as usual	Woke many times, poor night's sleep	Could not sleep at all
Clinical assessment				
Change in mental status	No change	Lethargy/lassitude	Disoriented/confused	Stupor/semi-consciousness
Ataxia (heel-to-toe walking)	No ataxia	Maneuvers to maintain balance	Steps off line	Falls down (NB: There is a 4 for "Can't stand")
Peripheral edema	No edema	One location	Two or more locations	

Each criterion is rated from 0 (no symptoms) to 3 (severe to incapacitating). For the standard scale, acute mountain sickness is diagnosed with the presence of headache and a total score of 4 or more. The clinical scale can be used as a supplement to medical assessment.

changes in mental status, ataxia, and peripheral edema. The scale has been adopted internationally and translated into many different languages, including phonetically into the Nepali language. Variations of the Lake Louise scale have also been adapted for use with children and infants, revolving around fussy behaviors not associated with specific causes like teething. While these questionnaires are an invaluable tool for field diagnosis, it is important to keep in mind their subjective nature. Therefore, another major thrust in the use of such tools must be in education and increasing self-awareness about AMS and its potentially lethal consequences.

High-Altitude Cerebral Edema

If AMS remains undiagnosed and preventive measures are not taken, the progression of symptoms can magnify in range and intensity. One potentially fatal endpoint of AMS is high-altitude cerebral edema (HACE). Ideally, therefore, AMS should never progress to this stage! As implied by the name, the major symptoms are severe headaches from brain swelling. As a consequence, specific symptoms for HACE revolve around the central nervous system, including potentially severe ataxia, impaired mental functioning, loss of consciousness, and retinal hemorrhage (Hackett and

Roach 2004). If untreated by pharmacological means, supplemental oxygen, and most preferably by descent to a lower altitude, HACE can easily cause permanent damage or death.

The exact mechanisms for AMS and HACE remain open to debate, but likely derive from swelling within the central nervous system (Hackett and Roach 2004). The pain from a headache can be multifactorial in cause, from changes in both mechanical and chemical factors within the body, making the relative contributions of individual factors difficult to isolate. Some of the more attractive, though at this stage speculative, models (see figure 9.3) propose that the hypoxemia at altitude results in a compensatory greater cerebral blood flow. This mechanism is supported by a higher increase in the mean blood flow velocity of the middle cerebral arteries in individuals experiencing AMS compared to those without symptoms (Baumgartner et al. 1994). This elevated blood flow to the brain may in turn result in changes in blood–brain barrier permeability, ultimately elevating intracranial blood pressure and cerebrospinal swelling. Fluid accumulation within the brain may arise from swelling within the individual cells, possibly due to changes in osmolality within the cells or interstitial fluid (cytotoxic

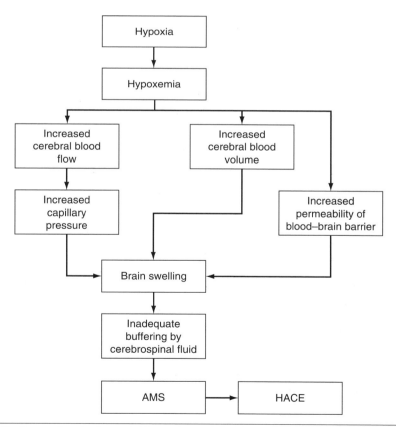

Figure 9.3 Proposed mechanism underlying acute mountain sickness (AMS) and high-altitude cerebral edema (HACE).

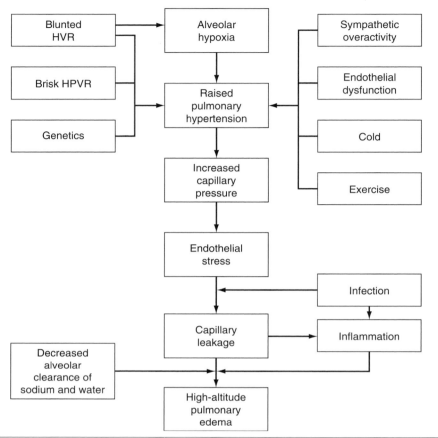

Figure 9.4   Proposed mechanism underlying high-altitude pulmonary edema (HAPE). HVR = hypoxic ventilatory response; HPVR = hypoxic pulmonary vascular response.

Reprinted from *The Lancet*, Vol. 361, B. Basnyat and D.R. Murdoch, "High-altitude illness," pgs. 1967-1974, Copyright 2003, with permission from Elsevier.

edema). Alternatively, changes in cerebrovascular pressure may cause additional leakage of proteins or fluids across the blood–brain barrier (vasogenic edema). Overall, it may be that a combination of brain and cerebrospinal swelling is responsible, as both would potentially affect neural functioning.

High-Altitude Pulmonary Edema

A clinical problem parallel to AMS and HACE, and one that shares many of the same predisposing risk factors, occurs in the lungs with high-altitude pulmonary edema (HAPE) (see figure 9.4). The time course for HAPE can be similar to that for AMS, generally beginning shortly upon arrival at altitude or after a rapid further increase in altitude. As evident in the name, the primary tissue of concern is the pulmonary system (Leon-Velarde et al. 2005). The dominant issue concerns inadequate oxygen transfer between the lungs and the cardiopulmonary circulation from altered blood flow and fluid buildup within the lungs. Therefore, initial symptoms include dyspnea

FIGHTING "SUMMIT FEVER"

Symptoms consistent with extreme fatigue and HACE, coupled with late return from a summit attempt, were prime predictive factors for subsequent death in a retrospective study of death on Everest between 1921 and 2006 (Firth et al. 2008). One factor working against diagnosis of AMS and other high-altitude illnesses is the inherent excitement of expeditions, whereby climbers may subconsciously downplay symptoms or else voluntarily withhold information about symptoms from other team members to avoid being left behind on summit attempts. Therefore, a strong self-awareness and resistance to "summit fever" often constitute one of the prime survival tools for mountaineers. As an example of a sound survival mentality, Ed Viesturs, the first American to summit all fourteen 8,000 m (26,246 ft) peaks (and all without supplemental oxygen), states as his slogan, "Getting up is optional, getting back down is mandatory" (Viesturs and Roberts 2006). At the same time, a strong command structure and objective medical analysis can help to protect people in expeditions from pushing each other or themselves too hard.

and difficulty breathing both during rest and with exertion. This can also be expressed as tachycardia and tachypnea as a physiological response to reduce hypoxemia. Similar to and sometimes misdiagnosed as pneumonia in the early days of altitude medicine, the buildup of fluid within the lungs will also be evident as crackles heard with chest examination, along with the coughing of sputum or blood. Another obvious physical symptom stemming from impaired oxygenation is cyanosis, marked by blueness in the lips and extremities. Diagnosis of HAPE can be made initially in the field with use of subjective criteria similar to those of the Lake Louise AMS scales, and AMS and HACE symptoms can often coincide with HAPE.

Similar to the situation with AMS and HACE, increased blood flow and pressure, in this case to the pulmonary tissues and arising from a variety of mechanisms, likely underlie the pathophysiology of HAPE (Dehnert et al. 2007). For example, the hypertension may be due to hypoxia-induced pulmonary vasoconstriction that forces fluid into the interstitial fluid. Endothelial dysfunction may also contribute to the increase in leakage across the pulmonary capillaries. In turn, an elevated sympathetic neural drive from hypoxic stress can also contribute to the vasoconstriction. Existing evidence of pulmonary hypertension or other cardiopulmonary circulatory issues may therefore be a predictor of individual susceptibility (Eldridge et al. 1996). Thus it may be useful to use these initial clinical histories, along with tests of pulmonary arterial pressures during normobaric hypoxic exposure, as prescreening tools of individuals prior to altitude deployment (Rodway et al. 2003, 2004).

Countermeasures

The primary initiating mechanism of reduced PO_2 for high-altitude illnesses suggests that beyond any other countermeasure, the dominant preventive and clinical countermeasure should be a gradual acclimatization to altitude, coupled with a rapid descent to lower altitudes at the first signs of distress or diagnosis of disease (Houston and Dickinson 1975; Leon-Velarde et al. 2005) (see table 9.3). Acute motion sickness, when mild, generally decreases in intensity with prolonged time at altitude. Nevertheless, it is important to continuously monitor the victim to ensure that symptoms do not increase and progress into the more severe cases of cerebral and pulmonary edema. Contingency plans should be made to evacuate the individual if necessary, and under no circumstances should anyone suffering from AMS be permitted to continue climbing to higher elevations until symptoms subside (Luks 2008). In essence, in this situation it is *never* inappropriate to descend, while it is always inappropriate to continue to ascend, leading to the mountaineering axiom "Never go up until symptoms go down." At the same time, the victim must avoid moderate to strenuous exercise, which will exacerbate symptoms by increasing physiological stress and risk of dehydration. In a mountaineering context, the problem with altitude illnesses is that victims can readily become incapacitated, requiring an enormous mobilization of people and resources to provide aid or perform evacuation. In turn, this places many other people, already operating near the limits of their functional capacity, in increased danger along with jeopardizing the expedition itself.

Where AMS risk is high, the most common prophylaxis is the prescription of acetazolamide (Diamox). Acetazolamide acts by blocking the action of carbonic anhydrase, resulting in a buildup of bicarbonate and subsequently its increased excretion from the body (Leaf and Goldfarb 2007). In turn, this decreases blood pH and stimulates ventilation. For AMS, it has been proposed that this pathway enhances the hypoxic hyperventilatory response and can therefore be used to help stimulate or

Table 9.3   Common Methods for Relieving Symptoms of Acute Mountain Sicknesses (AMS)

Method	Physiological rationale	Recommendations
Rest and descent	↑ PO_2 and ↓ hypoxic stress	*Always* the top option and priority in any severe case of AMS. Under no circumstances should activity or ascent continue until AMS symptoms have subsided.
Gamow bags	Sealed bag using manual pump to artificially increase pressure and PO_2 around victim	• ↓ equivalent altitude by ~1500 m. • Used as emergency aid in situations of severe AMS where descent is not immediately feasible.
NSAIDs (aspirin, acetaminophen, ibuprofen)	Pain blockers	Should be avoided at altitude, as they can mask the discomfort and headaches that are a primary symptom of AMS.
Acetazolamide (Diamox)	Carbonic anhydrase inhibitor used to ↓ blood pH and stimulate hyperventilation	• Can be used prophylactically or with AMS onset. May be useful for treating mild cases of AMS. • ↑ diuresis and risk of dehydration. • Does not replace rest and descent as first method of treatment.
Dexamethasone (Decadron)	Steroidal anti-inflammatory drug to reduce inflammation	• Can be used prophylactically or with AMS onset. Used to treat HACE and HAPE. • Does not replace rest and descent as 1° treatment.

accelerate the rate of acclimatization (Leaf and Goldfarb 2007). However, it absolutely should not be used to replace a well-planned period of natural altitude acclimatization. One major side effect is the increased diuresis and urination, which should also be accommodated with an emphasis on adequate fluid intake. Another drug that can be employed either prophylactically or upon AMS occurrence is the corticosteroid dexamethasone (Decadron), which helps to reduce inflammation. The two drugs are also sometimes used in combination (Wright et al. 2008).

Mechanical aids for short-term management of high-altitude illnesses include portable hyperbaric chambers, which are generally termed Gamow bags after the original designer (Kasic et al. 1989). At its most basic, the Gamow bag is an airtight bag that is inflatable with a foot pump. By pumping air into the Gamow bag, the effective elevation inside can be decreased by approximately 1500 m (4920 ft) or more. Most variations of the Gamow bag involve zipping the subject completely inside the unit, which has a clear window to permit monitoring; some include a CO_2 scrubbing unit while others rely on continual pumping to provide circulation of fresh air inside. Victims may be placed inside the bag for an hour or more, then reassessed, and the procedure can be repeated as necessary. In some cases, the effects of this brief repressurization may be sufficient to permit subjects to descend under their own power. However, some versions are also designed to be incorporated with evacuation devices such as stretchers for victims who remain nonambulatory. People who are unconscious and victims who are at risk for the cessation of breathing should not be placed inside Gamow bags, as there is a time delay for assisted ventilation once the person is inside. However, some of the larger tent systems may permit a medical attendant to be inside with the victim. Overall, as noted at the beginning of this section, it must be reemphasized that such devices are for the temporary management of altitude illness and that planning for evacuation and descent must be a concurrent priority.

Prediction of Altitude Illnesses

With the difficulties posed by high-altitude illnesses and wide individual variability in susceptibility, one obvious research and clinical goal is an understanding of the mechanisms of AMS and the potential prediction of response to hypoxia. Overall, as with predicting susceptibility to space motion sickness in astronauts (see chapter 10), such investigations have been difficult and largely unsuccessful in pinpointing any key individual physiological or genetic factors (Leon-Velarde and Mejia 2008). Along these lines, an epidemiological analysis of AMS prevalence highlighted that the primary predisposing factors were mainly environmental and situational, including past history of AMS, existing acclimatization to altitude (e.g., days above 3000 m [9840 ft] in the past two months), and the rate of ascent (Schneider et al. 2002). Furthermore, exertion at altitude can exacerbate and magnify the intensity of symptoms. Beyond these primary factors, other considerations, such as sex, age, training status, smoking history, and alcohol consumption, played minimal predictive roles. Therefore, as discussed in chapter 8 on altitude training, one ergogenic use of hypoxic houses or tents may be to preadapt to altitude. Such uses may greatly eliminate two of the three major risk factors just listed, thus minimizing the amount of time required to acclimatize prior to expeditions. Similar large-scale setups for passive hypoxic exposure (e.g., barracks set up for hypoxic control) may also prove useful for military units that might require rapid deployment for high-altitude missions.

One prominent area of focus for the physiological prediction of AMS has been the intensity of the hyperventilatory response to hypoxia or to hypercapnia, on the hypothesis that a lower ventilatory response would lead to lower oxygen level or breathing stimulus, resulting in less cerebral oxygenation and a greater risk of AMS (Podolsky et al. 1996). However, the association between ventilatory response to hypoxia or hypercapnia with AMS susceptibility is very weak or nonsignificant overall across a range of studies (Moore 2000). This is indirectly supported by the history of AMS in the six elite mountaineers studied by Oelz and colleagues (1986). While a slight hyperventilatory response was noted in the mountaineers, two had a history of light headaches during previous high-altitude exposures, and one had previously had severe symptoms and also HACE.

As the primary criterion for AMS involves moderate to severe headaches, another basis for AMS susceptibility may be slight anatomical variation in central nervous system architecture, such as the integrity of the blood–brain barrier. Another consideration may be the volume of the intracranial and intraspinal cerebrospinal fluid, and especially its relation to brain volume and ability to absorb or tolerate slight changes in volume and subsequently pressure. For example, it is theorized that a large brain-to-skull volume ratio, coupled with a high intracranial fluid volume, may leave little space available to accommodate further intracranial or brain swelling upon exposure to altitude (Krasney 1994). Such a hypothesis has some support from studies using neural imaging on AMS patients with differing severity of symptoms (Kallenberg et al. 2008). However, even if this idea is valid, the difficulty lies in employing neural imaging as a predictive and preventive tool for individuals to employ prior to exposure.

As with many other topics within the field of environmental and exercise physiology, understanding the mechanisms and susceptibility to high-altitude illnesses may benefit from recent advances in genetic research. In turn, pinpointing causative genetic factors may also lead to improved treatments for people suffering from respiratory conditions such as chronic obstructive pulmonary disease and pulmonary hypertension. Currently, the main genetic candidate for both potential performance at altitude and susceptibility to altitude illnesses has been polymorphisms in the angiotensin-converting enzyme (ACE) gene (Tsianos et al. 2005). However, others argue that a clear and direct association between these genes and susceptibility to altitude illnesses remains unconfirmed (Dehnert et al. 2002). Overall, it would appear that one problem with isolating genetic factors lies in the fact that AMS likely occurs due to a multitude of causes rather than a single determinant.

SUMMARY

High and extreme altitudes form a unique research opportunity by limiting the amount of oxygen available to the body. As with exercise training, thermal conditions, and other stressors, high altitude provides a method for understanding physiological limitations and the relationships between various systems. Therefore, such an environment is also a useful research analog for many clinical conditions, especially those involving the cardiorespiratory system. One major avenue for research has been experiments performed during expeditions to extreme altitudes, both in the lab and in the mountains. Such experiments allow investigation of the acute and chronic adaptations to hypoxia. Another research approach has been the physiological comparison between high-altitude and lowland natives. Work will likely continue in the coming decades with both approaches, supplemented by advances in genetic techniques to allow further understanding of the role of developmental factors in altitude response. Overall, it is clear that a multiplicity of interconnecting factors, rather than a single overriding determinant, confers enhanced tolerance and exercise performance at high altitudes.

Microgravity and Spaceflight

O n October 4, 1957, the flight of *Sputnik*—the first artificial satellite put into orbit by the former Soviet Union—launched the Space Age and ignited an intense international race for space supremacy between the USSR and the United States of America in all areas of science and technology. Fiery words, including President John Kennedy's famous speech pledging to put an American on the moon by the end of the 1960s, were matched by inspiring deeds. The 1960s began with Colonel Yuri Gagarin's becoming the first human in space on April 12, 1961, and culminated with Neil Armstrong's famous words ("That's one small step for [a] man, one giant leap for mankind") broadcast around the world from the surface of the moon on July 20, 1969. In between were many notable human milestones (see table 10.1), including the first spacewalks or "extravehicular activities" (EVAs). Apart from the *Apollo* lunar missions, human space exploration has been confined to Low Earth Orbit (LEO, generally categorized as ranging from altitudes of 200 to 2000 km [124 to 1243 miles]) with the space stations *Skylab, Salyut,* and *Mir;* the Space Shuttle; and the International Space Station (ISS). Such missions range from 10 to 16 days in the Shuttle to approximately three to six months on the ISS, and on to Dr. Valeri Polyakov's current record of 438 consecutive days on *Mir.* With the Shuttle flight of 1999, John Glenn at age 77 became the oldest astronaut to have flown in space.*

Even the relatively brief duration of Shuttle missions is enough to elicit substantial changes to the human body (Buckey 2006). Physiologically, the dominant effects of spaceflight, caused by the removal of gravitational gradients, are a redistribution of body fluids and alterations in neurovestibular balance. The physical sensation of weightlessness and altered cues from the neurovestibular system can result in space motion sickness even in individuals highly trained in ground-based analogs, causing delays in mission objectives. The removal of gravity means that blood volume is no

---

*Each of the three nations with direct-launch capabilities (United States, Russia, and China) has its own terms for space-farers (astro-, cosmo-, and taikonauts, respectively). For the sake of simplicity, this text uses only the term astronaut.

Table 10.1 Major Milestones in Human Spaceflight

Year	Event
1961	First man in space: Yuri Gagarin (USSR). Alan Shepard of the USA flew a suborbital flight in 1961, and John Glenn flew an orbital mission in 1962.
1963	First woman in space: Valentina Tereshkova (USSR). It was 19 years before the USSR launched its second female cosmonaut.
1965	First spacewalk: Alexei Leonov (USSR). Followed closely by Ed White of the USA.
1969	First lunar landing: Neil Armstrong and Buzz Aldrin, with Michael Collins in the command module. Currently only 12 humans on six *Apollo* missions have landed on the moon.
1971	*Salyut 1* enters orbit; this was the first of many long-duration space stations launched by the USSR.
1984	Bruce McCandless (USA) makes the first untethered spacewalk from the Space Shuttle using the Manned Maneuvering Unit. The picture of McCandless floating above the earth becomes one of the most famous and iconic of space images.
1986	The *Salyut* series is succeeded by the *Mir* space station, which was almost continuously inhabited until it was deorbited in 2001.
1994-1995	Valeri Polyakov (Russia) sets the record for continuous spaceflight duration at 437 days aboard *Mir*. Having also performed a 240-day mission in 1988-1989, he easily holds the record for longest time spent in space.
1998	John Glenn returns to space, becoming the oldest astronaut at 77 years of age.
1998	On-orbit assembly for the International Space Station begins.
?	Lunar base and manned mission to Mars.

longer predominantly pooled in the lower body, and the heart no longer has to work to pump blood against gravity. This redistribution and subsequent decreased workload can have a profound impact on the cardiovascular system in the initial phase of return to Earth and the sudden reintroduction of gravity. The resulting high incidence of orthostatic intolerance can be especially problematic in case of emergency egress following landing. At the same time, weightlessness alters the stresses imposed on the muscles and bones. Specifically, the body in space is no longer forced to hold itself upright and resist the forces of gravity. This leads to a reduced stress and also stimulus on postural muscles and the skeletal system, resulting in a progressive atrophy of these systems.

Previous space missions in LEO have greatly advanced our understanding of biological systems and human physiology. However, the United States again has bold plans in place for a lunar base and ultimately a manned mission to Mars, requiring a renewed focus on understanding the effects of the space environment on human physiology. Many physiological, medical, and psychological questions about prolonged spaceflight remain unanswered and are sources of active research, including the effects of radiation, the potential need for artificial gravity, the interpersonal dynamics of close confinement, and how to deal with medical emergencies far away from Earth.

The purpose of this chapter is to provide an overview of the effects of the microgravity environment on different human physiological systems, including the neurovestibular, cardiovascular, bone, and muscular systems. In addition, we will examine countermeasures designed to minimize such effects both during and following spaceflight. A section is also devoted to the human engineering of spacesuits and EVAs. Finally, we will explore some of the major outstanding physiological questions that need to be addressed to enable long-term human space exploration. The reader is also referred to other comprehensive texts on space physiology and medicine (Buckey 2006; Clément 2007, 2005; Kawaguchi et al. 1999; Nicogossian et al. 1994).

THE MICROGRAVITY ENVIRONMENT

The defining characteristic of spaceflight is the microgravity environment. While an object has the same overall mass in space as it did on Earth, the gravitational forces on it are reduced to nearly zero, producing a sensation of weightlessness. A useful analogy is an elevator whose cable has snapped so that the elevator is in freefall. The occupants are falling and accelerating toward the ground at the same rate as the elevator and therefore perceive themselves to be weightless. Indeed, tall drop towers with controlled deceleration at the bottom are one means of simulating microgravity on Earth. In a similar fashion, astronauts in LEO at the altitudes of the Shuttle and space stations (typically 200-300 km [124-186 miles]) remain subject to gravitational pull from the earth. While the pull is reduced compared to that experienced on the surface, the spacecraft would fall toward the earth without a counterbalancing force. That force is provided by forward movement of the spacecraft orbiting around the earth, with a tangential velocity that supplies a centrifugal (outward) force equal to the gravitational force. In essence, astronauts are in a weightless state of continuous freefall *around* the earth, which produces a microgravity environment (see figure 10.1).

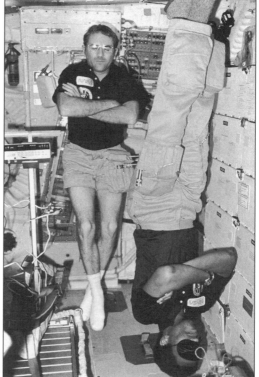

From NASA: nasaimages.org

Figure 10.1 Without gravitational cues, astronauts in microgravity can move and sleep in a range of postures and orientations. Such unfamiliar environments and the associated alterations in visual and neurovestibular cues can lead to space motion sickness in the initial period of spaceflight. Note the pointed toes (plantarflexion) and slight hip and knee flexion common to a natural posture in microgravity.

Microgravity Simulations

Due to the limited sample sizes, costs, and logistics related to performing human physiological research in space, much of what we know about physiological responses to the microgravity environment is derived from either cell or animal models in space or in Earth-bound simulations. Alternatively, much of the research on astronauts involves pre- and postmission testing following spaceflight, ideally with periodic monitoring and testing during a mission. The following are some of the primary Earth-based biological simulations for microgravity (Adams et al. 2003; Sonnenfeld 2003):

- **Hindlimb suspension or immobilization.** The primary animal simulation of microgravity involves prolonged hindlimb suspension. The suspension of the tail in a harness lifts the hindlimbs off the ground, removing weight-bearing forces on the hindlimb muscles and also producing fluid redistribution toward the head. In this way, both muscle and bone atrophy can be induced, along with the loss of tactile and other sensory input from the immobilized regions. Some studies have also employed casting to immobilize limbs in order to investigate the process of atrophy.

- **Human immobilization.** In a similar fashion, human injuries such as broken bones requiring prolonged immobilization, along with spinal cord injuries, can provide a scientific model of neural, muscular, cardiovascular, metabolic, and bone atrophy. Temporary casting is especially valuable for investigating the process of rehabilitation of these systems and developing training programs for returning astronauts.

- **Water immersion.** The water environment provides buoyancy support that mimics the weightlessness of spaceflight. While of minimal use as a physiological research tool, water immersion is of immense applied value in simulating extravehicular activities. As such, full-scale underwater mock-ups of the Shuttle and the ISS are used by astronauts to plan and rehearse both general and mission-specific EVA tasks prior to the actual mission. In addition, underwater simulations are prime environments for testing the ergonomics and usability of tools intended for EVAs.

- **Confinement.** While not a direct physiological analog, short- and long-term confinement of a small group in close proximity is one of the best simulations of the psychological issues arising from prolonged space missions. For long-term missions such as a lunar base or a Mars mission, the appropriate age, gender, and cultural composition of the crew, along with psychological makeup, training, and monitoring, remain major issues of debate. The psychological stresses involved with confinement can also provide insight into physiological stress markers such as cortisol, as well as the resultant immunological effects of spaceflight.

- **Parabolic flights.** Alternating rapid and steep ascents and descents, coupled with "pushovers" at the apex (the plane belly remains facing the ground rather than rolling over as seen in stunt flying), makes it possible to simulate microgravity for 20 to 40 s during the apex of these parabolic arcs. Specially reinforced airplanes, commonly nicknamed "Vomit Comets" for their effects on passengers, can fly these parabolic arcs repeatedly during a single flight, providing a means for astronauts to obtain training in the sensations of microgravity. As with water immersions and EVAs, parabolic flights also offer an invaluable platform for the testing of equipment that astronauts will be utilizing during space missions. However, the brief duration of microgravity and the constantly changing environment make sustained physiological research using parabolic flights largely impractical.

Bed Rest Simulations

While microgravity simulations such as drop zones and parabolic flights may be highly effective for materials and engineering testing, the physiological effects of chronic microgravity exposure are much more difficult to replicate on Earth. One of the best human models consists of bed rest, with its removal of head-to-toe gravity gradient and weight-bearing forces (Adams et al. 2003). As an added benefit, there is a direct transfer from such research to the prolonged bed confinement common in many clinical situations, along with rehabilitation and spinal cord injuries.

The actual protocol for bed rest simulations depends on the physiological system of interest. For example, acute bed rest simulations of 2 to 4 h, coupled with rapid tilting, can replicate some of the fluid redistribution and orthostatic tolerance challenges of spaceflight. To further maximize the fluid shifts found in microgravity, bed rest studies typically employ 6° to 12° head-down tilt rather than a horizontal supine posture. As deconditioning appears to be continual and progressive, the duration of bed rest studies becomes an important parameter, though it must be balanced against logistics and subject recruitment. The majority of bed rest simulations employ durations of 14 to 35 days. The National Aeronautics and Space Administration (NASA) continually conducts 90-day bed rest protocols in a full-time bed rest simulation facility in Galveston, Texas. Amazingly, some of the longest bed rest studies have utilized 6 to 12 months of confinement! Some emerging issues include the most appropriate bed rest simulation for the reduced gravity (one-sixth to one-third Earth gravity or G) of the moon and Mars.

Bed rest simulation studies have uncovered a wealth of knowledge concerning the effects of physiological deconditioning, and NASA has mandated that all space-based countermeasures must first be validated in a bed rest model. Yet the issues involved in executing bed rest studies are legion. From a strict simulation perspective, the main advantage of bed rest is that it can closely simulate the unloading of postural muscles and cause a redistribution of body fluids. However, one must keep in mind that the situation for astronauts is not analogous to bed rest in that they remain highly active during spaceflight, with a metabolic rate and energy demand similar to or above those on Earth. This amount of daily activity is difficult to simulate with bed rest and likely affects the extrapolation of results.

By far the largest hurdles to bed rest microgravity simulations are logistical, with the need to recruit subjects who are willing to be immobile and isolated from their normal environments for a prolonged period of time. While this is the normal reality for astronauts in space, it is a decidedly abnormal sociological environment on Earth, and one must question whether the baseline conditions of bed rest subjects (age, health, fitness, psychological status, etc.) accurately reflect those of the intended target audience of astronauts. In addition, as with much of space-based research, the costs associated with bed rest studies can be prohibitive. Apart from subject recruitment and payment, a full facility resembling a hospital ward must be maintained for the duration of the study. Such expenses often necessitate multiple researchers and teams combining forces and funding to address multiple research questions at the same time. In turn, this can lead to difficult management and research design issues to ensure that the various studies, each with its own manipulations, tests, and equipment, do not interfere with and confound one another. Overall, one can consider bed rest simulations as still in their infancy stages, as evidenced by the fact that it was only a few years into this century that the first large-scale female-specific bed rest study was performed in France by an international research team (Beavers et al. 2007; Smith et al. 2008).

MEDICAL REQUIREMENTS FOR SPACE TOURISTS?

With the awarding of the Ansari X-Prize in 2004 for a successful private launch and relaunch to an orbit of at least 100 km (62 miles), the era of space as the exclusive domain of governmental agencies and professional astronauts may have been replaced by an emerging era of the "space tourist." Without strict control on who can or cannot join the astronaut corps, an important consideration with this shift becomes the medical qualifications for nonprofessionals entering space. In this context, the various space agencies have developed and adopted a common medical screening policy for such nonoperational spaceflight participants (SFP) traveling to the ISS (Bogomolov et al. 2007). While basic operational imperatives such as the ability to participate in typical flight operations, utilize noncritical flight equipment, and perform emergency egress remain, the medical standards are substantially relaxed in comparison to those for the primary flight crew. Within this context, exclusion standards based on greater acceptance rates still must be balanced against increased risks of major medical emergencies in orbit. A perusal of the SFP ISS document demonstrates that medical standards, while relaxed, remain extremely strict. However, it is important to realize that this is the case only for ISS- and agency-based missions and that the privatization of space will likely not face such safety pressures. While the risk for medical incidents in space may become higher with this new focus on space tourism, one potential benefit is that more physiological research may actually be possible, especially on individuals with medical conditions that would ordinarily disqualify them from professional space missions.

PHYSIOLOGICAL RESPONSES TO MICROGRAVITY

Human beings have evolved in adaptation to the 1-G environment on the surface of the earth. However, the human body also appears to be an incredibly adaptable system in responding to the forces and stress (e.g., exercise, altitude) imposed upon it. Microgravity results in an extreme environmental change, and few human physiological systems appear resistant to effects from this change. Apart from long-term health concerns for astronauts upon return to Earth, the operational imperative for mission planning is to ensure that astronauts have the functional capacity to fulfill mission goals following long-term spaceflight. For example, the moon and Mars have ~1/6- and 1/3-G, respectively. Even if astronauts fly in a microgravity environment with relatively minimal operational demands during the prolonged transit to Mars, they still must maintain sufficient strength and functional physical capacity to perform work in the higher-gravity environments of the Martian surface upon arrival.

It should be noted that all of the existing human data from spaceflight have, from a scientific research design perspective, been "confounded" by the exercise protocols and other standard safety protocols (e.g., saline loading prior to reentry) performed by astronauts. Therefore, it becomes difficult to truly isolate the effects of microgravity. The separate effects of any one potential countermeasure are also difficult to tease out from the complex interactions among a number of countermeasures (e.g., exercise, nutrition, launch and reentry suits), many of which have been added in response to a specific issue without full testing of their separate and interactive effects on the entire human system. This places great emphasis on Earth-based simulations, such as bed rest studies, for understanding the physiological effects of microgravity and also for developing evidence-based health standards and protocols for space missions. At the same time, the fact that such physiological deconditioning continues to occur despite the existing interventions clearly demonstrates that current strategies can and need to be refined and improved.

Neurovestibular System

The immediate sensation of weightlessness in microgravity is a sensation that, apart from brief periods during parabolic flights, is difficult to replicate prior to a mission. One common occurrence with spaceflight is space motion sickness (SMS), marked by nausea and vomiting, with secondary symptoms of drowsiness and lethargy (Lackner and Dizio 2006). The incidence of SMS is high, affecting approximately 70% of first-time astronauts, though experienced astronauts seem to have reduced susceptibility (Davis et al. 1988). Space motion sickness is largely transient upon arrival in orbit or landing on Earth, strongly supporting the prevailing sensory conflict model (conflicting sensory input from visual and tactile systems, along with the otoliths and semicircular canals of the vestibular system) of motion sickness (Lackner and Dizio 2006). Apart from impairment of mental and physical functioning with SMS, vomiting can contribute to dehydration and loss of body fluids, further exacerbating fluid balance and cardiovascular issues with spaceflight.

Once in microgravity, the otolith organs provide minimal feedback concerning tilt and acceleration of the body, as the otoliths are not "weighted" down by gravity. Visually, the astronaut is free to float through space, such that the "normal" up or down environment of Earth is no longer relevant, producing conflict between the visual system and the otoliths. Further conflict arises in that the feet and muscles no longer provide muscular or tactile feedback of pressure on the floor. With head movements, the acceleration is perceived by vision and the semicircular canals, but not by the otoliths. Eventually, the brain adapts to this new and altered sensory input with microgravity, and the risk of SMS fades. Problems may arise again upon return to Earth and a reestablishment of 1-G sensory input.

Unfortunately, motion sickness tends to be environment specific. For example, susceptibility to Earth-based motion sickness (e.g., at sea [seasickness]; with parabolic flights or high-G training in centrifuges) has a relatively low correlation with SMS prevalence, making it difficult to predict which astronauts are susceptible to SMS (Oman et al. 1986). Currently, a multitude of different antimotion sickness drugs are employed by astronauts; intramuscular injections of promethazine or scopolamine are a common therapy during both parabolic flight training and spaceflight itself (Putcha et al. 1999). One potential area of concern is that the main pharmacological countermeasures to motion sickness tend to have sedative characteristics. This can in itself impair physical and mental functioning, though astronaut experience has indicated that, contrary to findings from land-based studies, intramuscular promethazine induced minimal sedation during spaceflight itself (Bagian and Ward 1994). Research to date shows that autogenic feedback training, in which individuals are taught to monitor various physiological factors (e.g., heart rate, respiration) and consciously attempt to control these responses, has had minimal efficacy with SMS—hence the continued reliance on pharmacological countermeasures (Lackner and Dizio 2006).

Despite high levels of inherent motivation in astronauts, the risk of SMS makes significant mission milestones and tasks difficult to accomplish during the initial days of a spaceflight (e.g., EVAs are not scheduled during the first three days of a Shuttle flight). This has been especially problematic with the relatively short overall mission lengths (7-16 days) for the Shuttle, greatly reducing the available operational time.

Cardiovascular System

On Earth, a hydrostatic gradient is established by gravity, whereby the majority of blood volume pools within the venous beds of the legs. Of the typical 5 L of total blood volume in the body, approximately 3 L is located in the legs during standing at rest. This also establishes different baseline blood pressure in the vascular system, with higher mean arterial pressure and vascular distension in the lower body. In contrast, the heart must work against gravity to maintain blood flow to the head, resulting in lower mean arterial pressures in the cerebral regions.

Along with neurovestibular changes, cardiovascular changes occur almost immediately upon microgravity exposure. The removal of gravitational forces results in a redistribution of fluid throughout the body. The initial symptoms observed by astronauts, nicknamed "puffy face and bird legs," are seen because the blood volume is no longer forced down into the legs and redistributed relatively evenly throughout the body. The resultant increase in cerebral blood pressure and fluid distribution can cause headaches, congested sinuses, and possibly altered sensory (taste, olfactory) perceptions as well as contribute to SMS (Aubert et al. 2005). The change in leg volume due simply to fluid redistribution must also be accounted for when alterations in body composition and muscle mass are monitored during spaceflight. For example, the use of calf girth to monitor muscle mass may be meaningless in the initial phases or with short-duration spaceflight, but might become more suitable with prolonged spaceflight after the initial adaptation has plateaued.

The overall fluid redistribution and the effects of the reduced gravity gradient influence the cardiovascular system dramatically. The primary clinical concern is the high prevalence of orthostatic intolerance in astronauts upon return to Earth and a 1-G environment (Buckey et al. 1996b). Operationally, this can become extremely problematic if astronauts have to egress rapidly from the spacecraft upon landing or perform other emergency maneuvers requiring movement or exercise (see table 10.2 for a summary of major countermeasures).

Illustrating the interconnection between different effects of cardiovascular deconditioning, briefly summarized in figure 10.2, are some of the contributing factors predisposing astronauts to orthostatic intolerance:

- The total blood volume available within the cardiovascular system is decreased. This occurs because redistribution during spaceflight increases central blood volume, inducing a decrease in renal water reabsorption and an increase in urinary excretion to reduce the central pressure. In addition, the cardiovascular deconditioning and reduced aerobic capacity may cause a downregulation of blood volume (Norsk 2005).

- Deconditioning of the cardiac muscles may result in decreased pumping capacity of the heart and arterial pressure (Perhonen et al. 2001).

- The muscular deconditioning during spaceflight, especially of the lower body, results in less muscle mass and subsequently less venous return via the skeletal muscle pump. However, calf compliance changes seem to remodel quickly to the onset of

Table 10.2 Common and Potential Countermeasures for Orthostatic Intolerance in Astronauts

Method	Physiological rationale	Notes
Aerobic exercise	Maintains cardiovascular conditioning or minimizes deconditioning.	Muscle strength work also serves to enhance skeletal muscle pump and venous return.
Lower body negative pressure (LBNP)	Negative pressure around legs or torso elicits migration of vascular fluid to lower body, in turn forcing cardiovascular adjustments and training the body to maintain systemic blood pressure.	Prototypes are being developed that incorporate a treadmill inside a LBNP device to combine exercise and orthostatic stress in flight.
Self-powered human centrifuge	Cycle ergometer spins astronaut along head–toe axis; combines aerobic exercise with gravitational loading.	Space-efficient method of providing artificial gravity for prolonged missions.
Saline loading	Hyperhydration with an isotonic fluid (15 mL · kg^{-1}) or other fluids prior to reentry temporarily elevates plasma volume.	Current NASA guideline; not extremely popular due to diuresis and potential nausea or vomiting from large volume.
Liquid-cooling garment	Chilled water minimizes heat stress and blood distribution to peripheral vasculature during reentry, maintaining venous return and central pressure.	Current NASA guideline.
G-suit	Increased pressure around lower body and abdomen decreases gravity-induced blood distribution to the lower body, maintaining venous return and central pressure.	Current NASA guideline.
Postflight recovery	Decreased activity minimizes cardiovascular and orthostatic stress.	Whenever possible, activity is minimized 30 min postlanding to permit initial physiological adaptation to gravity.

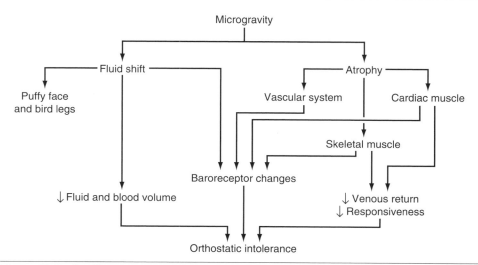

Figure 10.2 Simplified schematic of the major cardiovascular effects of microgravity exposure leading to increased risk of orthostatic intolerance upon return to 1-G Earth atmosphere. The shifts in fluid balance are generally acute upon initial exposure, and the magnitude of decrement plateaus. In contrast, the process of atrophy is more chronic and continues to occur throughout spaceflight.

microgravity with an increase in vasoconstriction, with preflight levels returning within one or two days postflight (Watenpaugh et al. 2001). To avoid the acute changes in blood pressure, one countermeasure is to wear an anti-G suit underneath the reentry suit to increase peripheral pressure and venous return.

• Impaired vascular compliance and baroreceptor response due to deconditioning of the arteries may result in a less rapid response to pressure changes (Buckey et al. 1996a). This is compounded as the reentry suit possibly contributes to the heat load. One countermeasure that has demonstrated effectiveness against orthostatic intolerance in ground-based tilt studies, in both normo- and hyperthermia, is the use of liquid-cooling garments that circulate chilled water next to the skin (Wilson et al. 2002). This appears to aid the vasoconstriction of the vascular skin beds, thus increasing venous return, and coolant is run through the flight suit within the reentry suit.

• Blood and plasma volume changes upon exposure to microgravity, with a decrease in plasma volume in the initial days of spaceflight (Alfrey et al. 1996b). This is concurrent with a decrease in hemoglobin and red blood cell mass, primarily through a destruction of newly released red blood cells or those soon to be released from the bone marrow, along with a decrease in erythropoietin activity (Alfrey et al. 1996a). Upon return from space, this decreased venous return contributes to orthostatic intolerance, and a period of space-induced anemia is evident before plasma volume and ultimately EPO production are upregulated (Alfrey et al. 1996b). To compensate for the decreased plasma volume, a mandated countermeasure for Shuttle flights is to hyperhydrate with an isotonic solution prior to reentry.

Bone

Bone is a living tissue that, like muscles and the cardiovascular system, constantly adapts itself to the forces imposed upon it. Bone tissue is also the primary reservoir of calcium in the body, and calcium levels in the body are a complex function of bone reabsorption or construction, along with calcium absorption through diet. In turn, each of these parameters is affected by physiological, dietary, and environmental factors. The removal of weight-bearing stress in microgravity greatly decreases the gravitational stress imposed on bone, leading to its remodeling to suit the lower strength demands (see table 10.3). As a result, bone density decreases and calcium loss occurs relatively rapidly and progressively with initial and continued spaceflight (Smith and Heer 2002). One additional concern with the calcium release from bone becomes a potentially greater risk of kidney stone development because the kidneys have to process and excrete greater loads of calcium from the body. Such a serious medical threat could derail an interplanetary mission, thus necessitating research on bone loss prevention and calcium metabolism in microgravity.

Due to both different strength demands and circulatory changes in microgravity, bone density may decrease at different rates throughout the body depending on location. Overall, *Skylab* data demonstrated that approximately 0.3% of total body calcium was lost monthly over the one- to three-month missions (Whedon et al. 1976). With decreased demands on the trunk and legs for support and locomotion, the greatest rate of bone mass decrease appears to be from the lower body and hips, along with the spine. In contrast, much of the propulsion in space is from the upper body, and bone mass loss there is minimal or occurs at a lower rate (Buckey 2006). Skull bone density may actually increase with prolonged microgravity, possibly due to

Table 10.3   Bone Response to Microgravity

	Preflight	Postflight	% change during flight	1-year recovery
vBMD (g/cm³)	0.332	0.297	−10.40*	0.309
Bone mass (g)	35.182	30.772	−11.10*	34.397
Volume (cm³)	106.322	104.242	−0.70	112.008

vBMD = volumetric bone mineral density (total mean femur)

*Indicates significance at $p < 0.001$.

One year of recovery to Earth's gravity involved an increase in bone size, but an incomplete recovery in BMD and mass. Data from Lang et al. 2006.

altered cardiovascular system dynamics and increased blood pressure to the cerebral region. Bone and calcium loss appears to continue progressively with increased durations (Lang et al. 2004). No clear data exist on any upper ceiling for bone mass loss, although data can be inferred from individuals experiencing spinal cord injury and loss of limb function. One area of concern is that the rate of recovery upon return to Earth seems to be much slower than that of loss during spaceflight, despite exposure to a 1-G environment and also active rehabilitation programs. Besides its implications for rehabilitation, this raises the concern that a "tipping point" may be crossed such that the body's metabolic ability to regenerate and recover may become chronically or irreversibly impaired (Payne et al. 2007).

In addition to the removal of weight-bearing stress upon microgravity exposure, factors influencing bone physiology during spaceflight include the following:

• Vitamin D is important in proper calcium absorption, while parathyroid hormone serves to enhance bone resorption to maintain blood calcium levels. These two substances may both be reduced during spaceflight, resulting in disruptions in normal calcium metabolism and an uncoupling in the matching of bone breakdown and formation (Caillot-Augusseau et al. 1998). Without sunlight exposure, natural vitamin D production within the body is greatly reduced, and absorption of dietary calcium may become impaired. As a result, supplementation of vitamin D, calcium, parathyroid hormone, or some combination of these may be required during prolonged spaceflight. The optimal form of supplementation, as well as the pharmacological and physiological interactions in microgravity and the effect on bone health, remains open to Earth- and space-based investigation.

• The rate of bone and calcium loss increases with aging, resulting in a greater risk of osteoporosis in older individuals. This is further magnified in postmenopausal females due to the decreased levels of estrogens. Therefore, the functional capacity of older astronauts and their ability to recover upon return to Earth may be affected to a greater degree following prolonged microgravity and may affect crew selection for long-duration missions.

• To decrease demands on the life support system, the Shuttle, ISS, and other spacecraft typically maintain a higher level of carbon dioxide (CO_2) of 0.7% to 1% as compared to the 0.03% composition on Earth. While this moderate hypercapnia generally does not affect health or work performance, there may be a long-term respiratory hyperventilation and disruption in acid–base balance that may in turn have a chronic effect on calcium balance and metabolism (Drummer et al. 1998).

• While individuals vary widely in their levels of bone density, the genetic bases for such variability, and also potentially for resistance to atrophy from microgravity, remain largely unknown (Buckey 2006).

• Drugs that may have a preventive role in minimizing the risks of kidney stones or may increase bone formation may need to be included as supplements, especially in astronauts screened to be at higher risk (Cavanagh et al. 2005). For example, thiazide diuretics decrease the rate of urinary calcium excretion and are clinically employed for kidney stone prevention.

Muscular System

As with bone, muscle is used both for locomotion and for maintaining posture against gravitational forces, and therefore faces similar problems with disuse and atrophy. Muscle is also the primary reservoir for protein within the body. The muscles involved in postural stability are primarily the large muscles on the posterior side of the leg because the center of gravity is slightly forward of the hamstrings in the frontal plane. These muscles include the gastrocnemius, soleus, and hamstrings. Furthermore, the large muscles of the back, such as the erector spinae, work to keep the spine in an upright posture. Therefore, microgravity causes the most rapid rates of muscle mass and strength loss in these muscles (Buckey 2006). As with bone, the upper body and arm muscles are apparently more resistant to atrophy due to their greater role in movement.

Because of the selective disuse of particular muscles and also the altered movement patterns in microgravity, atrophy of the various muscle fiber types may occur at different rates. Within the antigravity muscles, the removal of continuous low-intensity contractions means that type I, or aerobic-specific, muscle fibers likely atrophy at a greater rate than type II, or "fast-twitch" fibers. Although some muscle biopsies have been performed on astronauts pre- and postflight, the exact cellular changes within muscle fibers with prolonged microgravity are unclear. As with bone, such changes can be a result not only of the altered activity patterns due to microgravity, but also of a complex interplay between altered circulation, nutrition, and hormones during spaceflight, each of which can have significant influences on muscle tissue formation or catabolism.

Another postural effect of microgravity is that a curving of the back and a "hunched" posture are promoted, and compressive forces on the intervertebral disks within the spine are also removed. This results in slight expansion of the cartilaginous disks and an increase in height that can range from 1 to 5 cm (0.4 to 1.9 in.). Both the hunched posture and the stretching of the spine and musculature can cause mild to severe back pain emanating from muscle or from pinching and pulling of nerves (Sayson and Hargens 2008). To maintain a chronic compression on the spine and postural muscles, the Russian space program pioneered the use of "penguin suits," coveralls with bungee-like straps built in. Wearing these garments during normal activities, astronauts can continually apply and increase the tightness of the straps, causing the body to curl forward. Consequently, spinal stretching is reduced and the postural muscles are constantly engaged in low-level contraction in order to maintain a straight posture, with an estimate of approximately 70% of normal 1-G resting muscle activation (Convertino and Sandler 1995).

   More issues relevant to muscle function in microgravity are presented in the sections on exercise countermeasures, EVA physiology, and spacesuit design. The following are some other areas for consideration:

• Due to the difference in "baseline" posture in microgravity, tools and workstations need to be ergonomically designed for this altered posture to ensure that unnecessary strain is avoided. For example, the ankle tends to adopt a plantarflexed posture in microgravity, and footrests for stability need to be designed to reflect this (Coblentz et al. 1988; Wichman and Donaldson 1996).

• The need for or the efficacy of protein supplementation with prolonged spaceflight, as with many issues concerning nutrition, requires further investigation. Beyond nutritional supplementation, a potentially controversial approach to aiding muscle status is the use of pharmacological interventions. Growth hormone, synthetic testosterone, and testosterone precursors can stimulate muscle formation and thereby minimize atrophy during prolonged spaceflights. Long-term tracking of the levels of these naturally occurring hormones in astronauts has not been performed, and the rate of production and cellular sensitivity are also unclear. However, any consideration of anabolic steroid supplementation is counterbalanced by the well-documented side effects, along with the huge societal stigma relating to the use of anabolic steroids and synthetic testosterone in athletes. Therefore, even if this type of supplementation is ultimately deemed a medically appropriate countermeasure, a public relations backlash may preclude the use of such interventions.

• As noted in other sections within this chapter, standards for "acceptable" decay of muscular capacity and fitness need to be developed for prolonged spaceflight. One step toward creating such standards is to develop methods of tracking body composition, fitness, functional capacity, and activity levels that are simple, consistent, and validated. For example, ultrasound may provide tracking of body composition (muscle, bone) changes in space but will require specific training and standardized testing sites (Ma et al. 2007).

EXERCISE COUNTERMEASURES

As a consequence of the potential operational and long-term health implications of physiological deconditioning in microgravity, astronauts undergo an intensive exercise program during space missions. The challenges of exercising in space are many, requiring an interdisciplinary approach involving physiologists and engineers. This includes ergonomically engineered treadmills and equipment for working in a buoyant and weightless environment, where gravity does not pull the astronaut back to the treadmill floor. Equipment must adhere to mass and space constraints because of the high cost of launch and the limited volume inside the spacecraft, and also utilize damping to minimize vibration and magnetic disturbance from operation. Physiologically, heat dissipation is impaired because sweat beads on the skin rather than evaporating; this is compounded by the potential problems from sweat that floats through the spacecraft and into sensitive equipment.

   This section outlines three of the major available or proposed exercise countermeasures for astronauts (Convertino and Sandler 1995) and also deals with the potential for artificial gravity. Other forms of exercise or muscle stimulus, including electrical muscle stimulation and vibration training devices, either have been explored with minimal success or are continually under testing and development.

Aerobic Exercise

As in many fitness clubs, the primary aerobic exercise equipment on the Shuttle and the ISS consists of treadmills, along with cycling and rowing ergometers. Additionally, an arm ergometer can be used to provide upper body exercise. The choice between the different machines usually revolves around both scheduling and personal preference. As already noted, one major practical consideration is the need for astronauts to brace themselves against their pushing off the ergometer or treadmill. The treadmill can include a harness system for this purpose, or the astronaut can grab onto handholds while cycling. Each of these exercise modalities is performed with a lower gross efficiency than on Earth (mechanical power vs. metabolic cost of exercise), largely due to the added instability from weightlessness. Conversely, the extra muscle contractions required for stabilization may provide a valuable stimulus to the antigravity postural muscles.

Aerobic exercise in spaceflight is typically performed at moderate intensities for relatively long periods or intervals. One issue to explore is whether equal aerobic benefit can be obtained with a regimen of short efforts of very high intensity. Recent ground-based research suggested that a bout of only four to six supramaximal cycling sprints, performed three times per week for two weeks (totaling <18 min), was just as effective at improving $\dot{V}O_{2max}$ and aerobic performance as exercising at a steady submaximal effort for 90 min three times weekly for two weeks (Gibala et al. 2006). Of special importance was the observation that the gross aerobic performance measures were supported by equal increases in citrate synthase, a key marker of aerobic metabolic activity. Similar patterns of improvement have been observed in subjects who are already relatively fit and trained, and this approach has been adapted successfully for cardiac rehabilitation patients. If such high-intensity training protocols can be developed successfully for astronauts, then it is possible that less total time can be devoted daily to exercise, increasing active operational time. Conversely, greater training stimulus may be provided by a given investment of time.

Resistance Exercise

Resistance training is an important complement to aerobic training in order to maximize muscle stimulation, and has proven efficacious in controlled bed rest studies. In one study, 14 days of bed rest in healthy subjects significantly decreased the rate of muscle protein synthesis, but both the rate of muscle protein synthesis and absolute muscle strength (1-repetition maximum) were maintained at baseline levels in a group performing three to five sets of 10 to 12 isotonic knee extensions every other day (Ferrando et al. 1997). Therefore, it appears that concentric contractions can be targeted with a relatively simple resistance program focusing on large muscles and multijoint movements. However, postural muscles may prove harder to target with resistance training programs during spaceflight, as the gravitational impact forces and need for antigravity bracing are decreased in microgravity. These eccentric muscle contractions take on secondary importance, and the ability of muscles to perform such contractions may need to be a focus during resistance training (Kirby et al. 1992). Specific strength training can be isometric or dynamic, and resistance devices include bungee straps and machines for both upper (e.g., bench press) and lower body (e.g., squats) training (Convertino and Sandler 1995).

Lower Body Negative Pressure

An approach currently undergoing extensive ground testing and development is the use of lower body negative pressure (LBNP) (see figure 10.3). On Earth, placing the lower body in a sealed chamber, then decreasing the pressure within, applies a negative pressure gradient toward the legs. This draws blood downward; the effect is analogous to the orthostatic stress of rapidly tilting from a supine to an upright

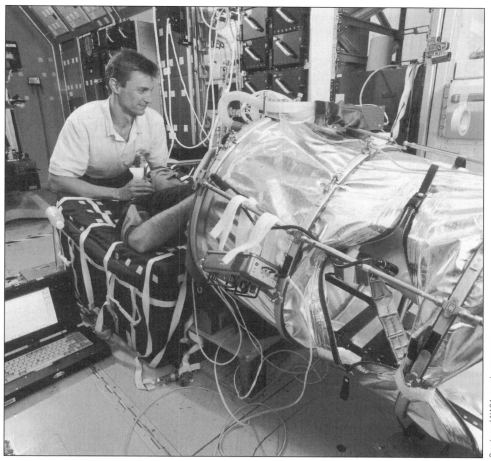

Figure 10.3 Lower body negative pressure (LBNP) devices can be used to simulate gravity pulling blood to the lower body, forcing the cardiovascular system to strengthen its resistance to orthostatic challenge. Furthermore, the combination of a supine treadmill inside an LBNP device has been proposed to enable simultaneous exercise and orthostatic stress. Potentially such a combination may closely simulate the effects of upright exercise in a gravity environment and may be an optimal countermeasure for both muscular and cardiovascular deconditioning during spaceflight.

posture. Periodic training with LBNP in space may therefore force the cardiovascular system to adapt by increasing its ability to return blood to the heart from the limbs. Lower body negative pressure units were employed on *Skylab* missions in the 1970s to investigate cardiovascular responses to microgravity. The continual stimulation of the cardiovascular system by LBNP may train the cardiovascular system and baroreceptor response, decreasing the effects of cardiovascular deconditioning discussed previously. The National Aeronautics and Space Administration is currently refining a LBNP chamber large enough to house a treadmill or cycle ergometer (Smith et al. 2008). Exercise within a LBNP environment may potentially maximize cardiovascular stimulus during aerobic exercise.

Artificial Gravity

The countermeasures just discussed provide exercise and environmental stimuli to the body, but they remain imperfect analogs of the stresses imposed by gravity. The creation of an entire spacecraft (or even just a small compartment) that has artificial gravity would be an engineering challenge of enormous scale, and such a spacecraft would likely be prohibitive to build and launch. One potential modality for providing gravitational stress during spaceflight might be via exercise on a human-powered centrifuge (Kreitenberg et al. 1998; Yang et al. 2007). In such a device, consisting possibly of paired cycle ergometers linked together on a fixed shaft, the pedaling forces would power the rotation of the centrifuge (see figure 10.4). With the astronaut positioned with his or her head closer to the axis of rotation, sustained angular acceleration would provide centrifugal force that pulls blood down toward the legs. An elegant feature of this concept is that the degree of gravitational force is dependent on the individual rather than relying on valuable external power sources (although external motors may also supplement human power). Physiologically, this simulation provides additional gravitational stimulus

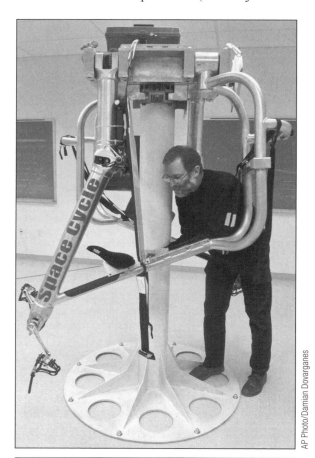

AP Photo/Damian Dovarganes

Figure 10.4 For long-duration spaceflight that might require artificial gravity, one efficient option may be the "Space Cycle." The rotation and gravity are self-generated by the cyclist, supplementing aerobic exercise with orthostatic stress from the centrifugal force generated as the bike spins.

not only to the cardiovascular system, but also to the muscular and bone tissues. An obvious disadvantage is the large space requirement for a rotational system. In addition, though designs intended for the ISS have been developed, many unknowns remain on the path to implementation. Specifically, the optimal combination of exercise and gravitational stress, including magnitude or intensity, frequency, and duration for both short (e.g., three- to six-month ISS) and long (e.g., Mars) missions, needs to be investigated before such a device can be usefully implemented.

— REHABILITATION FROM MICROGRAVITY EXPOSURE —

As we have seen, exposure to microgravity produces a host of physiological changes within the body. The changes generally intensify in magnitude with increasing mission duration, and exercise and other countermeasures appear to be only partially effective at limiting the changes. These responses can cause not only maladaptations such as orthostatic intolerance and anemia immediately upon return to the 1-G Earth environment, but also a prolonged period of impaired performance after spaceflight. Therefore, one of the major research and clinical thrusts of space agencies is longitudinal tracking of astronauts to understand the process of microgravity recovery and the role of physical rehabilitation (Payne et al. 2007). This is complicated by the nonlinear progression and also the pattern of recovery. Specifically, microgravity appears to induce systematic changes such that the matter is not one of simply reverting back to preflight status. For example, bone scans one year following four- to six-month-long ISS missions showed that overall bone mass recovered to near preflight levels, but that the architecture of bone reformation was altered—bone mineral density was lower and bone volume was greater than before flight (Lang et al. 2006).

Short-term, the NASA astronauts on ISS missions begin a medically supervised rehabilitation program immediately upon return to Earth regardless of landing location, and this is continued for approximately 2 h daily during the first 45 days postflight. While customized to the individual, rehabilitation can consist of the following phases (Payne et al. 2007):

- Phase 0: Landing and egress. From landing until return to Houston, the primary focus is on treating any symptoms arising from orthostatic intolerance and vestibular problems.
- Phase 1: Acute readaptation. Upon return to Houston, the focus for the first three days includes reuniting the astronauts with their families, along with overall physical and psychological monitoring.
- Phase 2: Incremental physical conditioning. From 3 to 14 days postreturn, the astronaut engages in a progressive program to restore strength and endurance. Customization of the rehabilitation program is a strong requirement during this phase because of high individual variability in the response to microgravity and also the high incidence of fatigue during rehabilitation training.
- Phase 3: Return to baseline and return-to-flight readiness. The intensity of physical training increases, with the overall goal of returning the astronaut to flight readiness by 45 days following return to Earth.

EXTRAVEHICULAR ACTIVITY PHYSIOLOGY

Although videos of EVAs make it appear that astronauts float easily and gracefully through space, the actual physiological requirements during spacewalks can be very intense. The frequency and overall hours of EVA were relatively low (ranging from 0 to 75 h per year but averaging 20 h per year) during the first three decades of space-walks, from 1965 to 1996. However, the need for EVAs has picked up dramatically with the ISS; it is estimated that >120 h per year is required during its operational life for both construction and maintenance (Cowell et al. 2002). With the proposed lunar base and Mars missions, EVAs will become even more critical for human exploration and at the same time range more freely from a space station or habitat. Despite this heavy emphasis on EVAs, much remains unknown about the actual physiological responses to exercise in microgravity and the limits of work capacity for astronauts.

Extravehicular activities are excellent case studies in integrative physiology and ergonomics, owing to the multiple external and internal stressors to which the astro-naut is exposed. An example is seen in the muscular demands of EVA. Extravehicular activity maneuvers require a combination of dynamic and isometric contractions (Cowell et al. 2002). For example, most of the dynamic movement is performed by the upper body muscles and occurs primarily at a sustained aerobic intensity. In contrast, the emphasis on manual dexterity and force generation through the hands (e.g., holding and operating tools) results in a high degree of isometric strain on hand and forearm gripping muscles to keep the fingers in a flexed position against the resis-tance of the suit. Furthermore, bracing against the buoyant effects of space requires clamping into footrests and isometric contractions of the lower body. As discussed throughout this chapter, the microgravity environment leads to a continual and progressive process of atrophy across nearly every physiological system. Therefore, reasonable work capacity and expectations during EVAs may need to incorporate models of physiological atrophy, and the development of proper physical training and exercise interventions during flight to maintain EVA-specific muscular capacity may become a critical component of mission planning. This is even more important with the ISS compared to the brief Shuttle flights, as the prolonged mission duration provides a greater potential for physiological atrophy prior to an EVA. Based on the specific demands of EVA and limited time available for exercise, it may eventually become most efficient for long-term missions to customize in-flight exercise programs specifically to target the muscles and the way they are employed during EVAs.

In an overall positive review of Russian experiences aboard *Mir,* minimal medical risks or complications during EVAs were reported (Katuntsev et al. 2004). While the physical demands may approach those of steady hard exercise, with peak metabolic rates of 9.9 to 13 kcal · min^{-1} and heart rates of 150 to 174 beats · min^{-1}, a review of 78 EVAs indicated no medical complications or premature termination due to medical issues during any of the EVAs. The primary medical issues during *Mir* EVAs were general muscle pain and fatigue from exertion, especially in the upper body. Some tachycardia and arrhythmias were also reported during moments of stress.

Despite this favorable perspective, much clearly remains unknown about the physiological requirements and responses to EVA exercise. Some of the many areas requiring research and development to maximize EVA efficiency and safety are the following (Barratt 1999; Gontcharov et al. 2005):

• Many ground-based exercise physiology studies simulating spaceflight have employed predominantly lower body exercise (e.g., cycling, running). In contrast, much EVA work places high levels of strain on the upper body musculature, making direct extrapolation of such studies difficult and likely inappropriate. The gross efficiency of upper and lower body exercise in microgravity may differ. Existing studies using bed rest simulations on the upper body have primarily employed tests of strength rather than the aerobic work more typical of EVAs. Ground simulation studies of EVA should also employ a supine rather than a sitting posture to more closely replicate the fluid distribution and cardiovascular responses to microgravity. In prolonged spaceflight, the rates of atrophy of the upper and lower body may also differ, and need to be incorporated into models of work capacity.

• Another issue is the accurate and safe assessment of work intensity during EVAs. The primary physiological monitoring during EVA consists of heart rate, which is recognized as an indirect measure of physiological strain. An accurate modeling of heart rate to oxygen uptake or actual muscle force production during typical EVA work, along with understanding how these relationships may change with prolonged microgravity exposure (e.g., cardiovascular and muscle atrophy), may help to develop more efficient and accurate EVA time lines for mission planning.

• As discussed in the next section, the hypobaric and pure-O_2 EVA environment inside a spacesuit is dramatically different than the normobaria in the main spacecraft. For example, the effects of repeated decompression–compression cycles from multiple exposures to EVAs, and their potential effects on the incidence of bubble formation and decompression sickness, require further investigation to better establish safe EVA schedules.

• Nutritional and drug countermeasures to offset oxidative stress from radiation and free radicals are another area requiring research. As detailed later, one major mechanism of cellular damage from space radiation may be through the elevated production of free radicals. Research needs to be conducted to determine whether elevated doses of antioxidants, via either ingestion or skin creams, can minimize the effects of increased radiation exposure during EVAs.

SPACESUIT DESIGN

A spacesuit must serve as a self-contained life support system (see figure 10.5), requiring climate control along with oxygen supply and purification of expired gases. In addition, as with the spacecraft itself, the spacesuit must provide protection from extremely harsh external conditions. These include extreme temperature gradients, radiation, and impact from microparticles. The dominant consideration in spacesuit design is energy efficiency, as power supply for the various support systems can compromise the duration of EVAs. The spacesuit is also a prime example of the complex interplay involved in human engineering, as ergonomic and physiological requirements affect engineering demands and vice versa.

The following sections outline how some engineering parameters affect the physiological and ergonomic functioning of the astronaut and form some of the important considerations in the optimal design and functioning of spacesuits.

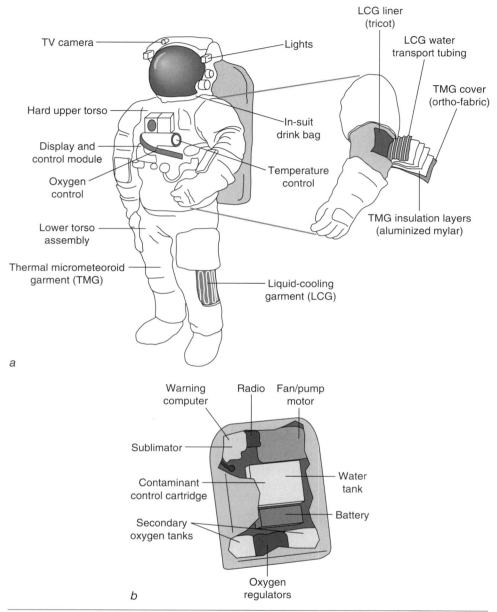

Figure 10.5 Illustration of *(a)* an astronaut in a spacesuit, highlighting the functional demands of extravehicular activities and *(b)* the main components of the life support system.
Courtesy of NASA: http://news.bbc.co.uk/1/hi/sci/tech/5120294.stm

Mobility and Suit Pressure

In the weightless environment of space, the mass of the spacesuit (>113 kg [250 lb] for those employed on the ISS) is not an important determinant of the metabolic costs of exercise. However, unlike most protective garments on Earth, where some air permeability and exchange is permissible, spacesuits must be completely sealed from the external environment, leading to a constant volume within the suit. Therefore, a high pressure within the suit compared to the minimal pressure of space would make the suit extremely stiff and difficult to maneuver and operate. For example, extra muscular force would be required to keep the arms in a flexed posture against the resistance of the spacesuit limbs. In turn, this would lead to local muscular fatigue and increased metabolic demands on the astronaut, further taxing the life support capabilities (e.g., oxygen supply, heat dissipation). Therefore, the advantages of a lower suit pressure lie in the lower resistance and therefore greater ease in operating the suit. Currently, the ISS operates at a cabin pressure of 748 mmHg, or nearly that at sea level on Earth, while spacesuits typically operate at an internal pressure of 220 to 295 mmHg pure oxygen to simulate the partial pressure of oxygen at Earth's surface (Cowell et al. 2002). The benefits of a lower suit pressure are balanced against the greater risk of decompression sickness, resulting from changes in pressure when the astronaut moves from the spacecraft environment to the spacesuit. However, an overly low pressure inside the suit would result in reduced inspired as well as alveolar partial pressure of oxygen—increasing respiratory stress, reducing work capacity, and even potentially inducing altitude sickness.

Decompression and Prebreathing Maneuvers

To safely equilibrate to the lower pressures within spacesuits, astronauts undergoing EVA first have to perform a carefully managed decompression protocol analogous to that for a diver ascending to the surface. Two broad approaches are theoretically possible: to gradually decrease the overall pressure within the entire spacecraft over a period of time prior to the EVA and affect everyone, or else to extend the preparation time of an EVA by having the astronaut undergo decompression in an airlock. As preparing a spacesuit for an EVA can be a lengthy process, it may be generally more efficient overall for astronauts to decompress at the same time they are donning the spacesuit. The clinical effectiveness of O_2 prebreathing has been clearly demonstrated in military aviator trainees undergoing hypoxia familiarization training; incidences of decompression sickness were significantly lower in both students and inside observers who underwent a 30 min O_2 prebreathe compared to no prebreathe prior to exposure to 7620 or 10,670 m (25,000 or 35,000 ft) in a hypobaric chamber (Rice et al. 2003).

The potential threat of a severe decompression sickness incident during space-flight is a major concern for mission planning. The optimal duration and protocol for prebreathing protocols prior to EVAs are constantly being debated and refined, with the ultimate aim of minimizing impact on valuable operational time and also non-EVA crew. Russian experience onboard *Mir* over 78 EVAs lasting up to 7 h 14 min (normobaric cabin pressure and Orlan suit pressure of 300 mmHg) suggests that a brief (30 min) pure-O_2 prebreathing period was sufficient for safe denitrogenation, with no reported DCS incidents (Katuntsev et al. 2004). In contrast, American decompression protocols are far more conservative, involving much longer pure-O_2 prebreathe durations of 90 to 240 min. During Shuttle missions, the pressure in the entire cabin is reduced for up to a day before an EVA in order to begin the denitrogenation process. In the ISS, the current protocol has the EVA crew begin decompression by first "camping out" overnight in an airlock at an intermediate (~530 mmHg) pressure during their normal sleep time, thus decreasing the "active" time required for final denitrogenation. Another method for reducing the risk of decompression sickness may eventually be to screen astronauts for patent foramen ovale (see chapter 7 for more detail), though this may be considered overly conservative and restrictive.

EXERCISE DURING PREBREATHE

Mild exercise during the prebreathe period may be an effective supplement for accelerating denitrogenation and lowering the risk of decompression sickness. Theoretically, the increased blood flow and tissue perfusion would accelerate the removal of nitrogen from the tissues and its elimination through the respiratory tract. The majority of ground-based simulations utilizing hypobaric chambers demonstrate that mild to strenuous lower or upper body exercise during 100% O_2 prebreathing can decrease the risks of bubble formation to that for passive 100% O_2 prebreathing lasting twice the duration or more. However, at least one research group has reported that the briefer the rest interval between a bout of moderate lower body exercise and decompression (chamber pressure simulating 6700 m [21,980 ft] altitude), the higher the incidence of decompression-induced bubble formation (Dervay et al. 2002). From a practical perspective, space and time limitations within the Shuttle or airlock can make it difficult to schedule exercise, such that creativity is needed to design appropriate exercises that will neither require a lot of time nor unduly fatigue astronauts prior to the actual EVA. For example, brief bouts of high-intensity exercise might be able to replace longer bouts of moderate exercise during prebreathe, and exercise using the larger lower body muscles may cause less overall fatigue than upper body exercise. Also, a degree of physical exertion is inevitable when the astronaut is donning and preparing the spacesuit for an EVA and should be accounted for in any exercise considerations.

Thermal Control and Fluid Balance During EVA

The influence of hypobaria and microgravity on thermoregulation remains largely unknown. It is possible that the hypobaria may enhance peripheral vasodilation and heat loss, though this may be counteracted by the negative effects of a lowered total blood volume. With high rates of heat production during EVA, there is a potential for both heat storage and hyperthermia. In addition, significant dehydration can occur via sweating to dissipate heat. While ~1 L of fluid and an energy bar are available for consumption within the spacesuit, most EVA crew prefer to "fly dry" and minimize the need for urination. To aid in temperature control, NASA pioneered the use of liquid-cooling garments (LCG), consisting of a stretchable undergarment embedded with a network of thin and flexible tubing through which coolant is circulated (see figure 3.4 for photo of a LCG). In this way, heat can be removed via conduction. The current LCG configuration consists of a single-zone coverage of all of the body surface except for the hands, feet, and head, with the cooling rate manually controlled by the astronaut. (See chapter 3 for a more detailed look at heat stress and cooling garments.)

Ironically, the most common thermoregulatory complaint during EVA, rather than feeling hot or experiencing heat stress, has been sensations of overcooling, especially during resting in shadow after a period of demanding physical work (Cowell et al. 2002; Katuntsev et al. 2004). This suggests that the thermal control systems, while able to sustain thermal balance and comfort during light to moderate metabolic rates, were not sufficiently responsive to rapid changes in metabolic heat generation due to intermittent or high-intensity exercise or both. Alternately, it is possible that thermoregulatory perception may be impaired in microgravity or during EVAs. All in all, the result is inefficient cooling and power utilization, and it is important to continue developing automated thermal control systems based on one or more physiological markers, along with possibly multiple thermal zones, in order to maximize comfort and efficiency (Flouris and Cheung 2006).

RADIATION EXPOSURE

Of all of the aspects of a microgravity environment and prolonged spaceflight, quite possibly the issue that we know least about and is most difficult to study—but that poses the most serious potential risk—concerns the acute and chronic effect of radiation exposure on astronaut functioning and health. Nuclear weapons and accidents like Chernobyl have taught horrific lessons about the threat from radiation. However, due to obvious ethical considerations, systematic testing on humans is impossible, and information can only be indirectly derived from cell culture and some animal models. In addition to effects on human physiology, the effects of space radiation on food nutrition and the efficacy of medicines are also unknown; these need to be understood in greater detail to ensure the sustainability of long-duration space missions.

To date, the majority of astronauts have been protected from radiation hazards largely by the earth's magnetosphere because they have traveled in LEO; the only exceptions were the *Apollo* astronauts. Formed from the magnetic properties of the earth and its atmosphere, the Van Allen radiation belt was discovered by the unmanned Explorer satellite missions in the late 1950s and named after the lead scientist. Close to a torus in shape and constantly changing due to Earth's rotation and the solar wind, the Van Allen belt consists of inner (~700-10,000 km [435-6214 miles]) and outer (31,000-65,000 km [19,260-40,390 miles]) fields of high-energy protons and electrons. The magnetic strength of this field largely shields the earth and astronauts in LEO from solar radiation and cosmic rays, though major solar flares can still break through the belt and both disrupt electronics and increase radiation exposure. Travel beyond LEO to the moon and Mars will require transit and return across the magnetosphere, further contributing to the astronaut's radiation exposure. Once beyond the Van Allen belt, astronauts will be exposed to the full brunt of galactic cosmic radiation (GCR), and such conditions are nearly impossible to replicate in any Earth-based simulation.

The actual biological effects of radiation on human tissue are complex to model because they are influenced by many different factors having to do with the nature of the radiation itself. Some of the major determining factors influencing the biological effect of radiation are outlined in table 10.4. The mechanisms of radiation damage are primarily via direct or indirect damage to the DNA molecules. Radiation greatly elevates the production of free radicals within cells via collision and ionization of water or oxygen molecules. Large concentrations of these highly reactive molecules, such as hydroxyl ion (OH^-), hydrogen peroxide ($H_2O_2^-$), and superoxide (O_2^-) with their unpaired electrons, become highly reactive with organic molecules, leading to disruption of proper cellular function and indirectly affecting DNA function or integrity. Alternately, radiation may directly ionize and damage the DNA molecule itself. In turn, damage to a tissue's reproductive capacity will affect both its normal functioning and its capacity to repair itself from further damage, while mutations within the DNA may stimulate cancer. Finally, high-energy radiation may cause cell death directly via impact.

Table 10.4 Factors Affecting the Relative Biologic Effect of Space Radiation

Term	Units	General definition	Example	Notes
Radiation type	Relative quality	Amount of damage to cells per given dose; depends on size and type of radiation	• X rays (photons); quality factor = 1 • Solar flares (mainly protons) • Galactic cosmic radiation or GCR (protons, neutrons from helium and heavier atoms); quality factor for iron nuclei = 20	X rays are diffuse and traverse all exposed cells; protons and neutrons collide with high energy on limited number of cells in exposed area, potentially causing much more damage to cell and also secondary radiation from collision with other nuclei.
Energy	Electron volts (eV)	The "speed" at which radiation is traveling	• Chest X rays ~40 to 60 KeV • GCR ~300 to 3000 MeV	Higher-energy radiation causes much more damage upon collision with cell tissue.
Fluence	particles · $cm^2 \cdot s^{-1}$	Number of particles that pass through given area per unit time	• GCR = low • X rays = high • Solar flares = very high	↑ fluence may minimize cell's capacity to recover from damage.
Absorbed dose	1 gray (Gy) = 1 j · kg^{-1}	Actual dose absorbed by the body; function of energy and fluence		
Dose	Sieverts (Sv)	Function of absorbed dose and quality factor of radiation		Permissible exposure limits also affected by tissue of interest (e.g., cataract formation may have lower threshold).
Dose rate		Rate at which radiation dose is accumulated		Same overall dose that is safe over lifetime may be fatal if acute.

SUMMARY

The current emphasis on Shuttle and ISS missions, along with ground-based simulations, has provided critical insight into the acute and short-term issues in spaceflight. In turn, space-based research has strong transfer to Earth-based physiology and clinical conditions, including prolonged bed rest, spinal cord injury, and rehabilitation medicine. However, with the declared goal of a lunar base and ultimately a manned mission to Mars in the coming two decades, thorough investigation of the human physiological responses to prolonged microgravity will become a critical and prominent branch of environmental physiology. A quantum advance in extrapolating what we know about short-term microgravity exposure to spaceflight durations lasting three years or more will become one priority. This needs to be coupled with determining

SPACECRAFT-SHIELDING STRATEGIES

Obviously, undue radiation exposure of astronauts is unacceptable, as it constitutes a serious and potentially fatal health hazard; unlike other effects such as physical deconditioning, radiation damage is permanent. Therefore, proper planning against radiation risks is essential in any long-duration mission, especially on voyages to the moon and Mars beyond LEO and the Van Allen belt. Protection strategies must be developed against both acute high doses, as during solar flares, and chronic low doses from galactic cosmic rays. Furthermore, the astronaut is at increased risk during EVAs because of the lower protective abilities of spacesuits and also difficulty in rapidly reentering the spacecraft.

Options for spacecraft shielding must be balanced against both acceptable risks and engineering constraints. Using current technology, safe shielding of the entire spacecraft to minimal radiation levels would require a weight prohibitive for launch from Earth. Another option is the creation of a smaller "safe room" that astronauts could retreat to during high-radiation events. The shielding material is also important, as different materials have different degrees of effectiveness against particular types of radiation. For example, high-energy GCR can affect lead shielding and cause secondary radiation of gamma rays or neutrons. Active shielding, whereby a magnetic field is maintained around the spacecraft, may be another technological approach to radiation protection but might have a prohibitive energy cost.

acceptable levels of physiological decline on such missions, along with targeted training during flight and rehabilitation upon return to Earth. In turn, translating such knowledge into sound and practical medical planning for space missions will rely on a highly interdisciplinary approach, especially in conjunction with ergonomics and engineering fields.

Exercise in Polluted Environments

T he 2008 Summer Olympics in Beijing, China, combined with the increased environmental awareness in Western society, helped to bring the issue of the effects of air pollution on human health to the forefront of scientific research. Indeed, while the host nation promised a "green" Games, the potential effects of air pollution on athletic performance was the dominant sport science question leading up to Beijing, much as the Mexico 1968 Summer Olympics served as a catalyst for altitude training research (see chapter 8). In the buildup to the 2008 Games, the men's marathon world record holder, Ethiopian Haile Gebrselassie, announced that he would not compete due to the pollution levels. This decision stemmed from his participation at the test event in late 2007, which exacerbated his exercise-induced asthma to the point that he was forced to quit the race. At the same time, the Canadian Olympic Committee arranged to base many of their athletes in Singapore prior to the Games, a country with levels of heat and humidity similar to those of Beijing and in a similar time zone. The plan consisted of conducting final training in Singapore and then flying athletes to Beijing as close to competition time as feasible in order to minimize exposure. Sport scientists from different countries also developed various masks for athletes to wear during rest, training, and sometimes even up to the point immediately before actual competition.

Such precautions and fears, imagined and real, cannot help but raise awareness and concern about the effects of air pollution on health, along with exercise capacity and long-term health for athletes of all abilities. Research in this topic is incredibly difficult due to a number of logistical and ethical considerations. First and foremost, because of the clear potential for health issues stemming from air pollutants, the idea of asking subjects to voluntarily increase their exposure levels makes gaining ethical approval a challenge. Then, even upon gaining clearance, there is the additional issue of recruiting willing subjects! The direct result of these challenges is that most pollutant research is based on animal models and requires extrapolation to humans. In turn, much of the existing research on humans is epidemiological. While this is certainly an important approach, the findings are clouded by the myriad and evolving array of pollutants in different regions and over time. This "cocktail" of multiple pollutants, each with different individual effects on the body, can also have major synergistic effects that influence both acute exercise and chronic health outcomes. Furthermore, the rapid acceleration of pollution levels can make it difficult to predict long-term effects given that environments will change in the coming years.

The purpose of the present chapter is to survey the current state of knowledge regarding the effects of air pollutants, within the specific context of athletes or workers during exercise. We will use the example of ozone to illustrate the multitude of factors influencing exercise capacity in polluted environments, along with long-term costs versus benefits. A final section explores current best practices and potential countermeasures when one is dealing with high-pollution environments.

AIR QUALITY INDICES

With the multitude of air pollutants, an air quality index needs to reflect the potentially different health hazards from each of these pollutants. Unlike the situation with the windchill index common to Canada and the United States, the two countries have their own definitions and models for determining an air quality index. Examining the derivations behind these two indices provides a good illustration of the different approaches to modeling air pollution, each with its own strengths and limitations.

United States

The U.S. Environmental Protection Agency established the AIRNow Web site to standardize air pollution reporting and provide health messages to the public. The U.S. version of the Air Quality Index (US-AQI) (see table 11.1) relies on monitoring sites throughout the country from a variety of jurisdictions and incorporates ground-level ozone, carbon monoxide (CO), particulate matter (PM), sulfur dioxide (SO_2), and nitrogen dioxide (NO_2). This US-AQI is a dimensionless value. For each pollutant, 100 corresponds to the national air quality standard as currently determined by the Environmental Protection Agency, and the highest individual value for any pollutant sets the US-AQI (e.g., if CO has a value of 120 and PM a value of 130, then the AQI is 130). For the US-AQI, a low value is preferable; progressively greater warnings begin at 51 for "moderate" and proceed to 151 for "unhealthy" *(Everyone may begin to experience health effects)* through to >301 for "hazardous" *(Health warnings of emergency conditions)*. The value of this dimensionless scale is that the range can easily be reset by changing the standard for any one pollutant. However, the fact that one pollutant sets the entire US-AQI negates any contribution from the others or possible interactions from multiple pollutants.

Canada

Environment Canada employs the CHRONOS (Canadian Hemispheric and Regional Ozone and NOx System) air quality model (NOx = nitrogen oxides) based on both known atmospheric chemistry of a range of pollutants and meteorological processes. As part of the output, ozone levels at different altitudes along with particulate levels ($PM_{2.5}$ and PM_{10}, with the subscript denoting the size of the particulate matter in microns) can be calculated over 48 h. From CHRONOS, the Canadian-AQI is calcu-

Table 11.1   The Air Quality Index (United States)

Air quality	Numerical value	Meaning
Good	0-50	Air quality is considered satisfactory, and air pollution poses little or no risk.
Moderate	51-100	Air quality is acceptable; however, for some pollutants there may be a moderate health concern for a very small number of people who are unusually sensitive to air pollution.
Unhealthy for sensitive groups	101-150	Members of sensitive groups may experience health effects. The general public is not likely to be affected.
Unhealthy	151-200	Everyone may begin to experience health effects; members of sensitive groups may experience more serious health effects.
Very unhealthy	201-300	Health alert: Everyone may experience more serious health effects.
Hazardous	>300	Health warnings of emergency conditions. The entire population is more likely to be affected.

The index is based on the highest concentration of one of six monitored air pollutants.
From http://airnow.gov/index.cfm?action=static.aqi.

lated, ranging from 0-25 (good) to 26-50 (fair), 51-75 (poor), and 76-100 (very poor). Similar to the American version, the Canadian-AQI normalizes the levels of six major air pollutants (sulfur dioxide, SO_2; ground-level ozone, O_3; nitrogen dioxide, NO_2; total reduced sulfur, TRS; carbon monoxide, CO; and fine particulate matter, $PM_{2.5}$) into a dimensionless numerical scale based on existing known health standards set by Environment Canada, and the highest single pollutant reading becomes the Canadian-AQI reading.

Canada rolled out an Air Quality Health Index (AQHI) beginning in 2007 and continuing over a period of four years across the country (see table 11.2). The AQHI is generally more conservative than the existing Canadian-AQI, predicting adverse health effects with lower pollutant levels. However, one significant advance over the existing US-AQI and Canadian-AQI is that the new index attempts to combine the potential impacts of ozone, particulate matter ($PM_{2.5}$ and PM_{10}), and NO_2 rather than simply letting the highest individual pollutant determine the index score. Another advance is that the AQHI actually interprets pollution levels by providing advice on managing outdoor activities, categorizing warnings for the general public and also for "at-risk" (children, elderly people, individuals with cardiopulmonary conditions) populations. Finally, the AQHI seeks to enhance public awareness and acceptance by standardizing to a scale of 1 to 10+, in line with the commonly employed UV index (Environment Canada, www.ec.gc.ca).

Major Pollutants

Table 11.3 briefly outlines the major air pollutants currently of interest in urban environments. Keep in mind that this is only a very small subset of the total range of compounds produced in modern society to which humans are exposed through air,

Table 11.2   Canada's Air Quality Health Index

Health risk	Air Quality Health Index	Health messages	
		At-risk population*	**General population**
Low	1-3	**Enjoy** your usual outdoor activities.	**Ideal** air quality for outdoor activities.
Moderate	4-6	**Consider reducing** or rescheduling strenuous activities outdoors if you are experiencing symptoms.	**No need to modify** your usual outdoor activities unless you experience symptoms such as coughing and throat irritation.
High	7-10	**Reduce** or reschedule strenuous activities outdoors. Children and the elderly should also take it easy.	**Consider reducing** or rescheduling strenuous activities outdoors if you experience symptoms such as coughing and throat irritation.
Very high	Above 10	**Avoid** strenuous activities outdoors. Children and the elderly should also avoid outdoor physical exertion.	**Reduce** or reschedule strenuous activities outdoors, especially if you experience symptoms such as coughing and throat irritation.

*People with heart or breathing problems are at greater risk. Follow your doctor's usual advice about exercising and managing your condition.

© Her Majesty the Queen in Right of Canada, 2009. The AQHI is a collaboration initiative of Environment Canada and Health Canada. Available: http://www.ec.gc.ca/cas-aqhi/default.asp?lang=En&n=79A8041B-1. Reproduced with the permission of the Minister of Public Works and Government Services, 2009.

Table 11.3   Major Categories of Airborne Pollutants

Pollutant	Source	Limits for danger to human health*	Primary effects
Particulate matter	• Solid particles from a wide variety of human and natural sources • High regional and temporal variability due to weather and geography • Dust, wood smoke, pollen • Tobacco smoke, fossil fuel combustion • Large (5-10 μm) Deposition in nasopharyngeal tract: inflammation, congestion • Fine (3-5 μm) Deposition in trachea and bronchi: bronchospasms, congestion, bronchitis • Ultrafine (<3 μm) Deposition in alveoli Inflammation	1000 μg · m^{-3}, 24 h average	• Effects dependent on particle size, mass, and composition, along with individual (e.g., mouth vs. nasal breathing, allergies) factors • Minimal to no controlled studies on exercise effects in humans

Pollutant	Source	Limits for danger to human health*	Primary effects
Sulfur oxides (SO_x)	• Fossil fuel combustion • Sulfur dioxide (SO_2) most common	2620 μg · m^{-3} (1.0 ppm), 24 h average	• Highly soluble gas • Respiratory irritation, bronchoconstriction, and spasms • Heightened sensitivity in asthmatics, especially with cold–dry air
Nitrogen oxides (NO_x)	• Fossil fuel combustion, cigarettes • Nitrogen dioxide (NO_2) most common	• 938 μg · m^{-3} (0.5 ppm), 24 h average • 3750 μg · m^{-3} (2.0 ppm), 1 h average	• Soluble gas that can be absorbed by nasopharyngeal tract • Pulmonary dysfunction, respiratory irritation • Minimal evidence for major exercise impairment at low levels • Acute high doses can induce prolonged pulmonary deficits
Carbon monoxide (CO)	Fossil fuel combustion	• 57.5 mg · m^{-3} (50 ppm), 8 h average • 86.3 mg · m^{-3} (75 ppm), 4 h average • 144 mg · m^{-3} (125 ppm), 1 h average	• High affinity (~230 times greater than that of O_2) for binding with hemoglobin, resulting in impaired oxygen-carrying capacity and cardiovascular function • Permanent cardiac and neural damage; ultimately fatal in high doses
Aerosols	Suspension of primary pollutants (e.g., fine particulates) in air or other (e.g., smoke, mist) gases; broad category of secondary pollutants from reactions of nitrogen oxides and sulfur oxides	• 800 μg · m^{-3} (0.4 ppm), 4 h average • 1200 μg · m^{-3} (0.6 ppm), 2 h average • 1400 μg · m^{-3} (0.7 ppm), 1 h average	Potentially wide range and variability of cardiovascular and respiratory responses, based on pollutant and allergy or asthma history
Ozone (O_3)	• Main component of smog • Secondary pollutant from interaction of automobile exhaust, hydrocarbons, and nitrogen oxides with sunlight and ultraviolet radiation • Highly susceptible to spikes during summers due to increased sunlight, heat, and local geography and weather	• 120 μg · m^{-3}, 8 h average (World Health Organization) • 0.075 ppm, 8 h average (EPA)	• Wide range of respiratory symptoms including irritation, bronchospasms, inflammation, dyspnea, and reduced pulmonary function • Intensity of impairment during exercise dependent on concentration and effective dosage from ventilation rate • Proposed threshold of 0.20 to 0.40 ppm for significant exercise impairment

*Environmental Protection Agency (EPA) of the United States unless noted.

water, and food. The table also does not include pollutants found in indoor environments or in specific regions or industries (e.g., radon, asbestos). The focus is on known major physiological effects and their potential impact on exercise performance. As can be seen from the table, different air pollutants can have a multitude of effects on different pulmonary responses, eliciting a range of symptoms from inflammation and bronchospasms (ozone, sulfur dioxide) to direct binding to hemoglobin and impaired oxygen-carrying capacity (carbon monoxide). This makes it difficult to generalize about the effects of air pollutants as a whole on human health and performance. However, when we study air pollutants, many consistent themes do emerge, from the concept of quantifying exposure levels and dose response to the risk from long-term exposure and the potential countermeasures. Therefore, in the following section, we will explore the effects of one pollutant, ozone (O_3), whose effect on exercise and health has been intensively studied over the past 40 years, and use it a template for understanding the effects of air pollutants on athletes and exercise in general.

OZONE EFFECTS ON HEALTH AND EXERCISE

Ozone is a gaseous pollutant formed from the photochemical reaction between sunlight and volatile organic compounds and nitrogen oxides. Of the major air pollutants, O_3 may be the one that is the most susceptible to high levels of variability due to its interaction with local environmental conditions like temperature, such that O_3 can spike to extremely high levels on sunny summer days. This is further exacerbated by increased automobile traffic, which generates more pollutants to react with sunlight (see figure 11.1), and also the greater risks of a temperature inversion in warm air, which traps the pollution and keeps it close to ground level. The World Health Organization sets safe O_3 thresholds at 120 $\mu g \cdot m^{-3}$ averaged over 8 h of exposure, and the U.S. Environmental Protection Agency translates this to 0.075 parts per million (ppm) averaged over 8 h.

Ozone is a respiratory irritant that can reduce pulmonary capacity during passive exposure. Along with irritation and bronchospasms, one of the most common symptoms of ozone exposure is tightness in the chest with breathing. Currently, one major theory is that the underlying mechanism behind O_3 impairment is through a stimulation of airway receptors or nerve endings, which in turn elicit a reflex central neural inhibition of inspiration and expiration (Hazucha et al. 1989). Another potential mechanism is oxidative damage and an inflammatory response within the lung tissue, with elevation in the level of neutrophil release and immune response upon ozone exposure (Devlin et al. 1997; Foster et al. 1996). The physiological responses to O_3 were first noted in the mid-19th century, and initial studies on exercising rats were performed in the 1950s (reviewed by Folinsbee 1993). In terms of humans and exercise performance, the initial scientific reports first appeared in the 1960s; these included observations of an inverse relationship between ozone levels and race times in high school runners (Wayne et al. 1967).

Ozone Effects on Exercise Capacity

It is difficult to ascertain whether it is possible or even desirable to become partially or fully adapted to air pollution in terms of exercise. While some research on repeated exposure to O_3 has been performed on both animals and humans, studies on the effects of repeated exposure on exercise capacity are almost nonexistent. Two studies

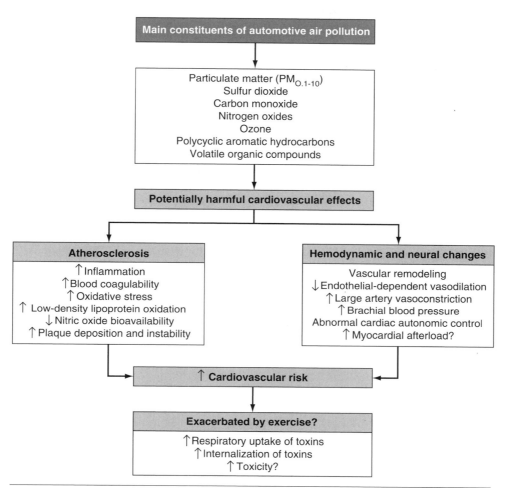

Figure 11.1   Schematic of auto pollution effects on the cardiovascular system.

J.E. Sharman, J.R. Cockcroft, and J.S. Coombes, "Cardiovascular implications of exposure to traffic air pollution during exercise," *QJM: An International Journal of Medicine,* 2004, Vol. 97, 637-643, by permission of Oxford University Press.

directly investigating the effects of acute and short-term repeated ozone exposure on aerobic capacity and subjective perceptions during exercise seem to have produced contrasting results.

In the first study, 11 subjects were exposed to 0.47 ppm O_3 for 2 h per day for four days with intermittent exercise (Linn et al. 1982). Forced expiratory volume and pulmonary symptoms showed moderate signs of adaptation over the four days of exposure. However, any evidence of adaptation was rapidly lost, with minimal retention after four days following the last exposure and no retention with weekly reexposure. In the second study, heavy exercise and maximal aerobic capacity were tested (Foxcroft and Adams 1986). Subjects performed submaximal exercise (60 L · min^{-1} ventilation rate) for four consecutive days while exposed to 0.35 ppm O_3 for 60 min, delivered via mouthpiece and therefore enhancing oral delivery and potentially dosage. At the beginning and end of this exposure period, $\dot{V}O_{2max}$ tests were performed. Maximal aerobic capacity decreased immediately with initial exposure to O_3 (3.62 L · min^{-1}) compared to filtered air (3.85 L · min^{-1}), but recovered after four days (3.79 L · min^{-1}). However, basic pulmonary functional measures remained impaired throughout exposure compared to baseline filtered air values.

Dose Response of Ozone

One of the difficulties with understanding the health effects of pollutants is quantifying and modeling the actual magnitude or pattern of exposure. Without a standardized methodology of quantifying and reporting, it can be very difficult to compare different experimental findings and consequently to generalize research. In addition, most pollutants seem to have a dose–response effect, with higher doses eliciting greater levels of physiological response and clinical symptoms (Hazucha 1987). From first principles, two environmental parameters common to all pollutants would appear to be the ambient concentration along with the duration of exposure. As we are ultimately dealing with humans living and exercising in such environments, a third parameter would logically be the actual amount breathed in by an individual, or the ventilation rate. One basic model of quantifying this effective dose (ED), therefore, is an unweighted product of these three parameters (Adams 1987):

$$\text{Effective dose (ppm · L)} =$$
$$\text{Concentration (ppm)} \times \text{Duration (min)} \times \text{Ventilation (L · min}^{-1})$$

Table 11.4 illustrates the different effective doses possible based on manipulation of each of these three parameters. Beyond this simple product, modeling can become much more complex, incorporating finer detail such as the following (Adams 1987):

• A relative weighting of the importance of each of the three parameters of concentration, duration, and ventilation. Adams and colleagues (1981) systematically investigated the relationship between the effective dosage of O_3 and pulmonary impairment by altering concentrations, durations, and ventilation rates (using mild through moderate continuous exercise) over 18 experimental trials. At the same time, the relative importance of each of these three input parameters in modeling effective dose was tested. This research design achieved similar ED values (e.g., five conditions had ED = 800 ppm · L) with different input values, and also tested a range of effective doses, from 0 to 1200 ppm · L. Greater decreases in spirometric measures such as forced vital capacity (FVC), forced expiratory volume over 1 s (FEV$_1$), and maximal midexpiratory flow rate were reported with increasing ED exposure. Of the three components modeled in effective dose (concentration, duration, and ventilation), the authors proposed that O_3 concentration remained the primary determinant of toxicity and pulmonary impairment, with ventilation rate in turn playing a more important role than exposure time. The general threshold for decreases in spirometric values compared to baseline values was between 0.20 and 0.30 ppm, although no finer resolution was possible because 0.10 ppm intervals were employed for the concentration manipulation (Adams et al. 1981).

• Finer resolution of pollutant concentration levels, including factors such as temporary or local spikes in concentration rather than overall average values. With intermittent moderate exercise during a prolonged 8 h exposure to 0.12 ppm O_3, differences in pulmonary impairment have been demonstrated between a square wave

Table 11.4   Calculating Effective Dose

O$_3$ concentration (ppm)	Exposure duration (min)	Ventilation rate	Effective dose (ppm · L)
0.1	120	50	600
0.1	60	100	600
0.2	60	50	600
0.2	120	50	1200
0.3	80	50	1200
0.3	40	100	1200

These values are based on an unweighted calculation of concentration, duration, and ventilation rate. More complex models may give relative weighting to the importance of each factor.

(i.e., constant 0.12 ppm) and a triangular (i.e., gradually increasing from 0 to 0.24 ppm over the first 4 h, decreasing back to 0 ppm over the final 4 h) exposure pattern (Hazucha et al. 1992). Specifically, greater impairment was observed during the later stages of a triangular exposure than during a square wave exposure.

- In addition to absolute levels of pollutants, the relative negative "quality" of exposure to pollutants within a given dosage is also likely higher in athletes than in others. For example, at resting or moderate exercise, most ventilation can occur through the nasal passage. However, higher exercise levels are associated with a greater requirement for oral ventilation or breathing through the mouth, thus removing the potential filtering of particulate matter by the nasal passage (Pierson 1989). Variations in breathing patterns even at the same average ventilation rate, such as those that occur with intermittent versus continuous exercise, may affect the deposition pattern of pollutants in the body. For example, even with a similar overall average ventilation rate, intermittent exercise may shift to greater oral breathing, providing less potential filtering by the nasal passage and increasing the level of pollutants actually reaching the lungs (Pierson 1989). Arguing against the importance of breathing patterns, however, Adams (2003) reported similar levels of pulmonary impairment in the same group of subjects whether they breathed through a mouthpiece (i.e., 100% oral breathing) or in a controlled pollution chamber (i.e., combination of oral and nasal breathing) with 2 h of intermittent exercise in 0.30 ppm ozone. Nevertheless, this difference may remain a concern for other pollutants such as sulfur dioxide, which is normally filtered relatively effectively as it goes through the nasal passage (Pierson 1989).

Exercising in Urban Environments

Analogous to the epidemiological studies demonstrating that elderly persons and those with existing cardiovascular conditions are the most at risk from heat waves (Yip et al. 2008), research suggests that air pollution may affect different subsets of the population more than others. From this perspective, athletes training and competing outdoors may be much more susceptible to air pollutants and have a much higher effective dose than the general population or athletes primarily training and competing indoors. Besides the higher ventilation rates during exercise, the first obvious difference is in the time spent outdoors by athletes, which often also coincides with the times of peak pollution levels. In urban settings, walkers, runners, and cyclists may also be exercising near major roadways, where dosage of pollutants may be much higher than average reported values (Sharman et al. 2004).

In urban environments, it appears essential that both recreational and elite athletes be properly educated about protecting themselves from the worst effects of high pollutants while exercising. As it is difficult to actually change the air that we breathe, most of the advice comes down to commonsense methods to reduce exposure. Chief among these are to exercise in quieter settings away from heavy traffic and busy roadways. If this is not possible, then exercise should be performed during nonpeak periods of traffic flow. For example, analysis of pollutants in Toronto, Canada demonstrated that different pollutants peak at different times throughout the day (Campbell et al. 2005). O_3, particulates, and SO_2 peaked during midday, while CO and NO_2 peaked during the morning rush hour. Overall, pollution levels were lowest prior to 0700 h and after 2000 h.

Lessons From the Olympic Games in Beijing

The hosting of the 2008 Summer Olympics in Beijing, one of the world's most polluted megacities, instigated a thrust within the sport science community to understand the potential effects of air pollution on exercise. The pollution levels for ozone, carbon dioxide, sulfur dioxide, nitrogen oxides, and particulate matter are very close to or

SECONDHAND SMOKE AND EXERCISE

Clear and direct evidence shows that secondhand smoke (SHS) has long-term health impacts and increases the risk for lung cancer. The prevalence of policies and legislation against indoor tobacco use in many jurisdictions throughout North America and parts of Europe has greatly reduced the amount of contact with passive tobacco smoke for non-smoking adults. Yet during the course of traveling for training and competitions, nonsmoking athletes can become acutely exposed to high levels of tobacco smoke, and it is possible that this could lead to acute decreases in functional capacity and performance. Direct research on the effects of passive smoking on exercise capacity in healthy nonsmoking adults or athletes is nonexistent. However, indirect evidence exists for a number of physiological changes with acute passive exposure that may relate to exercise capacity in athletes. Short-term exposure to SHS of 30 to 60 min in healthy nonsmokers can reduce coronary flow velocity (Otsuka et al. 2001) and increase thyroid hormone secretion (Metsios et al. 2007). Secondhand smoke can also reduce gonadal hormones such as testosterone, estradiol, and progesterone and appears to lead to responses in elevated blood pressure and interleukin-1 levels in males but not females (Flouris et al. 2008). Also missing from this discussion is information on the effects of chronic passive smoke exposure on exercise capacity during growth and development in children and adolescents. Nonetheless, the copious literature on the effects of SHS makes it highly likely that the long-term potential for athletic development would be compromised in some fashion.

exceed the standards set for long-term health in the general population by the U.S. Environmental Protection Agency (Qi et al. 2007), and the effects of pollution are potentially exacerbated by the high heat and humidity typical during August. The potential for safe athletic performance remained a concern for the Beijing Games Organizing Committee, the International Olympic Committee, and various nations through the years and months leading to the Games, accompanied by scientific speculation about the potential for achieving record performances (Lippi et al. 2008). In response, the Chinese government placed extra emphasis on reducing air pollution by closing factories and limiting vehicle traffic in the months prior to the Games, at the same time increasing meteorological monitoring (Wang et al. 2008) and implementing additional medical planning for dealing with allergies and asthma in tourists and athletes (Li et al. 2008).

The preparation and planning by sport agencies and athletes revolved primarily around avoiding arrival in Beijing until shortly prior to the actual competition times. For example, many teams traveled early to areas in or close to the same time zone as Beijing (e.g., Japan, South Korea, Singapore) for final training in the weeks before the Games. Such an approach avoided the deleterious effects of pollution but offered the opportunity to adjust the chronobiological clocks of the athletes (see chapter 12), along with providing temperatures and humidity similar to those that the athletes would be experiencing in Beijing. Many athletes then arrived very shortly (one to three days) prior to their competition to avoid prolonged pollution exposure. This practice was not universally adopted, however. The Swiss cyclist Fabian Cancellara arrived in Beijing a full two weeks before the road cycling events specifically in order to adapt himself to the local environment, training for 4+ h daily outdoors in the Beijing area. Ultimately, Cancellara earned a bronze in the 6+ h cycling road race and gold in the time trial (~1 h duration). Such an accomplishment, with ventilation rates exceeding $100 \text{ L} \cdot \text{min}^{-1}$, demonstrates the unique ability of humans to adapt to different environmental stressors. Another option provided to many athletes by their national agencies was various masks to wear when they were not competing or training. However, the high-profile negative publicity that occurred when some American athletes were photographed wearing masks upon arrival at the Beijing airport prior to the opening ceremonies likely contributed to a quick curtailment of their use by most athletes.

The ultimate question posed by sport scientists is the effect of air pollution on actual sport performance. Judging from a survey of Olympic and world records, the impact on performance is arguably mixed. With events indoors, where there was some degree of air filtering along with climate control, multiple world records in both swimming and track cycling were shattered in Beijing. This would suggest that the continual improvements in training and technology were effective in improving and maximizing human capabilities. However, in outdoor running competitions, it is perhaps notable that while Olympic records were broken in numerous events, very few world records were broken. Olympic records were set by men in the 100, 200, 5000, and 10,000 m events, along with the marathon and the 50 km race-walk. Of these, only the 100 and 200 m times were world records. On the women's side, Olympic records were set in the 3000 m steeplechase, the 10,000 m run, and the 20 km race-walk, with only the steeplechase breaking a world record. It should be noted that world record performances are reliant on many factors, not the least of which are race dynamics. However, given the combination of multiple world records set indoors and relatively few outdoors, it can be argued that despite optimal preparation, the environmental conditions in Beijing may have limited performance capacity—only slightly, but just enough to minimize the potential for world record performances in outdoor events.

Cost–Benefit of Exercise Versus Air Pollution Threats

The previous section demonstrated that with both acute and short-term exposure over a few days, elevated levels of ozone can decrease pulmonary capacity along with impairing exercise capacity during both moderate- and high-intensity exercise, and that these impairments appear to be progressive with increasing effective doses. Another aspect to consider when we discuss air pollution is the long-term impact of living and exercising in polluted environments for both sedentary and active populations. Answers in this area are more difficult to obtain through experimental research. Instead, knowledge can be gained through an epidemiological approach, which is designed to tease out key insights from massive data sets of information on the long-term health of large populations.

With all the data available about the potential health threats of air pollution and its potential impact on exercise, it is only natural to pursue the question whether it is worthwhile for urban dwellers to be exercising at all. At a very basic level, this question can be posed as an issue of cost versus benefit over the course of a lifetime: *Do the potential negatives from increased exposure to air pollutants, acting primarily on the respiratory system but possibly increasing the risks of other illnesses (e.g., cancer), outweigh the positive health outcomes stemming from a physically active lifestyle?*

Having emigrated to Canada in 1975, I well remember revisiting my birthplace, the large urban metropolis of Hong Kong, for the first time in 1990. The very first morning after arrival, I woke up early to a seemingly never-ending convoy of garbage trucks spewing diesel exhaust outside my window. Later that day, my eyes were so irritated by the pollution that I couldn't comfortably wear my contact lenses any more and switched to glasses for the remainder of my visit. If this was my experience as a tourist engaging in nothing more strenuous than sightseeing, what might be the long-term effects of exercising in such an environment for huge quantities of time over the course of my life?

Coincidentally, one of the largest such cost–benefit analyses was conducted in Hong Kong (Wong et al. 2007) as part of a massive project on the various causes of mortality. Previous reports on the cohort had separately demonstrated that exercise decreased mortality risks (Lam et al. 2004) and that higher air pollution levels elevated risks (Wong et al. 2001). Therefore, the project aimed to bridge these two reports by directly addressing the correlation between exercise levels and susceptibility to higher pollution levels. The database consisted of 24,053 individuals >30 years who had died in Hong Kong in 1998, representing nearly 80% of total deaths in this age group. From interviews with next of kin, exercise activity levels were obtained and broadly categorized into "never exercise" and "exercise once a month or more." Correlations between exercise levels and daily pollution levels over 1998 showed that the group older than 65 years that never exercised had a significantly higher risk of mortality, independent of socioeconomic status, smoking history, or health status. Interestingly, the majority of exercise in the elderly group consisted of walking and tai chi. These activities are generally moderate in intensity and often take place in parks or in early morning before traffic levels become heavy. Therefore, the effective dosage of pollutants may differ substantially from models based on peak or daily levels averaged over the entire region or even localized to areas of residence.

Within the limitations of the research, interpretation of the data would suggest that moderate exercise over the life span provides protection against mortality from transient spikes in pollution levels. The mechanisms underlying such benefits are obviously not open to investigation from such reports, but may potentially derive from greater respiratory clearance of pollutants with increased activity and fitness, enhanced immune function, and altered gene expression protecting against environmental damage (Wong et al. 2007).

An epidemiological approach may ultimately be best suited to teasing out the long-term health effects of outdoor pollution on athletes. As an example, one possibility may be to use sedentary and active adults within a large urban area as a cohort and from there categorize the athletic group into outdoor (e.g., runners, cyclists) and indoor (e.g., masters swimmers) subgroups to explore whether one group experiences increased risk. Comparing the health outcomes and morbidity and mortality rates

INDOOR POLLUTION

While this chapter primarily focuses on outdoor exercise and pollution, one significant venue for indoor air pollution with exercise is ice arenas because of the use of ice resurfacing machines. The use of combustion engines in enclosed spaces with potentially poor ventilation can lead to very high levels of particulates and CO that remain in place and build up over time. For example, indoor arena levels of ultrafine and fine (0.2-1.0 μm diameter) particulate matter can be approximately 30 times higher than levels outside the arena, largely due to the use of gasoline- or propane-powered ice resurfacing and edging machines (Rundell 2003). The process of ice resurfacing can also transiently raise particulate matter levels fourfold. Ice arenas and other indoor athletic venues, such as sport complexes and swimming facilities, may also be susceptible to molds and fungi due to the combination of active exercise, humidity, and poor ventilation.

Potential problems stemming from faulty ice surfacing machines include chronic coughing, which occurred within days of 3 h of exposure in a group of 16 male hockey players (Kahan et al. 2007). Additionally, these athletes had high incidences of dyspnea, chest pain, and hemoptysis, to the point that some required hospitalization and a high proportion were treated with steroids or bronchodilators. A major concern also exists regarding chronic effects from even such relatively brief exposure, as symptoms of dyspnea or coughing upon exertion remained in 46% and 54% of subjects, respectively, upon reexamination at six months. Overall, such a study highlights the potential for long-term health and performance issues for athletes even when they compete only for a short time in a polluted environment. It also demonstrates the importance of proper monitoring, equipment maintenance, and attention to pollution indices. One avenue for future research on the chronic effects of indoor air pollution on exercise may be the long-term tracking of youth and adult hockey players, figure skaters, or speed skaters in ice arenas, especially as the air quality in such a confined space can be closely localized and monitored.

between the sedentary and the combined active groups would provide information on whether exercise confers a protective health benefit despite the added pollutant exposure. Similarly, if outdoor exercise poses an additional health risk compared to indoor exercise, then it would be anticipated that runners and cyclists would experience greater morbidity and mortality rates than swimmers.

MANAGING AIR POLLUTION FOR ATHLETES

It is clear that, certainly with acute exposure, air pollution can pose a challenge to the respiratory and circulatory systems and thus present a major obstacle for peak athletic performance. National and world championships, along with large multisport competitions, will remain tied to major urban metropolises largely because of logistics and marketing, so air pollution is likely to remain an issue for many elite athletes in the coming years. During smog alert days or periods when air quality is low, extra planning is required for athletes during both training and competition. The aim in this section is to outline some major considerations and avenues for planning and research by athletes and sport scientists.

Preparing and Planning for Pollution

Sport scientists can plan ahead by performing sophisticated modeling of the pollution patterns of a competition site, as occurred prior to the Athens 2004 Olympics (Flouris 2006). This analysis, based on existing meteorological data for the greater Athens area from 1984 to 2003, was able to accurately predict O_3 and PM_{10} as the most problematic pollutants during the period before and throughout the Games. Furthermore, the model was able to break down pollution levels both throughout the 24 h cycle and for different regions of the city, including the northern region where the Village was located and the downtown area for many of the venues. Models of particulate matter concentrations, distribution, and emission sources were also developed prior to the Beijing 2008 Olympics (Wang et al. 2008). Such models can be used by athletes and teams to plan the timing, location, and intensity of training sessions or decide whether to train off-site completely and arrive shortly before competition.

To minimize pollutant exposure, athletes may consider moving some training sessions indoors, preferably to an air-conditioned facility in the summer so that temperature and humidity are controlled. Ideally, the incoming air is filtered to screen out particulates. This idea is obviously based on the assumption that the air quality of the indoor facility is adequate and does not present a different set of pollutants! However, research on the relative merits of exercise indoors versus outdoors in terms of total pollutant load, during both smog alert and non-alert days, remains lacking. Periods of high pollution levels or the requirement to exercise outdoors may necessitate the use of masks to filter out pollutants. These may range from simple gauze masks through to sophisticated designs aimed at minimizing flow resistance or particular pollutants.

While it may make little sense to expose athletes unnecessarily to air pollution long-term, some short-term adaptation over one to four days prior to competition may help to alleviate some of the major inflammatory and respiratory responses with acute exposure to O_3 and other pollutants, along with psychologically habituating the athlete to the potential discomfort posed by the pollution. Keep in mind that any adaptation is brief in duration, so such preexposure must happen immediately prior to competition. It should be possible to design this environmental acclimatization into an overall tapering regimen. For example, athletes might arrive on-site three to seven days prior to competition to begin their tapering. During this time, high-intensity training, with its high ventilation rates, can be performed, if possible, in a controlled environment. At the same time, passive exposure to the ambient environment during rest and recovery phases, or active exposure during lower-intensity training sessions, may be used for acclimatization purposes. However, it must be noted that three days of 2 h passive exposure to 0.20 ppm ozone did not provide any protective effect from acute exposure to higher (0.42 or 0.50 ppm) ozone compared to no preexposure, with the high ozone doses eliciting similar levels of spirometric impairment (e.g., FEV_1) with or without preexposure (Gliner et al. 1983). Therefore, passive acclimatization to low effective doses may not ultimately translate to competition at high exercise intensities.

Antioxidant Supplementation

It is evident from this chapter that O_3 can impair pulmonary function and potentially exercise capacity. It may also appear that pollutants are a systemic problem whose effects can only be minimized rather than neutralized. However, besides the avoidance and management of exposure, one potential area of interest is the use of nutritional countermeasures such as vitamins and antioxidants. As one of the proposed pathways for pollutant damage is through inflammation within the cells of the lungs and respiratory pathways, it has been proposed that antioxidants may minimize oxidative stress. The exact biochemical mechanism for antioxidant protection from pollution remains unclear, but may revolve around attenuating the ozone-induced production of arachidonic acid, which may in turn also decrease central neural inhibition of ventilation.

Grievink and colleagues (1998) studied the effects of three months of β-carotene and vitamins C and E versus no supplementation on a group of amateur Dutch racing cyclists. Lung function measures were taken posttraining and postcompetition on 4 to 14 occasions per subject over this period, with the results regressed over the previous 8 h average ozone level, which in turn averaged 101 $\mu g \cdot m^{-3}$ across the two groups. Pulmonary function, including FVC, FEV_1, and peak expiratory flow, decreased with increasing O_3 levels in the control group. This was significantly different from findings for the supplementation group, which experienced no reduction in pulmonary function with increasing O_3 levels. In a subsequent study, the same research group examined a placebo versus vitamin C and E supplementation (Grievink et al. 1999) in another cohort of amateur racing cyclists. Similar results were obtained, with progressive decrement in FEV_1 and FVC during exposure to ozone levels of 100 $\mu g \cdot m^{-3}$ in the placebo group but no decrements in the supplementation group.

Antioxidant supplementation may also prove beneficial as short-term protection for nonathletes exposed to high levels of ozone. Mexico City street workers, tested in a placebo-crossover study, took a supplement consisting of β-carotene and vitamins C and E or a placebo (Romieu et al. 1998). Protective effects similar to those previously discussed were observed in the supplement group versus the placebo group. Interestingly, after crossing over into the second phase of the study, the group that had received the supplementation first showed lower lung function decrements than the group receiving the placebo first had shown during the first phase. This suggests that antioxidants have a washout period, resulting in a slight residual protective effect.

Subjects in both of the Dutch cycling studies stopped taking vitamin and mineral supplements prior to the study and abstained throughout the study; however, the typical dietary intake of the subjects was not reported, so existing diet could potentially have confounded the results. However, the magnitude of supplementation was relatively small at 500 to 650 mg and 75 to 100 mg for vitamins C and E, respectively, and the utility of such conclusions is that these may be less than the amounts commonly available in many supplements. Overall, it appears that moderate vitamin intake, either via alterations in diet or through supplementation, may prove valuable for athletes preparing for competition in polluted regions and also for individuals chronically exposed to elevated ozone levels.

SUMMARY

While it is almost entirely man-made, airborne pollution has become an increasingly important focus within the field of environmental exercise physiology. This has come about due to a confluence of increasing urbanization, environmental awareness, and exercise participation, along with high-profile events such as the 2008 Beijing Summer Olympic Games. Five important pollutants that have been studied include ozone, sulfur dioxide, nitrogen oxides, particulate matter, and carbon monoxide. With the exception of carbon monoxide, which directly impairs oxygen transport through direct binding onto hemoglobin, the effects of pollutants are concentrated in the respiratory organs, as the primary method of entry into the body is through breathing during rest and exercise. This chapter has detailed the effects of one of these pollutants, ozone, as it is a primary constituent of smog and its concentration can spike rapidly to very high levels due to environmental factors such as temperature and weather patterns. At the same time, studying ozone brings out many commonalities in the field of air pollution, including effective dosage and managing exposure in athletes. Unfortunately, most countermeasures to directly mitigate the impact of high pollutants seem to be only marginally effective (e.g., antioxidants) or nondesirable (e.g., acclimatization), such that planning for most athletes revolves around adjusting training and competition preparation to minimize pollution exposure. One important avenue for future research is to gain a better understanding of the potential synergistic effects from multiple pollutants in the ambient air that we breathe. This may help us to develop better models and greater understanding of the mechanisms for the physiological impact of pollutants on the human organism during both acute and chronic exposure.

Chronobiological Rhythms and Exercise Performance

Sleep deprivation is possibly one of the most insidious health issues facing modern society as a whole. With advances in technology enabling increased work demands in and out of the office along with family and social responsibilities, sleep often becomes the first casualty in time management for busy professionals and recreational athletes. Sleep deprivation becomes extremely problematic in occupational settings due to the commonality of shift work in many professions, and contributes to health impairment along with social and work–home conflicts (Demerouti et al. 2004). Furthermore, the risk of accidents occurring due to sleep deprivation-related fatigue can pose severe hazards beyond the worker; Chernobyl and the *Exxon Valdez* were major disasters in which fatigue from sleep deprivation contributed to inappropriate, and ultimately catastrophic, responses (Folkard and Lombardi 2006; Mitler et al. 1988).

Elite and professional athletes face constant travel demands that can play havoc with both sleep patterns and performance (Reilly et al. 2005). In North America, players in the National Basketball Association and the National Hockey League compete roughly three times a week in different cities during road trips, with each league spread over three time zones. Promotional exhibition and league games in Europe or the Far East have also recently been staged for many North American leagues. This globalization trend has long been common in many Olympic sports; World Cup events occur across the globe over each season. The fatigue and jet lag associated with such prolonged travel can lead to a multitude of problems for athletes, ranging from simple discomfort and fatigue to reduced exercise capacity and recovery through to suppressed immune systems and elevated risks of infections (Waterhouse et al. 2007b). Therefore, appropriate strategies to rapidly and safely synchronize circadian rhythms can help to mitigate the effects of travel.

The purpose of this chapter is to review current knowledge concerning the effects of sleep deprivation and shift work on exercise performance. The chapter also covers the effects of transmeridian travel on exercise capacity, as well as current best practices to minimize disruptions among shift workers and athletes.

SLEEP–WAKE CYCLE BASICS

Sleep, which is similar to fatigue as a multifactorial phenomenon, is not triggered by a single hormone or mechanism but rather by a host of physiological and environmental factors. This chapter mainly focuses on the sleep–wake cycle and its related circadian rhythms, particularly those involving body temperature and major hormones such as melatonin. However, the field of biological rhythms is much broader. Circadian rhythms are generally within a period of 19 to 28 h; but other physiological cycles, such as the menstrual cycle in human females and hibernation in animals, may have periodicities based more closely on lunar and annual cycles.

Most adult humans require approximately 6 to 10 h of sleep daily for proper functioning and health, though the primary purpose of sleep remains open to speculation. Two peaks in sleepiness are typical over a circadian cycle, with a large spike in the early morning (~0200-0700 h) and a smaller spike in the afternoon (~1400-1700 h) (Mitler et al. 1988). Such a pattern of increased sleepiness and sleep tendency, including decreased alertness and the presence of micro-sleeps, appears to be consistent regardless of whether sleep patterns are normal or disrupted. Interestingly, this bimodal circadian pattern in sleepiness is strongly associated with the temporal pattern for heart attacks and death from all causes (Mitler et al. 1987) and also with performance errors as well as vehicular and industrial accidents across a wide range of occupations (Mitler et al. 1988). Such findings highlight the importance for human health and performance of understanding the physiological process of sleep.

As with other autonomic functions, the brain appears to be the major control site for sleep regulation, specifically the suprachiasmatic nucleus (SCN) within the hypothalamus. The location of the SCN is close to the primary thermal integration site within the posterior hypothalamus, and may be a link to the close relationship between body temperature and sleep. The SCN also appears receptive to light (photic) stimulation, which forms possibly the strongest environmental regulator of the sleep–wake cycle. The ventrolateral SCN receives input from the retinohypothalamic tract, with neurons traveling from the retina to the hypothalamus.

Temperature

While humans can maintain body temperature within a narrow range over the course of a lifetime, fluctuations in deep body temperature of approximately 1 °C are one of the most characteristic features of circadian rhythms. Irrespective of metabolic heat generation from exercise, a trough during nighttime and a gradual rise occurring over the course of the day are typical simply with basal metabolism (see figure 12.1).

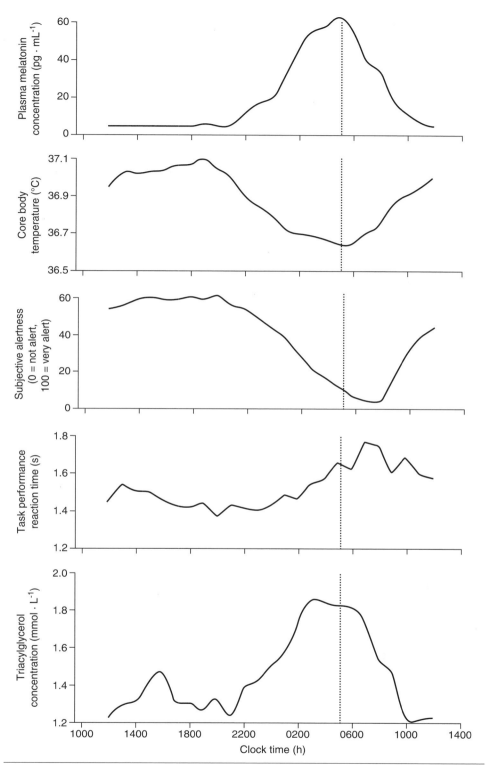

Figure 12.1 Plasma melatonin over the course of a 24 h cycle, along with the circadian patterns of body temperature and mental functioning measures.

Reprinted from *The Lancet*, Vol. 358, S.M.W. Rajaratnam and J. Arendt, "Health in a 24-h society," pgs. 999-1005, Copyright 2001, with permission from Elsevier.

Temperature patterns appear to be intimately related to sleep, falling by approximately 0.3 °C over the initial hours of sleep onset compared to times when wakefulness has been maintained (Barrett et al. 1993). This is likely mediated by an increase in peripheral vasodilation and a consequent elevation in heat dissipation (Krauchi et al. 2006). Indeed, during moderate exercise in the evening (2300 h), as body temperature begins its circadian decrease, forearm blood flow and sweating rates have been found to be higher than in the morning (1100 h) during the circadian rise in body temperature. Complementing this enhanced heat dissipation, rectal temperature also decreased more rapidly following evening exercise (Waterhouse et al. 2007a). With this change in baseline body temperature throughout the day, an obvious question is whether thermoregulatory responses and tolerance to cold or heat are affected by the time of day of exposure or exercise.

Cold responses appear to vary on an absolute basis but are stable relative to core temperature over the circadian cycle. When resting subjects were tested at 0700 and 1500 h, baseline rectal temperature was 0.4 °C higher at 1500 h (Castellani et al. 1999). Despite the change in absolute body temperature across the two testing times, responses to moderate cold (20 °C) immersion exhibited minimal differences. Both the rate of heat production and heat exchange were similar over the course of the two immersion trials. Specifically, neither the vasoconstriction nor shivering thresholds, when expressed relative to initial starting core temperature, differed between 0700 and 1500 h.

Responses to heat also seem to follow a consistent circadian pattern based on core temperature, though the underlying relationships between the effector responses may be more complex than a simple shift (Stephenson et al. 1984). Over six experimental sessions on different days, subjects were tested at evenly spaced time points. The nadir and peak of esophageal temperature occurred at 0400 and 1600 h, respectively, with a range of 0.7 °C. Both vasodilation and sweating thresholds tracked closely relative to core temperature in response to moderate exercise in a warm environment. However, the sensitivity of vasodilatory responses increased between 0400 and 2400 h, while sweating sensitivity remained stable throughout, possibly due to circadian variation in autonomic cardiovascular control.

Melatonin

One hormone with a significant, though not necessarily dominant, role in circadian rhythms of temperature and the sleep–wake cycle is melatonin. Produced in the pineal gland of the brain, melatonin begins to be secreted in the evening hours and is strongly suppressed by photic (light) stimulation during the day (Cajochen 2007). The complete pharmacological basis for melatonin remains to be resolved, but in relation to the sleep–wake cycle it can be viewed as preparatory or facilitative to initiating sleep in humans rather than as a direct sedative or trigger (Atkinson et al. 2003). Mild direct interaction occurs between melatonin and body temperature, with ingestion provoking a temperature decrease of 0.1 to 0.3 °C in bed rest situations. However, in keeping with the lack of circadian effects on thermoregulatory thresholds already discussed, melatonin ingestion does not provide any ergogenic benefit in prolonging exercise tolerance in hot conditions (McLellan et al. 2000). While specific research on melatonin and cold exposure has not been done, it is unlikely that responses to cold exposure are affected by melatonin.

CIRCADIAN PATTERN IN EXERCISE CAPACITY

With circadian patterns in many physiological and hormonal systems, an interesting initial question in chronobiology is whether a circadian cycle exists in exercise capacity (Drust et al. 2005). If a pattern exists whereby exercise capacity peaks at specific points, this may lead athletes and sport organizations to time competitions or world record attempts to maximize athletic potential. Current evidence suggests that a circadian pattern of improved performance capacity in the afternoon and evening does exist for tasks involving both peak muscular force and skill. With both dart throwing and the badminton serve, which emphasize motor control and technical skill, improved performance over the course of a day has been reported (Edwards et al. 2007, 2005b). Female subjects were tested at both 0600 and 1800 h for maximal voluntary and evoked muscle force during knee extension and flexion, and voluntary maximal isokinetic torque was 4% to 6% higher at 1800 h (Bambaeichi et al. 2005). Though other muscle measures did not exhibit diurnal variation, the knee flexion increase was consistent whether the subjects had received normal sleep or had been sleep deprived for 24 h. An evening increase was also observed in power output over a 30 s supramaximal Wingate cycling test, with an approximately 8% increase in peak and mean power outputs along with a lower decrease in power over the 30 s (Souissi et al. 2007). Though the Wingate test is heavily reliant on anaerobic metabolism for energy, a significant contribution is provided by aerobic metabolism. No differences were observed in blood lactate profiles across the two trials, suggesting similar rates of anaerobic metabolism. However, oxygen uptake was higher in the evening trial; thus an enhanced aerobic metabolic capacity, possibly related to faster oxygen uptake kinetics, likely contributed to the improved performance. Indeed, during submaximal exercise, oxygen uptake kinetics were demonstrated to be faster and net efficiency was higher in the evening compared to morning (Brisswalter et al. 2007).

A related question is the timing of final training sessions leading up to the competition. Prior to a 16 km (9.94-mile) cycling time trial test performed at 0700 h, the effects of moderate training (60% $\dot{V}O_{2peak}$) on the previous day at either 0700 or 1200 h were tested (Edwards et al. 2005a). While no muscle biopsies were performed to measure muscle glycogen levels, it was theorized that the shorter recovery time between the 1200 h training and 0700 h testing was not a confounding factor, due to the relatively low training load and duration along with lack of changes in physiological parameters at rest prior to the time trial. Power output and time trial performance improved by a significant 2% when training occurred at 0700 h, which translates to a mean improvement of 34 s. With no other significant differences in physiological responses across the two experimental conditions, the authors attributed the improvement to familiarization with a morning schedule, though the additional 5 h of recovery may have been more critical.

SLEEP DEPRIVATION

The first obvious question about sleep deprivation is whether it has a practically relevant effect on overall exercise performance in field settings. Most athletic situations are discrete events such that sleep and recovery are possible even if the competition occurs over multiple days, as in the Tour de France. The effects of sleep deprivation and accumulated sleep debt on exercise performance are most relevant with ultra-endurance competitions, ranging from the Race Across America (RAAM) bicycle race to ultramarathons and multiday adventure races. Military training is another field model for sleep deprivation research, especially sustained operations training performed by special forces in which sleep is reduced to minimal levels for days or weeks on end. While such events are excellent applications for sleep deprivation research, using them as experimental models is difficult because of the intertwined nature of extreme and continuous exercise, nutritional deficits, and sleep deprivation. Therefore, they should be viewed as indirect approaches to studying sleep deprivation and its effects on exercise.

The capacity for tolerating short-term sleep deprivation is supported by results from simulated sustained military operations featuring severe daily caloric deficit (~1600 kcal) and sleep deprivation (2 h of sleep daily) (Nindl et al. 2002). Some laboratory-based performance measures, such as squat jump power, decreased over the 72 h simulation. However, no decrements were observed in bench press power. Furthermore, performance on operationally relevant tasks such as marksmanship and grenade throw was not impaired. As the overall duration of the simulation was only 72 h, it may have been too brief to maximize impairment from accumulated sleep debt. In a field study, both sleep deprivation (4 h of sleep daily) and caloric imbalance (−850 kcal) were milder, but the duration was greatly prolonged to the full 61 days of real-life U.S. Army Ranger training in a group of qualified military candidates (Young et al. 1998). All eight subjects successfully completed the full training program, as do approximately 30% to 40% of candidates for such special forces, demonstrating that prolonged exercise capacity can be maintained even in extreme situations of sleep and energy deficit for many fit, well-trained, and highly motivated individuals. One needs to consider, however, that the high motivation levels typical of military special forces candidates may not be directly relevant or applicable to athletic or other nonmilitary situations, no matter how strenuous the setting.

A more direct study on exercise capacity and the physiological strain across multiple systems in athletes was provided by Lucas and colleagues (2008) in the sport of adventure racing. The Southern Traverse adventure race in New Zealand features >400 km (250 miles) and 120 h of competition involving mountain biking, kayaking, orienteering, and coasteering. During the 2003 edition, competitors demonstrated a significant drop-off in self-paced exercise intensity, from an average of 64% maximal heart rate (HR_{max}) in the first 12 h of the race down to an average of 41% HR_{max} from 24 h through the rest of the race. This was true in male and female members of both the winning and the last-place teams, suggesting that the downregulation in exercise intensity was not a by-product of race placement. It can be argued that this decrease was due to general pacing errors, as an overly high initial pacing is common to most time trials of any distance. In support of this contention, heart rate responses did not differ during laboratory exercise tests performed pre- and postcompetition, suggesting no autonomic impairment in cardiovascular regulation. However, subjective ratings of exercise intensity were significantly higher postcompetition, likely due to changes in psychological motivation following such a strenuous event. Although no laboratory tests were performed, core temperature during competition remained within normal values despite wide-ranging environmental conditions. Immediately following this competition, upper respiratory problems and skin wounds, along with gastrointestinal problems, were more prevalent than before the competition and fairly common (Anglem et al. 2008). Mood alterations were also common, but these as well as the physical symptoms generally resolved themselves within two weeks. As it is difficult to separate and isolate the effects from sleep deprivation in such field studies, it remains impossible to generalize about the contribution of sleep deprivation to performance in sustained exercise.

In summary, it appears that, at least in highly motivated and trained individuals, ultra-endurance exercise and sustained operations with significantly reduced sleep time and quality can be tolerated with minimal decrements in exercise and operational capacity, and that any physiological or psychological impairments resolve relatively rapidly with appropriate recovery. One issue that must be raised, though, is the level of motivation and the psychological characteristics of subjects in such environments, such that extrapolation to populations with lower inherent motivation may not be valid. Relatively minimal work has focused on the direct effect of fitness on sensitivity or response to sleep deprivation. However, in one study on 14 university-aged males categorized across a range of self-reported physical activity levels, no correlation was found between activity and thermoregulatory, psychological, or exercise response to sleep deprivation (Meney et al. 1998).

In contrast to indirect conclusions on sustained exercise drawn from field studies, controlled scientific experiments have provided direct data on the relationship between short-term (24 h) sleep deprivation and strength and anaerobic capacity. Twenty-four hours of sleep deprivation did not impair peak torque with either isometric or isokinetic knee extension and flexion in eumenorrheic women (Bambaeichi et al. 2005). National-caliber weightlifting athletes were tested for sleepiness and on a battery of psychological tests before and after a night of normal sleep or before and after overnight sleep deprivation. Following the sleep manipulation, strength tests were performed; testosterone and cortisol were measured immediately pre- and post-testing, as well as 1 h posttesting. Subjective sensations were significantly impaired by sleep deprivation, with decreased vigor and elevated confusion, sleepiness, and mood disturbance. The stress marker cortisol was attenuated immediately after the strength test and 1 h later in the subjects who had been assigned to normal sleep. However, despite the impaired psychological profile, no differences were reported across sleep conditions in any of the strength measures, including maximal lifts along with individual exercise and total work volume. Therefore, the authors concluded that in these elite athletes, a single bout of sleep deprivation did not negatively affect strength or exercise capacity (Bambaeichi et al. 2005).

This discussion has focused on acute bouts of sleep deprivation or else forced sleep loss in the course of sustained activity. What cannot be determined from the current literature, however, are the effects of sustained sleep deprivation on long-term training and athlete development. Daily suboptimal sleep quantity or quality can lead to a sustained sleep debt for many athletes, especially given the large daily time requirement for training in addition to work, school, and family responsibilities. In a survey on 24 athletes from a Bobsleigh Canada Skeleton team, 78% scored above the threshold for poor sleep quality in the Pittsburgh Sleep Quality Index, a validated self-reported questionnaire for use in adults (Samuels 2009). Poor sleep quality and accumulated sleep debt may be especially problematic in the long-term development of athletes during adolescence, with the additional physiological stress from physical maturation superimposed on training. This is highlighted by findings that a similarly high percentage (85%) of 44 adolescents at an elite Canadian sport-centered high school scored above the Pittsburgh threshold for poor sleep, with most of the athletes obtaining less than 8 h of sleep nightly (Samuels 2009).

SHIFT WORK

Apart from its effects on athletes, chronobiology plays an important role in occupational health and safety. Shift work involves an alteration of the typical circadian rhythm, similar to that seen with transmeridian travel and jet lag as discussed later in this chapter. As with such travel, the circadian period can be phase-advanced via moving from a normal daytime schedule to a nighttime schedule, or phase-delayed via moving from a daytime to an evening schedule. Working in shifts that differ from the traditional daylight working hours is common in occupations ranging from health care (nursing), emergency response (firefighting), utilities (work in nuclear generators), and transport (shipping, air traffic control). Each of these examples represents major potential for accidents from sleep deprivation that provokes physical fatigue or cognitive impairment.

Unlike transmeridian travel, shift work retains a permanent mismatch in the sleep–wake cycle between endogenous rhythms and environmental cues and also social constraints. This makes it difficult, if not impossible, to fully acclimatize to shift work at night (Driscoll et al. 2007). Circadian impairment is further exacerbated by the continued rotation of working shift times in many of these occupations, with the occurrence of only a few shifts before another work schedule is imposed and readaptation is again required. Therefore, it is unsurprising that shift workers have a high prevalence of insomnia and other sleep disorders (Akerstedt 2005). Symptoms of nonadaptation to shift work include decreased cognitive performance across a range of tasks, along with decreased attention and vigor (see table 12.1). Problematically, such cognitive impairments appear to progressively worsen with chronic exposure to shift work. In contrast, removal from a shift work regimen seems to provide some degree of recovery (Rouch et al. 2005). Such data imply that experience on the job does not seem to confer an adaption that protects against cognitive impairment, but rather that the effect of shift work is cumulative. Initial exposure to shift work also elicits many of the cardiovascular and neuroendocrine changes observed in clinical conditions such as chronic fatigue syndrome.

Other hidden societal costs of shift work include increased risk of depression, addictions, and mental health issues. For example, nursing students exposed to shift work for the first time exhibited marked increases in psychological indices typical of depression, such as disturbed sleep and appetite, along with lethargy and poor concentration (Healy et al. 1993). Such findings were above and beyond the slight elevation in baseline values in control subjects due to the stress of nursing training itself. Furthermore, the exposure to shift work induced feelings of helplessness, eroded perceptions of social support, and increased feelings that one is under criticism as well as psychosomatic complaints. The presence of shift work during weekends, when familial and societal expectations for interaction are often elevated, may prove especially problematic.

From this discussion, it appears that response to shift work is a matter of coping rather than adapting. Temporary strategies should accommodate what is required to minimize the effects of this shift in cycles. An important coping strategy is the

Table 12.1 Selected Studies on Cognitive Functioning Following Partial Sleep Loss and Sleep Deprivation

Study	Experimental design	Main findings
Partial sleep loss		
Van Dongen et al. (2003)	Subjects ($n = 48$) randomized to one of the following sleep doses: 4 h, 6 h, or 8 h per night for 14 days, or 0 h for 3 days. Psychomotor vigilance and working memory performance were tested every 2 h throughout each day.	• Vigilance (lapses in attention) and working memory were significantly impaired in a dose-dependent, cumulative manner following sleep restriction of 4 h and 6 h per night. • Sleep restriction of 4 and 6 h per night resulted in cognitive deficits equivalent to those with two nights of total sleep deprivation.
Balkin et al. (2004), Belenky et al. (2003)	Commercial truck drivers ($n = 66$) were randomized to seven nights of 3, 5, 7, or 9 h of sleep per night. Cognitive and behavioral performance tests, including a psychomotor vigilance task, a simulated driving task, and the Walter Reed Performance Assessment Battery[a], were administered four times per day.	• Vigilance performance (speed and number of lapses) was significantly impaired in the 3 and 5 h groups. • Tests that demanded constant attention or were monotonous (i.e., psychomotor vigilance and driving a simulator) were most sensitive to sleep restriction.
Sleep deprivation		
Glenville et al. (1978)	Repeated-measures design with subjects ($n = 8$) undergoing performance tests following one night of total sleep loss and one night of regular sleep. Performance battery tested auditory vigilance, choice reaction time (RT), simple RT, short-term memory, and simple motor skill.	• Vigilance performance was most affected by sleep loss, with choice and simple RT also significantly impaired. • Neither memory nor simple motor skill was affected by sleep loss.
Landrigan et al. (2004)	Randomized study of interns working a traditional schedule (≥ 24 h work shifts every third night) or an intervention schedule (no extended work shifts; decreased number of hours per week).	More serious medical errors (35.9% more) were made during traditional schedule versus the intervention schedule.
Akerstedt et al. (2005)	Shift workers ($n = 10$) performed a simulated driving task following a normal night sleep (7.6 ± 0.32 h) and a night of regular work.	Night shifts resulted in more incidents, decreased time to accidents, increased variability in lateral position, and increased eye closure durations during simulated driving.

[a]The Walter Reed Performance Assessment Battery includes the following: Stanford Sleepiness Scale, Profile of Mood States, code substitution, serial addition/subtraction, grammatical (logical) reasoning, running memory, time estimation, 10-choice reaction time, Stroop color naming, and delayed recall.

scheduling of shift work, both in its timing and in relation to rotating patterns. In a large cohort of Dutch military police, Demerouti and colleagues (2004) demonstrated that the presence of shift work was a strong predictor of work–home conflict, with a greater increase when shift work took place during weekends. Differentially, a rotating shift schedule was highly predictive of unfavorable job attitudes and job satisfaction. If rotation is required, a forward progression (morning, afternoon, night shifts) may be the best option, possibly because it provides a gradual phase delay rather than phase advance (Driscoll et al. 2007). Individual situations and also tolerance to shift work and rotation can be highly variable, such that it would appear useful to develop methods of assessing response and tolerance to variable schedules and then ideally customizing schedules where feasible. In that vein, the hormone cortisol, which also exhibits a circadian rhythm, has been proposed as a potential monitoring tool for adaptability to shift work in general and also for the individualized customization of shift work (Hennig et al. 1998).

JET LAG

From an athletic perspective, the greatest acute impact of chronobiology can occur due to the physiological and psychological disruptions from transmeridian travel for training or competition (Reilly et al. 2005) (see figure 12.2 for a general time line for readjustment). As with the insomnia and other sleep-related issues discussed in the context of shift work, jet lag can induce feelings of disorientation, fatigue, irritability, light-headedness, impatience, lack of energy, and problems with appetite and bowel movements. Individual variability exists in tolerance, but a threshold of three or more time zones appears to be required to evoke significant symptoms in most individuals (Waterhouse et al. 2002). Proper planning can help to minimize, though not necessarily eliminate, the impairment in exercise performance. The first priority in athletic and occupational settings should be to provide work schedules and environments that minimize problems with sleep deprivation to begin with. The following approaches may be useful for both athletes and shift workers in assisting with adaptation.

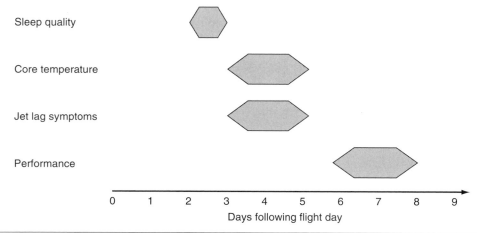

Figure 12.2 Time frame for readjustment of key circadian rhythms following air travel. Sleep quality is attained relatively quickly following travel, whereas performance variables can take up to eight days to return to normal values.

Travel Planning

A survey study on a group of 85 athletes, support staff, and academics traveling 10 time zones eastward from the United Kingdom to Australia highlights the important potential for proper planning to influence the rate of adaptation to a new time zone, along with the individual factors that may influence the magnitude of jet lag (Waterhouse et al. 2002). Various jet lag symptoms were reported by the large majority of subjects throughout the study period (the first six days upon arrival); but age, somewhat surprisingly, was a predictor for lower intensity of jet lag while fitness, flexibility in sleep habits, and chronotype were not significant factors. The importance of travel logistics and scheduling was highlighted in the results. One group left the United Kingdom in the morning, then had a layover in Singapore before arriving in Australia in the late afternoon. Much of the actual flying on both legs of the journey was during local daytime, resulting in only 1.5 h of sleep on the first leg. The other group departed in the evening from the United Kingdom and slept for 5.5 h and also flew the second leg in local nighttime, arriving in Australia in the morning. The first group, the "daytime fliers," had greatly reduced jet lag symptoms compared to the "nighttime" group despite their sleeping less during the actual journey. The authors attribute this to an overall shorter period of time between sleeps in an actual bed in the daytime fliers (~30 h), as they flew shortly after a full sleep and arrived close to local nighttime. In contrast, the nighttime fliers began their journey after having been awake for a full day and arrived during local morning, resulting in approximately 50 h between sleeps in a bed (Waterhouse et al. 2002).

Bright Light Exposure

With the strong linkage between light exposure and melatonin levels and alertness (Cajochen 2007), bright light exposure (BLE) has been investigated as a means to produce shifting of the circadian rhythms. Used by itself in laboratory studies without travel, BLE appears to be capable of adjusting circadian melatonin and temperature rhythms. This phase adjustment can be achieved with a single dose of very bright (12,000 lux compared to typical room illumination levels of 400-600 lux) light exposure for 4 h, but the exact pattern of phase adjustment is dependent on the timing of exposure (Dawson et al. 1993). Specifically, when applied during the evening just prior to habitual sleep time and the initiation of melatonin release and temperature decline, BLE created a 2.39 h phase delay (i.e., delayed the onset of melatonin release and temperature decline). In contrast, BLE in the morning at the time of habitual wake-up had the opposite effect, advancing the circadian cycle by 1.49 h (Dawson et al. 1993). Furthermore, in a healthy population, eight weeks of moderate exercise increased mood levels and decreased atypical depressive episodes, but subjects exercising in bright light (2500-4000 lux) conditions had a stronger positive response than those exercising in normal light levels.

While interindividual variability in response to bright lights exists, current data strongly suggest that BLE can be used as a tool by athletes in adapting to transmeridian travel. In such applications, BLE may be able to facilitate entraining circadian rhythms to a new time zone both prior to and following travel, along with improving mood—thus increasing sleep quality upon arrival and minimizing jet lag symptoms. Bright light exposure can possibly begin the week prior to travel across multiple (more than six) time zones in order to initiate acclimatization to the new time zone, and then be intensively used upon arrival to maximize the rate of circadian adjustment.

An extension of the research on BLE and also the time of day in relation to exercise performance is the concept of whether there is an optimal level of illumination for exercise performance. This may be of special importance in world record attempts, whether indoors where illumination can be controlled or in outdoor world record competitions, which could be scheduled to coincide with optimal sunlight availability. One study had subjects perform a 45 s supramaximal Wingate cycling test following 90 min of exposure to either dim (50 lux) or bright (5000 lux) light; all subjects performed the test itself in normal (500 lux) light levels (Ohkuwa et al. 2001). Bright light exposure preexercise decreased the level of plasma epinephrine but did not appear to influence supramaximal exercise capacity; no differences were reported in power outputs throughout the 45 s test, nor were there differences in blood lactate, ammonia, or norepinephrine following the sprint. Similar lack of efficacy was reported with

BRIGHT LIGHT EXPOSURE AS A PRECOOLING MODALITY?

Apart from the context of transmeridian travel, some novel applications for BLE have been suggested for athletes. In one intriguing extension of the idea of adjusting circadian rhythms and body temperatures, the use of BLE has been proposed as a precooling modality prior to exercise in hot conditions (see chapter 3 for a discussion of heat stress and precooling). When trained runners were exposed to bright (10,000 lux) light in the late evening (2200-2300 h), melatonin release was suppressed, and the core temperature minimum was phase-delayed by 1.46 h (Atkinson et al. 2008). Compared to what occurred with a similar dose of BLE in the morning (0600-0700 h), core temperature decreased at 0700 h by 0.20 °C with evening BLE, and this lower temperature trended toward being maintained throughout 40 min of moderate running in a warm and humid (31 °C, 66% relative humidity) environment. A similar nonsignificant trend in lower perceived exertion was reported with evening BLE. One major limitation of the study is the small sample size, only six subjects; the low statistical power may partially explain why the perceived exertion and core temperature during exercise were close to, but did not achieve, statistical significance. In another study investigating light exposure immediately prior to exercise, untrained women with normal sleep the night before were exposed to 6 h of bright (5000 lux) or dim (50 lux) light immediately prior to exercise in a warm, humid (27 °C, 60% relative humidity) environment (Zhang and Tokura 1999). Despite no differences in resting temperature following the light manipulation, sweat rate was elevated and core temperature remained approximately 0.2 °C lower during the moderate exercise following BLE. These studies are intriguing because they suggest that chronobiological manipulations, whether by BLE or other modalities, may provide a practical method of precooling to minimize the risk of heat illnesses and may maximize performance in hot environments while avoiding the logistical constraints with precooling in the immediate precompetitive period.

prolonged submaximal exercise. Trained cyclists averaged similar power outputs (~270-274 W) over 20 min of self-paced maximal cycling whether exposed to 1320, 2640, or approximately 6000 lux; the ambient lighting was similar in the various conditions, but the illumination reaching the retina was adjusted using sunglasses of different filtration (O'Brien and O'Connor 2000). In addition, no differences were reported in physiological responses such as heart rate and oxygen uptake, nor in subjective responses of alertness, local perceived effort, or muscle pain. One caveat, however, is that all illumination conditions were much higher than the typical 400 to 600 lux room illumination, so there were no direct comparisons between bright light and normal illumination, and a threshold level for improvement may have existed in all conditions. Nevertheless, current data suggest that while BLE may be effective in adjusting circadian rhythms, it may have minimal direct ergogenic effect on exercise performance.

Exercise

Exercise itself could act as a catalyst for altering circadian rhythms and may help the body adjust to the rhythms of the local environment. However, the overall efficacy of exercise as a synchronizer, along with the underlying mechanisms, remains unclear and difficult to assess due to methodological problems (Atkinson et al. 2007). In one extensive study, eight subjects underwent 40 days of isolation and forced desynchronization of their normal schedules to a 20 h day rather than a 24 h day (i.e., 48 cycles of a 20 h day equals 40 actual 24 h days); three bouts of moderate exercise in dim light were evenly distributed over each 20 h day. As evidenced by plasma melatonin levels and core body temperatures, exercise did not contribute to any shifting of circadian rhythms (Cain et al. 2007). However, bouts of exercise in the evening, timed near the period of melatonin onset, induced a phase delay in melatonin release, with a greater magnitude of delay the closer the exercise was performed to the normal melatonin onset time (Barger et al. 2004). Many methodological issues need to be accounted for in investigations of the direct effects of exercise or other circadian adjusters, including the control of light, training status of the participants, individual susceptibility to phase shift, and the type and intensity of exercise manipulations (Atkinson et al. 2007). For example, nocturnal exercise can serve to phase-delay the circadian rhythm of melatonin levels, but appears to have no influence on the circadian body temperature rhythm. Furthermore, the lack of agreement between exercise and the phase-shifting effects of bright light suggests that exercise is not exerting its effects via photic entrainment pathways.

Sleep-Enhancing Substances

The reverse of the pharmacological approach to optimizing alertness and resistance to fatigue is the use of substances that may enhance people's ability to attain adequate quantity or quality of sleep (or both). The use of sleeping pills (benzodiazepines) and other sedatives to induce the onset of sleep may appear to be beneficial with travel or shift work. However, while these substances may promote sleep, the quality and ergogenic effect of such sleep is questionable; there is little evidence of advantageous effects on chronobiology and shifting circadian rhythms (Lemmer 2007). The potential for addiction to sedatives such as alcohol and barbiturates also forms a strong counterindication for their use in sleep enhancement except for clinical conditions.

PRACTICAL RECOMMENDATIONS FOR TRANSMERIDIAN TRAVEL

Susceptibility to jet lag appears to be highly individual. Therefore, rather than a general template, a customized approach to international travel may be required for each person. Overall, the goal should be to adjust circadian rhythms to the local time zone as rapidly as possible rather than anchoring circadian cycles to the original time zone, and a number of interventions may help to minimize jet lag effects and accelerate adaptation to the new environment. Recommendations include the following:

- Long flights should be scheduled so that the focus is not on maximizing sleep during the flight and any layovers but rather on minimizing the time between sleep periods in bed. For eastward travel, this may involve daytime rather than "red-eye" flights, and the reverse may apply for westward travel.

- Awareness of an issue not strictly related to chronobiology—blood pooling and potential deep vein thrombosis with prolonged sitting during flying—has increased in the past decade. While no research has involved athletes, data from clinical populations have suggested that commonly prescribed behaviors such as drinking nonalcoholic fluids, exercising, or wearing antiembolic stockings had minimal to no influence on decreasing the risk for developing thrombosis. However, choosing aisle seats compared to window seats decreased risk twofold, and flying business class also slightly decreased risk (Schreijer et al. 2009).

- Social contact may mask many symptoms but does not completely compensate or readjust circadian rhythms.

- Napping during the initial adjustment phase might not be appropriate, as it can help to anchor the individual back to the departure time zone and rhythms. People who need a nap should not allow it to become prolonged; this will avoid deep sleep and strong sleep inertia that may affect performance after the nap.

- Melatonin appears to have minimal impact on performance and can be used to facilitate a circadian phase advance. This may be especially effective for accelerating the onset of sleep tendency with eastward travel.

- Bright light exposure appears to be highly effective in adjusting circadian rhythms and can either phase-delay or phase-advance depending on the timing of exposure. Therefore, it appears that BLE can be utilized for both westward and eastward travel. For westward travel (phase delay required), 2 to 4 h of bright (>5000 lux) light exposure at the time of "normal" sleep (this will shift over the adjustment process, but as an example would be one's normal sleep time at home immediately upon arrival at the destination) can phase-delay circadian rhythm in melatonin release and temperature by 2 to 3 h. For eastward travel (phase advance required), BLE should be timed to coincide with the "normal" wake times for maximal effect.

- While exercise may be capable of assisting in adjusting circadian rhythms, scheduling and logistics may preclude training at optimal times for phase shifting.

As discussed earlier, melatonin is a natural precursor to a decrease in alertness and appears to prepare the body for sleep. Therefore, the use of exogenous melatonin may potentially be beneficial for athletes and shift workers in aiding to reset the circadian rhythm to a new sleep–wake cycle, though research on optimal dosing and clear evidence of efficacy in all users are unclear. Melatonin is categorized as a herb or supplement rather than a pharmaceutical by the U.S. Food and Drug Administration and therefore is available without a prescription but also without strict regulation of purity. Though it is considered legal by the World Anti-Doping Agency, an important consideration regarding melatonin is its potential effect, either negative or positive, on exercise capacity. Melatonin ingestion does not appear to either improve or impair physical performance. While alertness, short-term memory, and reaction times were impaired following 5 mg melatonin ingestion compared to a placebo, no differences were observed in a 4 km (~2.5-mile) cycling time trial either 1.5 or 4.5 h later (Atkinson et al. 2005). Similarly, 5 mg melatonin ingested prior to sleep did not influence grip strength or a 4 km cycling time trial in fit cyclists the following morning, and also did not improve sleep quality compared to placebo (Atkinson et al. 2001). Therefore, it would appear that melatonin is neutral in its effect on exercise capacity, and could constitute an important countermeasure in relation to circadian adjustment following transmeridian travel.

An additional caveat with melatonin is that the timing of its ingestion appears to be critical to its ultimate effect. Ingestion after the trough in circadian body temperature (i.e., in early morning with a normal sleep–wake cycle) appears to phase-delay circadian rhythm. In contrast, a phase advance occurs when melatonin is taken prior to the peak in body temperature (i.e., about 8 h following wake-up). Furthermore, ingestion during the normal sleep cycle (i.e., from bedtime through to wake-up) will aid in inducing drowsiness but does not appear to have any chronobiotic effect. Overall, the ergogenic effect of melatonin for jet lag recovery is equivocal and may be minimal, especially in individuals with busy schedules upon arrival. This was evidenced by the lack of any significant benefit from melatonin ingestion in British athletes traveling to Australia across 10 time zones; no improvement in any symptoms of jet lag compared to those in a placebo group was demonstrated (Edwards et al. 2000).

SUMMARY

Modern society and the globalization of sport impose great demands on athletes and workers throughout the 24 h day. Apart from the need for regular sleep, the body's physiological responses, including temperature and release of hormones like melatonin, fluctuate over the 24 h circadian cycle. Not surprisingly, therefore, cognitive and exercise capacity also exhibits a circadian pattern, with muscle strength and sprint performance higher in the afternoon than in the morning. A sudden desynchronization of the natural circadian rhythms from societal norms, as occurs with nighttime work and shift work, can lead to major problems with both mental and physical capacity, and this shift work–induced fatigue may be a major contributor to increased errors and the risk for accidents in many occupational settings. Athletes may be required to travel frequently over multiple time zones for training and competition, and the physiological and psychological stress of travel may impair performance for days following arrival, with these changes falling under the umbrella term of jet lag. The susceptibility to and intensity of jet lag are highly variable across individuals, and the improved fitness of athletes does not appear to provide general protection. Besides careful planning and scheduling of travel, taking melatonin and the use of BLE are two interventions that can accelerate the readjustment of circadian rhythms to local time zones following transmeridian travel.

Afterword

Future Frontiers

This volume forms a concise but certainly noncomprehensive summary of the major physiological issues facing humans exercising in a multitude of environmental conditions. One consistent theme throughout this volume has been performance in real-life situations under various environmental conditions. It is not enough to understand merely the effects of environmental stressors on human physiology; rather, the critical extension is how these responses may affect human functioning during exercise and work in applied situations. This is essential to designing safe work systems, appropriate countermeasures, and also safety guidelines. It should be evident that much work continues to be required both to expand physiological knowledge and to implement such knowledge into applications ranging from equipment design through to policy development.

Expanding on the major issues raised in chapter 1, the following are some closing thoughts on potential avenues for development within the field of environmental physiology.

INDIVIDUALITY

Perhaps the most perplexing question in environmental physiology, and indeed in human physiology in general, is the basis for individual variability. We have seen this in the chapters on altitude training, high altitude, and also microgravity. What makes one person highly susceptible to acute mountain sickness or to space motion sickness while another person seems to thrive during high-altitude mountaineering expeditions or rapidly adjust to the disorientation of spaceflight? Similarly, why do some people apparently show minimal physiological responses or performance changes with altitude training? Such differences are likely a synergistic combination of genetic and environmental factors during development, made further complex by sociocultural factors and differences in experience and training status. Such complexity makes it difficult to tackle environmental physiology from solely a reductionist perspective of isolating and investigating any single factor, though such research will continue to be essential. This is especially true for genetics and microcellular research. At the same time, systemic research on the responses of the human as a whole will continue to be critical to integrate such findings and also to broach further questions.

PREDICTION AND SCREENING

Hand in hand with a better understanding of individual factors in environmental response is potentially predicting those responses and screening for individuals who will perform best under such stressors. While a benign and noble approach in general, the selection or exclusion of particular individuals can also pose serious ethical and legal questions. This issue has consistently arisen in cases involving legal challenges to fitness and other criteria for occupations such as wildland firefighting and many military specializations. Therefore, with advancing research on the bases for individual variability in physiological response, scientists must be diligent in developing evidence-based prediction tools that are both scientifically valid and occupationally relevant. Only through this approach, along with close collaboration with the occupational group itself, can predictive tools be proactive and facilitative rather than judgmental and exclusionary.

ADVANCEMENT OF ENVIRONMENTAL PHYSIOLOGY AS A FIELD OF STUDY

One theme throughout this volume has been the interdisciplinary nature of research into physiology and performance in environmental extremes. The broad yet focused nature of this field emphasizes the need for people who can integrate knowledge from disparate fields and disciplines into a cohesive research and development program. Within the field of physiology itself, this requires integrating research approaches across the continuum from molecular and genetic research through to systems-based research on the entire body's response. I see undergraduate and graduate studies in kinesiology as the perfect entry and hub for such work, as students gain a strong foundation in basic and applied physiology, anatomy, and human factors. At the same time, significant contributions may come from cross-pollination and integration of knowledge encompassing fields as diverse as ergonomics, engineering, psychology, and health promotion. Therefore, significant contributions may also be achieved by students entering the field of environmental physiology from these disparate disciplines.

One challenge for the field of environmental physiology, however, is a lack of recognition on the part of industry and the general public of what kinesiologists and ergonomists are capable of. Unlike physiotherapy or occupational therapy, there is no distinct or defined job description for the field; students will enter highly diverse work, for example as personal trainers and coaches, as leaders of companies developing offshore survival systems, and as ergonomists designing cockpit and cabin instrumentation layouts. Yet this diversity is also a major strength for kinesiologists and ergonomists, as it enables a wide range of employment and opportunities. Therefore, one major advance for the field of environmental physiology in the future will likely stem from greater awareness of kinesiology and ergonomics as professions, along with the development of professional bodies and organizations.

Appendix

Resources for Further Information

The information contained within this text forms an initial foundation for understanding the disparate ideas and concepts within the large and burgeoning field of environmental physiology. Much more information on the effects of environmental stressors on exercise physiology and capacity can be obtained through a reading of textbooks focusing in detail on particular environments. At the same time, many scientific societies and government agencies perform research or advocacy work in enhancing safe exercise and work conditions across a range of environmental conditions. Listed below are further resources that may be of interest.

SCIENTIFIC SOCIETIES

The following is a listing of scientific societies with a significant or primary interest in environmental physiology.

- Aerospace Medical Association (AsMA): www.asma.org
- American College of Sports Medicine (ACSM): www.acsm.org
- American Physiological Society (APS): www.the-aps.org
- Canadian Society for Exercise Physiology (CSEP): www.csep.ca
- European College of Sport Science (ECSS): www.ecss.de
- International Society for Environmental Ergonomics (ICEE): www.environmental-ergonomics.org
- Undersea and Hyperbaric Medical Society (UHMS): www.uhms.org

AGENCIES

This is a listing of agencies that operate in extreme environments and also disseminate public information on human response in such environments.

- Canadian Space Agency (CSA): www.space.gc.ca
- Coast Guards: Canadian (www.ccg-gcc.gc.ca) and United States (www.uscg.mil)
- European Space Agency (ESA): www.esa.int
- Lifesaving Society of Canada: www.lifesaving.ca
- National Aeronautics and Space Agency (NASA): www.nasa.gov
- Red Cross Societies: Canadian (www.redcross.ca) and United States (www.redcross.org)
- The Royal Lifesaving Society UK: www.lifesavers.org.uk

TEXTBOOKS

This is a listing of major textbooks dealing with individual environments.

- Brubakk, A. and T. Neuman. 2003. *Bennett and Elliott's physiology and medicine of diving* (5th ed.). Amsterdam: Saunders.
- Buckey, J.C. 2006. *Space physiology.* New York: Oxford University Press.
- Burton, A.C. 1955. *Man in a cold environment: physiological and pathological effects of exposure to low temperatures.* Monographs of the Physiological Society. London: Arnold.
- Clément, G. 2005. *Fundamentals of space medicine.* Dordrecht, Netherlands: Springer.
- Hornbein, T. and R. Schoene. 2001. *High altitude: an exploration of human adaption.* London: Informa Healthcare.
- Nicogossian, A.E., C.L. Huntoon, and S.L. Pool, eds. 1994. *Space physiology and medicine* (3rd ed.). Philadelphia: Lea & Febiger.
- Parsons, K. 2003. *Human thermal environments: the effects of hot, moderate, and cold environments on human health, comfort and performance* (2nd ed.). London: Taylor & Francis.
- West, J.B., R. Schoene, and J. Milledge. 2007. *High altitude medicine and physiology* (4th ed.). Oxford: Oxford University Press.

OTHER BOOKS ON ENVIRONMENTAL PHYSIOLOGY

This is a listing of highly readable and practical books on environmental physiology aimed at the practitioner or the lay public.

- Armstrong, L.E. 2000. *Performing in extreme environments.* Champaign, IL: Human Kinetics.
- Armstrong, L.E. 2003. *Exertional heat illnesses.* Champaign, IL: Human Kinetics.
- Ashcroft, F. 2002. *Life at the extremes.* Berkeley: University of California Press.
- Golden, F. and Tipton, M.J. 2002. *Essentials of sea survival.* Champaign, IL: Human Kinetics.
- Wilber, R.L. 2003. *Altitude training and athletic performance.* Champaign, IL: Human Kinetics.

OTHER RESOURCES

This is a collection of other important references in environmental physiology.

- American College of Sports Medicine (ACSM) position stands (www.acsm-msse .org/pt/re/msse/positionstandards.htm). The largest and most influential scientific society for sports medicine and exercise science, ACSM publishes influential position stands on various topics, including (1) exertional heat illness during training and competition (2007), (2) exercise and fluid replacement (2007), and (3) prevention of cold injuries during exercise (2006).
- Environmental Protection Agency (EPA): www.epa.gov
- Environment Canada (www.ec.gc.ca)

- IUPS Thermal Commission. 2001. Glossary of terms for thermal physiology. *Japanese Journal of Physiology* 51: 245-280. Published by the Thermal Commission of the International Union of Physiological Societies, this is the current authoritative glossary of terms and definitions used within the field of temperature regulation research.

- National Athletic Trainers' Association (NATA: www.nata.org). This association of athletic trainers helps to regulate and promotes the industry. Its educational initiatives include publishing evidence-based information on a variety of topics involving exercise in different environments.

- National Oceanic and Atmospheric Administration's National Weather Service (www.nws.noaa.gov)

- World Anti-Doping Agency (WADA): www.wada-ama.org

References

Chapter 1

Anonymous. 2001. Glossary of terms for thermal physiology. *Japanese Journal of Physiology* 51: 245-280.

Enoka, R.M. and J. Duchateau. 2008. Muscle fatigue: what, why and how it influences muscle function. *Journal of Physiology* 586: 11-23.

Chapter 2

Bligh, J. 2006. A theoretical consideration of the means whereby the mammalian core temperature is defended at a null zone. *Journal of Applied Physiology* 100: 1332-1337.

Bligh, J. and K. Voigt. 1990. *Thermoreception and temperature regulation.* New York: Springer-Verlag.

Boulant, J.A. 2006. Neuronal basis of Hammel's model for set-point thermoregulation. *Journal of Applied Physiology* 100: 1347-1354.

Cabanac, M. 2006. Adjustable set point: to honor Harold T. Hammel. *Journal of Applied Physiology* 100: 1338-1346.

Cowell, S.A., J.M. Stocks, D.G. Evans, S.R. Simonson, and J.E. Greenleaf. 2002. The exercise and environmental physiology of extravehicular activity. *Aviation, Space, and Environmental Medicine* 73: 54-67.

Hammel, H.T., D.C. Jackson, J.A.J. Stolwijk, J.D. Hardy, and S.B. Stromme. 1963. Temperature regulation by hypothalamic proportional control with an adjustable set point. *Journal of Applied Physiology* 18: 1146-1154.

Hartung, G.H., L.G. Myhre, and S.A. Nunneley. 1980. Physiological effects of cold air inhalation during exercise. *Aviation, Space, and Environmental Medicine* 51: 591-594.

Havenith, G., I. Holmer, E.A. den Hartog, and K.C. Parsons. 1999. Clothing evaporative heat resistance—proposal for improved representation in standards and models. *Annals of Occupational Hygiene* 43: 339-346.

Kenny, G.P., P. Webb, M.B. Ducharme, F.D. Reardon, and O. Jay. 2008. Calorimetric measurement of postexercise net heat loss and residual body heat storage. *Medicine and Science in Sports and Exercise* 40: 1629-1636.

Mekjavic, I.B. and O. Eiken. 2006. Contribution of thermal and nonthermal factors to the regulation of body temperature in humans. *Journal of Applied Physiology* 100: 2065-2072.

Moran, D.S., J.W. Castellani, C. O'Brien, A.J. Young, and K.B. Pandolf. 1999. Evaluating physiological strain during cold exposure using a new cold strain index. *American Journal of Physiology* 277: R556-R564.

Moran, D.S., S.J. Montain, and K.B. Pandolf. 1998a. Evaluation of different levels of hydration using a new physiological strain index. *American Journal of Physiology* 44: R854-R860.

Moran, D.S., K.B. Pandolf, A. Laor, Y. Heled, W.T. Matthew, and R.R. Gonzalez. 2003. Evaluation and refinement of the environmental stress index for different climatic conditions. *Journal of Basic and Clinical Physiology and Pharmacology* 14: 1-15.

Moran, D.S., A. Shitzer, and K.B. Pandolf. 1998b. A physiological strain index to evaluate heat stress. *American Journal of Physiology* 44: R129-R134.

Nunneley, S.A. 1989. Heat stress in protective clothing: interactions among physical and physiological factors. *Scandinavian Journal of Work and Environmental Health* 15 (Suppl. 1): 52-57.

Nunneley, S.A. and R.F. Stribley. 1979. Fighter index of thermal stress (FITS): guidance for hot-weather aircraft operations. *Aviation, Space, and Environmental Medicine* 50: 639-642.

Parsons, K.C. 1993. *Human thermal environments.* London: Taylor & Francis.

Reardon, F.D., K.E. Leppik, R. Wegmann, P. Webb, M.B. Ducharme, and G.P. Kenny. 2006. The Snellen human calorimeter revisited, re-engineered and upgraded: design and performance characteristics. *Medical and Biological Engineering and Computing* 44: 721-728.

Siple, P.A. and C.F. Passel. 1999. Excerpts from: measurements of dry atmospheric cooling in subfreezing temperatures. 1945. *Wilderness and Environmental Medicine* 10: 176-182.

Sullivan, P.J., I.B. Mekjavic, and N. Kakitsuba. 1987. Determination of clothing microenvironment volume. *Ergonomics* 30: 1043-1052.

Taylor, N.A., N.K. Allsopp, and D.G. Parkes. 1995. Preferred room temperature of young vs aged males: the influence of thermal sensation, thermal comfort, and affect. *Journal of Gerontology A: Biological Science and Medical Science* 50: M216-221.

Tikuisis, P., M.B. Ducharme, and D. Brajkovic. 2007. Prediction of facial cooling while walking in cold wind. *Computers in Biology and Medicine* 37: 1225-1231.

Webb, P. 1995. The physiology of heat regulation. *American Journal of Physiology* 37: R838-R850.

Wright, H.E. and S.S. Cheung. 2006. Cranial-neck and inhalation rewarming failed to improve recovery from mild hypothermia. *Aviation, Space, and Environmental Medicine* 77: 398-403.

Yaglou, C.P. and D. Minard. 1957. Control of heat casualties at military training centers. *A.M.A. Archives of Industrial Health* 16: 302-316.

Chapter 3

Adolph, E.F. 1947. *Physiology of man in the desert.* New York: Interscience.

Armstrong, L.E., D.J. Casa, M. Millard-Stafford, D.S. Moran, S.W. Pyne, and W.O. Roberts. 2007. American College of Sports Medicine position stand. Exertional heat illness during training and competition. *Medicine and Science in Sports and Exercise* 39: 556-572.

Arngrimsson, S.A., D.S. Petitt, M.G. Stueck, D.K. Jorgensen, and K.J. Cureton. 2004. Cooling vest worn during active warm-up improves 5-km run performance in the heat. *Journal of Applied Physiology* 96: 1867-1874.

Ball, D., C. Burrows, and A.J. Sargeant. 1999. Human power output during repeated sprint cycle exercise: the influence of thermal stress. *European Journal of Applied Physiology* 79: 360-366.

Bar-Or, O. and B. Wilk. 1996. Water and electrolyte replenishment in the exercising child. *International Journal of Sport Nutrition* 6: 93-99.

Booth, J., F. Marino, and J.J. Ward. 1997. Improved running performance in hot humid conditions following whole body precooling. *Medicine and Science in Sports and Exercise* 29: 943-949.

Bouchama, A. 2004. The 2003 European heat wave. *Intensive Care Medicine* 30: 1-3.

Candas, V. and A. Dufour. 2005. Thermal comfort: multisensory interactions? *Journal of Physiological Anthropology and Applied Human Science* 24: 33-36.

Casa, D.J., L.E. Armstrong, S.K. Hillman, S.J. Montain, R.V. Reiff, B.S. Rich, W.O. Roberts, and J.A. Stone. 2000. National Athletic Trainers' Association position statement: fluid replacement for athletes. *Journal of Athletic Training* 35: 212-224.

Cheung, S.S. and T.M. McLellan. 1998. Influence of heat acclimation, aerobic fitness, and hydration effects on tolerance during uncompensable heat stress. *Journal of Applied Physiology* 84: 1731-1739.

Cheung, S.S., T.M. McLellan, and S. Tenaglia. 2000. The thermophysiology of uncompensable heat stress. Physiological manipulations and individual characteristics. *Sports Medicine* 29: 329-359.

Cheung, S.S. and A.M. Robinson. 2004. The influence of upper-body pre-cooling on repeated sprint performance in moderate ambient temperatures. *Journal of Sports Sciences* 22: 605-612.

Cheung, S.S. and G.G. Sleivert. 2004a. Lowering of skin temperature decreases isokinetic

maximal force production independent of core temperature. *European Journal of Applied Physiology* 91: 723-728.

Cheung, S.S. and G.G. Sleivert. 2004b. Multiple triggers for hyperthermic fatigue and exhaustion. *Exercise and Sport Sciences Reviews* 32: 100-106.

Cotter, J.D. and N.A.S. Taylor. 2005. The distribution of cutaneous sudomotor and alliesthesial thermosensitivity in mildly heat-stressed humans: an open-loop approach. *Journal of Physiology* 565: 335-345.

Davis, J.M. and S.P. Bailey. 1997. Possible mechanisms of central nervous system fatigue during exercise. *Medicine and Science in Sports and Exercise* 29: 45-57.

Dawson, B.T. 1994. Exercise training in sweat clothing in cool conditions to improve heat tolerance. *Sports Medicine* 17: 233-244.

Dennis, S.C. and T.D. Noakes. 1999. Advantages of a smaller bodymass in humans when distance-running in warm, humid conditions. *European Journal of Applied Physiology* 79: 280-284.

Dokladny, K., P.L. Moseley, and T.Y. Ma. 2006. Physiologically relevant increase in temperature causes an increase in intestinal epithelial tight junction permeability. *American Journal of Physiology* 290: G204-212.

Duffield, R., B. Dawson, D. Bishop, M. Fitzsimons, and S. Lawrence. 2003. Effect of wearing an ice cooling jacket on repeat sprint performance in warm/humid conditions. *British Journal of Sports Medicine* 37: 164-169.

Falk, B., O. Bar-Or, R. Calvert, and J.D. MacDougall. 1992. Sweat gland response to exercise in the heat among pre-, mid-, and late-pubertal boys. *Medicine and Science in Sports and Exercise* 24: 313-319.

Falk, B. and R. Dotan. 2008. Children's thermoregulation during exercise in the heat: a revisit. *Applied Physiology, Nutrition, and Metabolism* 33: 420-427.

Flouris, A.D. and S.S. Cheung. 2006. Design and control optimization of microclimate liquid cooling systems underneath protective clothing. *Annals of Biomedical Engineering* 34: 359-372.

Fox, R.H., R. Goldsmith, I.F.G. Hampton, and T.J. Hunt. 1967. Heat acclimatization by controlled hyperthermia in hot-dry and hot-wet climates. *Journal of Applied Physiology* 22: 39-46.

Frank, S.M., S.N. Raja, C.F. Bulcao, and D.S. Goldstein. 1999. Relative contribution of core and cutaneous temperatures to thermal comfort and autonomic responses in humans. *Journal of Applied Physiology* 86: 1588-1593.

Fuller, A., R.N. Carter, and D. Mitchell. 1998. Brain and abdominal temperatures at fatigue in rats exercising in the heat. *Journal of Applied Physiology* 84: 877-883.

Gagnon, D., O. Jay, F.D. Reardon, W.S. Journeay, and G.P. Kenny. 2008. Hyperthermia modifies the nonthermal contribution to postexercise heat loss responses. *Medicine and Science in Sports and Exercise* 40: 513-522.

Galloway, S.D. and R.J. Maughan. 1997. Effects of ambient temperature on the capacity to perform prolonged cycle exercise in man. *Medicine and Science in Sports and Exercise* 29: 1240-1249.

Gonzalez-Alonso, J., C. Teller, S.L. Andersen, F.B. Jensen, T. Hyldig, and B. Nielsen. 1999. Influence of body temperature on the development of fatigue during prolonged exercise in the heat. *Journal of Applied Physiology* 86: 1032-1039.

Grahn, D., J.G. Brock-Utne, D.E. Watenpaugh, and H.C. Heller. 1998. Recovery from mild hyperthermia can be accelerated by mechanically distending blood vessels in the hand. *Journal of Applied Physiology* 85: 1643-1648.

Griefahn, B. 1997. Acclimation to three different hot climates with equivalent globe temperatures. *Ergonomics* 40: 223-234.

Hocking, C., R.B. Silberstein, W.M. Lau, C. Stough, and W. Roberts. 2001. Evaluation of cognitive performance in the heat by functional brain imaging and psychometric testing. *Comparative Biochemistry and Physiology Part A: Molecular and Integrative Physiology* 128: 719-734.

Jay, O., L.M. Gariepy, F.D. Reardon, P. Webb, M.B. Ducharme, T. Ramsay, and G.P. Kenny. 2007. A three-compartment thermometry model for the improved estimation of changes in body heat content. *American Journal of Physiology* 292: R167-175.

Kenefick, R.W., S.N. Cheuvront, and M.N. Sawka. 2007. Thermoregulatory function during the marathon. *Sports Medicine* 37: 312-315.

Kenny, G.P., O. Jay, and W.S. Journeay. 2007. Disturbance of thermal homeostasis following dynamic exercise. *Applied Physiology, Nutrition, and Metabolism* 32: 818-831.

Khomenok, G.A., A. Hadid, O. Preiss-Bloom, R. Yanovich, T. Erlich, O. Ron-Tal, A. Peled, Y. Epstein, and D.S. Moran. 2008. Hand immersion in cold water alleviating physiological strain and increasing tolerance to uncompensable heat stress. *European Journal of Applied Physiology* 104: 303-309.

Kregel, K.C. 2002. Heat shock proteins: modifying factors in physiological stress responses and acquired thermotolerance. *Journal of Applied Physiology* 92: 2177-2186.

Lambert, G.P. 2004. Role of gastrointestinal permeability in exertional heatstroke. *Exercise and Sport Sciences Reviews* 32: 185-190.

Marino, F.E. 2002. Methods, advantages, and limitations of body cooling for exercise performance. *British Journal of Sports Medicine* 36: 89-94.

Marino, F.E., M.I. Lambert, and T.D. Noakes. 2004. Superior performance of African runners in warm humid but not in cool environmental conditions. *Journal of Applied Physiology* 96: 124-130.

Meeusen, R., P. Watson, H. Hasegawa, B. Roelands, and M.F. Piacentini. 2006. Central fatigue: the serotonin hypothesis and beyond. *Sports Medicine* 36: 881-909.

Montain, S.J., M.N. Sawka, B.S. Cadarette, M.D. Quigley, and J.M. McKay. 1994. Physiological tolerance to uncompensable heat stress: effects of exercise intensity, protective clothing, and climate. *Journal of Applied Physiology* 77: 216-222.

Morrison, S., G.G. Sleivert, and S.S. Cheung. 2004. Passive hyperthermia reduces voluntary activation and isometric force production. *European Journal of Applied Physiology* 91: 729-736.

Naito, H., S.K. Powers, H.A. Demirel, T. Sugiura, S.L. Dodd, and J. Aoki. 2000. Heat stress attenuates skeletal muscle atrophy in hindlimb-unweighted rats. *Journal of Applied Physiology* 88: 359-363.

Nakamura, M., H. Esaki, T. Yoda, S. Yasuhara, A. Kobayashi, A. Konishi, N. Osawa, K. Nagashima, L.I. Crawshaw, and K. Kanosue. 2006. A new system for the analysis of thermal judgments: multipoint measurements of skin temperatures and temperature-related sensations and their joint visualization. *Journal of Physiological Sciences* 56: 459-464.

Nielsen, B., T. Hyldig, F. Bidstrup, J. Gonzalez-Alonso, and G.R. Christoffersen. 2001. Brain activity and fatigue during prolonged exercise in the heat. *Pflugers Archive European Journal of Physiology* 442: 41-48.

Nunneley, S.A. 1989. Heat stress in protective clothing: interactions among physical and physiological factors. *Scandinavian Journal of Work and Environmental Health* 15 (Suppl. 1): 52-57.

Nunneley, S.A., C.C. Martin, J.W. Slauson, C.M. Hearon, L.D. Nickerson, and P.A. Mason. 2002. Changes in regional cerebral metabolism during systemic hyperthermia in humans. *Journal of Applied Physiology* 92: 846-851.

Nybo, L. 2007. Exercise and heat stress: cerebral challenges and consequences. *Progress in Brain Research* 162: 29-43.

Nybo, L., B. Nielsen, E. Blomstrand, K. Moller, and N. Secher. 2003. Neurohumoral responses during prolonged exercise in humans. *Journal of Applied Physiology* 95: 1125-1131.

Pandolf, K.B. 1997. Aging and human heat tolerance. *Experimental Aging Research* 23: 69-105.

Pandolf, K.B., R.L. Burse, and R.F. Goldman. 1977. Role of physical fitness in heat acclimatization, decay and reinduction. *Ergonomics* 20: 399-408.

Prosser, C., K. Stelwagen, R. Cummins, P. Guerin, N. Gill, and C. Milne. 2004. Reduction in heat-induced gastrointestinal hyperpermeability in rats by bovine colostrum and goat milk powders. *Journal of Applied Physiology* 96: 650-654.

Proulx, C.I., M.B. Ducharme, and G.P. Kenny. 2003. Effect of water temperature on cooling efficiency during hyperthermia in humans. *Journal of Applied Physiology* 94: 1317-1323.

Roberts, W.O. 2006. Exertional heat stroke during a cool weather marathon: a case study. *Medicine and Science in Sports and Exercise* 38: 1197-1203.

Saunders, A.G., J.P. Dugas, R. Tucker, M.I. Lambert, and T.D. Noakes. 2005. The effects of different air velocities on heat storage and body temperature in humans cycling in a hot, humid environment. *Acta Physiologica Scandinavica* 183: 241-255.

Sawka, M.N., A.J. Young, K.B. Pandolf, R.C. Dennis, and C.R. Valeri. 1992. Erythrocyte, plasma, and blood volume of healthy young men. *Medicine and Science in Sports and Exercise* 24: 447-453.

Selkirk, G.A. and T.M. McLellan. 2001. Influence of aerobic fitness and body fatness on tolerance to uncompensable heat stress. *Journal of Applied Physiology* 91: 2055-2063.

Selkirk, G.A., T.M. McLellan, and J. Wong. 2004. Active versus passive cooling during work in warm environments while wearing firefighting protective clothing. *Journal of Occupational and Environmental Hygiene* 1: 521-531.

Stephenson, L.A., C.R. Vernieuw, W. Leammukda, and M.A. Kolka. 2007. Skin temperature feedback optimizes microclimate cooling. *Aviation, Space, and Environmental Medicine* 78: 377-382.

Supinski, G., D. Nethery, T.M. Nosek, L.A. Callahan, D. Stofan, and A. DiMarco. 2000. Endotoxin administration alters the force vs. pCa relationship of skeletal muscle fibers. *American Journal of Physiology* 278: R891-896.

Taylor, N.A., J.N. Caldwell, A.M. Van Den Heuvel, and M.J. Patterson. 2008. To cool, but not too cool: that is the question—immersion cooling for hyperthermia. *Medicine and Science in Sports and Exercise* 40: 1962-1969.

Thomas, M.M., S.S. Cheung, G.C. Elder, and G.G. Sleivert. 2006. Voluntary muscle activation is impaired by core temperature rather than local muscle temperature. *Journal of Applied Physiology* 100: 1361-1369.

Tikuisis, P., T.M. McLellan, and G. Selkirk. 2002. Perceptual versus physiological heat strain during exercise-heat stress. *Medicine and Science in Sports and Exercise* 34: 1454-1461.

Tucker, R., T. Marle, E.V. Lambert, and T.D. Noakes. 2006. The rate of heat storage mediates an anticipatory reduction in exercise intensity during cycling at a fixed rating of perceived exertion. *Journal of Physiology (London)* 574: 905-915.

Walters, T.J., K.L. Ryan, L.M. Tate, and P.A. Mason. 2000. Exercise in the heat is limited by a critical internal temperature. *Journal of Applied Physiology* 89: 799-806.

Yip, F.Y., W.D. Flanders, A. Wolkin, D. Engelthaler, W. Humble, A. Neri, L. Lewis, L. Backer, and C. Rubin. 2008. The impact of excess heat events in Maricopa County, Arizona: 2000-2005. *International Journal of Biometeorology* 52: 765-772.

Chapter 4

Armstrong, L.E., R.W. Hubbard, B.H. Jones, and J.T. Daniels. 1986. Preparing Alberto Salazar for the heat of the 1984 Olympic marathon. *Physician and Sportsmedicine* 14: 73-81.

Boulze, D., P. Montastruc, and M. Cabanac. 1983. Water intake, pleasure and water temperature in humans. *Physiology and Behavior* 30: 97-102.

Caldwell, J.E., E. Ahonen, and U. Nousiainen. 1984. Differential effects of sauna-, diuretic-, and exercise-induced hypohydration. *Journal of Applied Physiology* 57: 1018-1023.

Carter, J.M., A.E. Jeukendrup, and D.A. Jones. 2004. The effect of carbohydrate mouth rinse on 1-h cycle time trial performance. *Medicine and Science in Sports and Exercise* 36: 2107-2111.

Casa, D.J., L.E. Armstrong, S.K. Hillman, S.J. Montain, R.V. Reiff, B.S. Rich, W.O. Roberts, and J.A. Stone. 2000. National Athletic Trainers' Association position statement: fluid replacement for athletes. *Journal of Athletic Training* 35: 212-224.

Casa, D.J., P.M. Clarkson, and W.O. Roberts. 2005. American College of Sports Medicine roundtable on hydration and physical activity: consensus statements. *Current Sports Medicine Reports* 4: 115-127.

Cheung, S.S. and T.M. McLellan. 1998. Influence of hydration status and fluid replacement on tolerance during uncompensable heat stress. *European Journal of Applied Physiology* 77: 139-148.

Cheung, S.S., T.M. McLellan, and S. Tenaglia. 2000. The thermophysiology of uncompensable heat stress. Physiological manipulations and individual characteristics. *Sports Medicine* 29: 329-359.

Cheuvront, S.N., R. Carter 3rd, J.W. Castellani, and M.N. Sawka. 2005. Hypohydration impairs endurance exercise performance in temperate but not cold air. *Journal of Applied Physiology* 99: 1972-1976.

Convertino, V.A., L.E. Armstrong, E.F. Coyle, G.W. Mack, M.N. Sawka, L.C. Senay, and W.M. Sherman. 1996. American College of Sports Medicine position stand. Exercise and fluid replacement. *Medicine and Science in Sports and Exercise* 28: i-vii.

Costill, D.L., R. Cote, and W. Fink. 1976. Muscle water and electrolytes following varied levels of dehydration in man. *Journal of Applied Physiology* 40: 6-11.

Eaton, D. and J. Pooler. 2009. *Vander's renal physiology.* New York: McGraw-Hill Medical.

Houmard, J.A., P.C. Egan, R.A. Johns, P.D. Neufer, T.C. Chenier, and R.G. Israel. 1991. Gastric emptying during 1 h of cycling and running at 75% VO_{2max}. *Medicine and Science in Sports and Exercise* 23: 320-325.

Hubbard, R.W., B.L. Sandick, W.T. Matthew, R.P. Francesconi, J.B. Sampson, M.J. Durkot,

O. Maller, and D.B. Engell. 1984. Voluntary dehydration and alliesthesia for water. *Journal of Applied Physiology* 57: 858-875.

Jacobs, I. 1980. The effects of thermal dehydration on performance of the Wingate anaerobic test. *International Journal of Sports Medicine* 1: 21-24.

Jeukendrup, A.E., S. Hopkins, L.F. Aragon-Vargas, and C. Hulston. 2008. No effect of carbohydrate feeding on 16 km cycling time trial performance. *European Journal of Applied Physiology* 104: 831-837.

Judelson, D.A., C.M. Maresh, J.M. Anderson, L.E. Armstrong, D.J. Casa, W.J. Kraemer, and J.S. Volek. 2007. Hydration and muscular performance: does fluid balance affect strength, power and high-intensity endurance? *Sports Medicine* 37: 907-921.

Lambert, G.P. 2004. Role of gastrointestinal permeability in exertional heatstroke. *Exercise and Sport Sciences Reviews* 32: 185-190.

Latzka, W.A., M.N. Sawka, S.J. Montain, G.S. Skrinar, R.A. Fielding, R.P. Matott, and K.B. Pandolf. 1997. Hyperhydration: thermoregulatory effects during compensable exercise-heat stress. *Journal of Applied Physiology* 83: 860-866.

Latzka, W.A., M.N. Sawka, S.J. Montain, G.S. Skrinar, R.A. Fielding, R.P. Matott, and K.B. Pandolf. 1998. Hyperhydration: tolerance and cardiovascular effects during uncompensable exercise-heat stress. *Journal of Applied Physiology* 84: 1858-1864.

Lee, J.K., S.M. Shirreffs, and R.J. Maughan. 2008. Cold drink ingestion improves exercise endurance capacity in the heat. *Medicine and Science in Sports and Exercise* 40: 1637-1644.

Maughan, R.J., J.B. Leiper, and S.M. Shirreffs. 1996. Restoration of fluid balance after exercise-induced dehydration: effects of food and fluid intake. *European Journal of Applied Physiology* 73: 317-325.

Maughan, R.J., J.H. Owen, S.M. Shirreffs, and J.B. Leiper. 1994. Post-exercise rehydration in man: effects of electrolyte addition to ingested fluids. *European Journal of Applied Physiology* 69: 209-215.

Maughan, R.J. and S.M. Shirreffs. 2008. Development of individual hydration strategies for athletes. *International Journal of Sport Nutrition and Exercise Metabolism* 18: 457-472.

Maughan, R.J., S.M. Shirreffs, and J.B. Leiper. 2007. Errors in the estimation of hydration status from changes in body mass. *Journal of Sports Sciences* 25: 797-804.

Mitchell, J.B., P.W. Grandjean, F.X. Pizza, R.D. Starling, and R.W. Holtz. 1994. The effect of volume ingested on rehydration and gastric emptying following exercise-induced dehydration. *Medicine and Science in Sports and Exercise* 26: 1135-1143.

Montain, S.J. and E.F. Coyle. 1992. Influence of graded dehydration on hyperthermia and cardiovascular drift during exercise. *Journal of Applied Physiology* 73: 1340-1350.

Montain, S.J. and E.F. Coyle. 1993. Influence of the timing of fluid ingestion on temperature regulation during exercise. *Journal of Applied Physiology* 75: 688-695.

Murray, R., G.L. Paul, J.G. Seifert, and D.E. Eddy. 1991. Responses to varying rates of carbohydrate ingestion during exercise. *Medicine and Science in Sports and Exercise* 23: 713-718.

Nelson, J.L. and R.A. Roberts. 2007. Exploring the potential ergogenic effects of glycerol hyperhydration. *Sports Medicine* 37: 981-1000.

Neufer, P.D., M.N. Sawka, A.J. Young, M.D. Quigley, W.A. Latzka, and L. Levine. 1991. Hypohydration does not impair skeletal muscle glycogen resynthesis after exercise. *Journal of Applied Physiology* 70: 1490-1494.

Neufer, P.D., A.J. Young, and M.N. Sawka. 1989. Gastric emptying during exercise: effects of heat stress and hypohydration. *European Journal of Applied Physiology* 58: 433-439.

Noakes, T.D. 2007. Drinking guidelines for exercise: what evidence is there that athletes should drink "as much as tolerable", "to replace the weight lost during exercise" or "ad libitum"? *Journal of Sports Sciences* 25: 781-796.

Oliver, S.J., S.J. Laing, S. Wilson, J.L. Bilzon, and N. Walsh. 2007. Endurance running performance after 48 h of restricted fluid and/or energy intake. *Medicine and Science in Sports and Exercise* 39: 316-322.

Rowell, L.B. 1974. Human cardiovascular adjustments to exercise and thermal stress. *Physiological Reviews* 54: 75-159.

Saunders, A.G., J.P. Dugas, R. Tucker, M.I. Lambert, and T.D. Noakes. 2005. The effects of different air velocities on heat storage and body temperature in humans cycling in a hot, humid environment. *Acta Physiologica Scandinavica* 183: 241-255.

Saunders, M.J., M.D. Kane, and M.K. Todd. 2004. Effects of a carbohydrate-protein beverage on cycling endurance and muscle damage. *Medicine and Science in Sports and Exercise* 36: 1233-1238.

Sawka, M.N., L.M. Burke, E.R. Eichner, R.J. Maughan, S.J. Montain, and N.S. Stachenfeld. 2007. American College of Sports Medicine position stand. Exercise and fluid replacement. *Medicine and Science in Sports and Exercise* 39: 377-390.

Sawka, M.N., S.J. Montain, and W.A. Latzka. 2001. Hydration effects on thermoregulation and performance in the heat. *Comparative Biochemistry and Physiology Part A: Molecular and Integrative Physiology* 128: 679-690.

Sawka, M.N., A.J. Young, R.P. Francesconi, S.R. Muza, and K.B. Pandolf. 1985. Thermoregulatory and blood responses during exercise at graded hypohydration levels. *Journal of Applied Physiology* 59: 1394-1401.

Sawka, M.N., A.J. Young, K.B. Pandolf, R.C. Dennis, and C.R. Valeri. 1992. Erythrocyte, plasma, and blood volume of healthy young men. *Medicine and Science in Sports and Exercise* 24: 447-453.

Shirreffs, S.M. and R.J. Maughan. 1997. Restoration of fluid balance after exercise-induced dehydration: effects of alcohol consumption. *Journal of Applied Physiology* 83: 1152-1158.

Shirreffs, S.M. and R.J. Maughan. 1998. Volume repletion after exercise-induced volume depletion in humans: replacement of water and sodium losses. *American Journal of Physiology* 274: F868-875.

Shirreffs, S.M., A.J. Taylor, J.B. Leiper, and R.J. Maughan. 1996. Post-exercise rehydration in man: effects of volume consumed and drink sodium content. *Medicine and Science in Sports and Exercise* 28: 1260-1271.

Slater, G.J., A.J. Rice, K. Sharpe, R. Tanner, D. Jenkins, C.J. Gore, and A.G. Hahn. 2005. Impact of acute weight loss and/or thermal stress on rowing ergometer performance. *Medicine and Science in Sports and Exercise* 37: 1387-1394.

Turlejska, E. and M.A. Baker. 1986. Elevated CSF osmolality inhibits thermoregulatory heat loss responses. *American Journal of Physiology* 251: R749-754.

van Essen, M. and M.J. Gibala. 2006. Failure of protein to improve time trial performance when added to a sports drink. *Medicine and Science in Sports and Exercise* 38: 1476-1483.

Watt, M.J., M.A. Febbraio, A.P. Garnham, and M. Hargreaves. 1999. Acute plasma volume expansion: effect on metabolism during submaximal exercise. *Journal of Applied Physiology* 87: 1202-1206.

Watt, M.J., A.P. Garnham, M.A. Febbraio, and M. Hargreaves. 2000. Effect of acute plasma volume expansion on thermoregulation and exercise performance in the heat. *Medicine and Science in Sports and Exercise* 32: 958-962.

Webster, S., R. Rutt, and A. Weltman. 1990. Physiological effects of a weight loss regimen practiced by college wrestlers. *Medicine and Science in Sports and Exercise* 22: 229-234.

Wilk, B. and O. Bar-Or. 1996. Effect of drink flavor and NaCl on voluntary drinking and hydration in boys exercising in the heat. *Journal of Applied Physiology* 80: 1112-1117.

Chapter 5

Adams, T. and R.E. Smith. 1962. Effect of chronic local cold exposure on finger temperature responses. *Journal of Applied Physiology* 17: 317-322.

Bell, D.G., P. Tikuisis, and I. Jacobs. 1992. Relative intensity of muscular contraction during shivering. *Journal of Applied Physiology* 72: 2336-2342.

Brajkovic, D. and M.B. Ducharme. 2003. Finger dexterity, skin temperature, and blood flow during auxiliary heating in the cold. *Journal of Applied Physiology* 95: 758-770.

Burton, A.C. and O.G. Edholm. 1955. *Man in a cold environment: physiological and pathological effects of exposure to low temperatures.* Monographs of the Physiological Society. London: Arnold.

Cain, J.B., S.D. Livingstone, R.W. Nolan, and A.A. Keefe. 1990. Respiratory heat loss during work at various ambient temperatures. *Respiration Physiology* 79: 145-150.

Castellani, J.W., A.J. Young, M.B. Ducharme, G.G. Giesbrecht, E. Glickman, and R.E. Sallis. 2006. American College of Sports Medicine position stand: prevention of cold injuries during exercise. *Medicine and Science in Sports and Exercise* 38: 2012-2029.

Castellani, J.W., A.J. Young, M.N. Sawka, and K.B. Pandolf. 1998. Human thermoregulatory responses during serial cold-water immersions. *Journal of Applied Physiology* 85: 204-209.

Costford, S., A. Gowing, and M.E. Harper. 2007. Mitochondrial uncoupling as a target in the treatment of obesity. *Current Opinion in Clinical Nutrition and Metabolic Care* 10: 671-678.

Daanen, H.A. 2003. Finger cold-induced vasodilation: a review. *European Journal of Applied Physiology* 89: 411-426.

Daanen, H.A. and M.B. Ducharme. 2000. Axon reflexes in human cold exposed fingers. *European Journal of Applied Physiology* 81: 240-244.

Daanen, H.A. and N.R. van der Struijs. 2005. Resistance Index of Frostbite as a predictor of cold injury in arctic operations. *Aviation, Space, and Environmental Medicine* 76: 1119-1122.

Daanen, H.A. and H.J. van Ruiten. 2000. Cold-induced peripheral vasodilation at high altitudes—a field study. *High Altitude Medicine and Biology* 1: 323-329.

Dulloo, A.G. and S. Samec. 2001. Uncoupling proteins: their roles in adaptive thermogenesis and substrate metabolism reconsidered. *British Journal of Nutrition* 86: 123-139.

Flouris, A.D., S.S. Cheung, J.R. Fowles, L.D. Kruisselbrink, D.A. Westwood, A.E. Carrillo, and R.J. Murphy. 2006. Influence of body heat content on hand function during prolonged cold exposures. *Journal of Applied Physiology* 101: 802-808.

Fox, R.H. and H.T. Wyatt. 1962. Cold-induced vasodilatation in various areas of the body surface of man. *Journal of Physiology* 162: 289-297.

Geurts, C.L., G.G. Sleivert, and S.S. Cheung. 2006. Central and peripheral factors in thermal, neuromuscular, and perceptual adaptation of the hand to repeated cold exposures. *Applied Physiology, Nutrition, and Metabolism* 31: 110-117.

Haman, F. 2006. Shivering in the cold: from mechanisms of fuel selection to survival. *Journal of Applied Physiology* 100: 1702-1708.

Hamlet, M.P. 1988. Human cold injuries. In *Human performance physiology and environmental medicine in terrestrial extremes*, ed. Pandolf, K.B., M.N. Sawka, and R.R. Gonzalez, 435-466. Indianapolis: Benchmark Press.

Hartung, G.H., L.G. Myhre, and S.A. Nunneley. 1980. Physiological effects of cold air inhalation during exercise. *Aviation, Space, and Environmental Medicine* 51: 591-594.

Jobe, J.B., R.F. Goldman, and W.P. Beetham Jr. 1985. Comparison of the hunting reaction in normals and individuals with Raynaud's disease. *Aviation, Space, and Environmental Medicine* 56: 568-571.

Kotaru, C., R.B. Hejal, J.H. Finigan, A.J. Coreno, M.E. Skowronski, L. Brianas, and E.R. McFadden Jr. 2003. Desiccation and hypertonicity of the airway surface fluid and thermally induced asthma. *Journal of Applied Physiology* 94: 227-233.

Krog, J., B. Folkow, R.H. Fox, and K.L. Anderson. 1960. Hand circulation in the cold of Lapps and North Norwegian fisherman. *Journal of Applied Physiology* 15: 654-658.

Layden, J.D., D. Malkova, and M.A. Nimmo. 2004a. During exercise in the cold increased availability of plasma nonesterified fatty acids does not affect the pattern of substrate oxidation. *Metabolism: Clinical and Experimental* 53: 203-208.

Layden, J.D., D. Malkova, and M.A. Nimmo. 2004b. Fat oxidation after acipimox-induced reduction in plasma nonesterified fatty acids during exercise at 0 degrees C and 20 degrees C. *Metabolism: Clinical and Experimental* 53: 1131-1135.

Lewis, T. 1930. Observations upon the reactions of the vessels of the human skin to cold. *Heart* 15: 177-208.

Livingstone, S.D. 1976. Changes in cold-induced vasodilation during Arctic exercises. *Journal of Applied Physiology* 40: 455-457.

Makinen, T.M., M. Mantysaari, T. Paakkonen, J. Jokelainen, L.A. Palinkas, J. Hassi, J. Leppaluoto, K. Tahvanainen, and H. Rintamaki. 2008. Autonomic nervous function during whole-body cold exposure before and after cold acclimation. *Aviation, Space, and Environmental Medicine* 79: 875-882.

Mekjavic, I.B., U. Dobnikar, S.N. Kounalakis, B. Musizza, and S.S. Cheung. 2008. The trainability and contralateral responses of cold-induced vasodilation in the fingers following repeated cold exposure. *European Journal of Applied Physiology* 104: 193-199.

Nelms, J.D. and D.J. Soper. 1962. Cold vasodilation and cold acclimatization in the hands of British fish filleters. *Journal of Applied Physiology* 17: 444-448.

Passias, T.C., G.S. Meneilly, and I.B. Mekjavic. 1996. Effect of hypoglycemia on thermoregulatory responses. *Journal of Applied Physiology* 80: 1021-1032.

Purkayastha, S.S., G. Ilavazhagan, U.S. Ray, and W. Selvamurthy. 1993. Responses of Arctic and tropical men to a standard cold test and peripheral vascular responses to local cold stress in the Arctic. *Aviation, Space, and Environmental Medicine* 64: 1113.

Purkayastha, S.S., W. Selvamurthy, and G. Ilavazhagan. 1992. Peripheral vascular response to local cold stress of tropical men during sojourn in the Arctic cold region. *Japanese Journal of Physiology* 42: 877-889.

Reynolds, L.F., I.B. Mekjavic, and S.S. Cheung. 2007. Cold-induced vasodilatation in the foot is not homogenous or trainable over repeated cold exposure. *European Journal of Applied Physiology* 102: 73-78.

Savourey, G., A.L. Vallerand, and J.H.M. Bittel. 1992. General and local cold adaptation after a ski journey in a severe arctic environment. *European Journal of Applied Physiology* 64: 99-105.

Scholander, P.F., H.T. Hammel, J.S. Hart, D.H. Lemessurier, and J. Steen. 1958. Cold adaptation in Australian aborigines. *Journal of Applied Physiology* 13: 211-218.

Shave, R., E. Dawson, G. Whyte, K. George, M. Nimmo, J. Layden, P. Collinson, and D. Gaze. 2004. The impact of prolonged exercise in a cold environment upon cardiac function. *Medicine and Science in Sports and Exercise* 36: 1522-1527.

Stensrud, T., S. Berntsen, and K.H. Carlsen. 2006. Humidity influences exercise capacity in subjects with exercise-induced bronchoconstriction (EIB). *Respiratory Medicine* 100: 1633-1641.

Stensrud, T., S. Berntsen, and K.H. Carlsen. 2007a. Exercise capacity and exercise-induced bronchoconstriction (EIB) in a cold environment. *Respiratory Medicine* 101: 1529-1536.

Stensrud, T., K.V. Mykland, K. Gabrielsen, and K.H. Carlsen. 2007b. Bronchial hyperresponsiveness in skiers: field test versus methacholine provocation? *Medicine and Science in Sports and Exercise* 39: 1681-1686.

Ternesten-Hasseus, E., E.L. Johansson, M. Bende, and E. Millqvist. 2008. Dyspnea from exercise in cold air is not always asthma. *Journal of Asthma* 45: 705-709.

Tikuisis, P., D.G. Bell, and I. Jacobs. 1991. Shivering onset, metabolic response, and convective heat transfer during cold air exposure. *Journal of Applied Physiology* 70: 1996-2002.

Tikuisis, P., M.B. Ducharme, D. Moroz, and I. Jacobs. 1999. Physiological responses of exercised-fatigued individuals exposed to wet-cold conditions. *Journal of Applied Physiology* 86: 1319-1328.

Tikuisis, P., D.A. Eyolfson, X. Xu, and G.G. Giesbrecht. 2002. Shivering endurance and fatigue during cold water immersion in humans. *European Journal of Applied Physiology* 87: 50-58.

Toner, M.M. and W.D. McArdle. 1988. Physiological adjustments of man to the cold. In *Human performance physiology and environmental medicine in terrestrial extremes,* ed. Pandolf, K.B., M.N. Sawka, and R.R. Gonzalez, 361-400. Indianapolis: Benchmark Press.

van Marken Lichtenbelt, W.D. and H.A. Daanen. 2003. Cold-induced metabolism. *Current Opinion in Clinical Nutrition and Metabolic Care* 6: 469-475.

Watanabe, M., T. Yamamoto, C. Mori, N. Okada, N. Yamazaki, K. Kajimoto, M. Kataoka, and Y. Shinohara. 2008. Cold-induced changes in gene expression in brown adipose tissue: implications for the activation of thermogenesis. *Biological and Pharmaceutical Bulletin* 31: 775-784.

Weber, J.M. and F. Haman. 2005. Fuel selection in shivering humans. *Acta Physiologica Scandinavica* 184: 319-329.

Whayne, T. and M. DeBuakey. 1958. Cold injury, ground type. U.S. Army Med. Dept. Washington, DC: Superintendent of Documents, U.S. Government Printing Office.

Yoshimura, H. and T. Iida. 1950. Studies on the reactivity of skin vessels to extreme cold Part 1. A point test on the resistance against frostbite. *Japanese Journal of Physiology* 1: 147-159.

Young, A.J., J.W. Castellani, C. O'Brien, R.L. Shippee, P. Tikuisis, L.G. Meyer, L.A. Blanchard, J.E. Kain, B.S. Cadarette, and M.N. Sawka. 1998. Exertional fatigue, sleep loss, and negative energy balance increase susceptibility to hypothermia. *Journal of Applied Physiology* 85: 1210-1217.

Chapter 6

Barwood, M.J., J. Dalzell, A.K. Datta, R.C. Thelwell, and M.J. Tipton. 2006. Breath-hold performance during cold water immersion: effects of psychological skills training. *Aviation, Space, and Environmental Medicine* 77: 1136-1142.

Brooks, C.J., C.V. MacDonald, L. Donati, and M.J. Taber. 2008. Civilian helicopter accidents into water: analysis of 46 cases, 1979-2006. *Aviation, Space, and Environmental Medicine* 79: 935-940.

Brooks, C.J., H.C. Muir, and P.N. Gibbs. 2001. The basis for the development of a fuselage evacuation time for a ditched helicopter. *Aviation, Space, and Environmental Medicine* 72: 553-561.

Canadian Red Cross. 2006. *Drownings and other water-related injuries in Canada,* 1991-2000. Ottawa: Canadian Red Cross Society.

Cheung, S.S., N.J. d'Eon, and C.J. Brooks. 2001. Breath-holding ability of offshore workers inadequate to ensure escape from ditched helicopters. *Aviation, Space, and Environmental Medicine* 72: 912-918.

Ducharme, M.B. and D.S. Lounsbury. 2007. Self-rescue swimming in cold water: the latest advice.

Applied Physiology, Nutrition, and Metabolism 32: 799-807.

Giesbrecht, G.G. 2000. Cold stress, near drowning and accidental hypothermia: a review. *Aviation, Space, and Environmental Medicine* 71: 733-752.

Golden, F.S. and G.R. Hervey. 1977. The mechanism of the after-drop following immersion hypothermia in pigs [proceedings]. *Journal of Physiology* 272: 26P-27P.

Golden, F.S., G.R. Hervey, and M.J. Tipton. 1991. Circum-rescue collapse: collapse, sometimes fatal, associated with rescue of immersion victims. *Journal of the Royal Navy Medical Service* 77: 139-149.

Golden, F. and M. Tipton. 2002. *Essentials of sea survival.* Champaign, IL: Human Kinetics.

Golden, F.S., M.J. Tipton, and R.C. Scott. 1997. Immersion, near-drowning and drowning. *British Journal of Anaesthesiology* 79: 214-225.

Hayward, J.S., J.D. Eckerson, and M.L. Collis. 1975. Effect of behavioral variables on cooling rate of man in cold water. *Journal of Applied Physiology* 38: 1073-1077.

Hayward, J.S., C. Hay, B.R. Matthews, C.H. Overweel, and D.D. Radford. 1984. Temperature effect on the human dive response in relation to cold water near-drowning. *Journal of Applied Physiology* 56: 202-206.

Johnson, D.G., J.S. Hayward, T.P. Jacobs, M.L. Collis, J.D. Eckerson, and R.H. Williams. 1977. Plasma norepinephrine responses of man in cold water. *Journal of Applied Physiology* 43: 216-220.

Lounsbury, D.S. 2004. Swimming performance and judgement in cold water. MSc thesis. University of Toronto.

Schagatay, E., H. Haughey, and J. Reimers. 2005. Speed of spleen volume changes evoked by serial apneas. *European Journal of Applied Physiology* 93: 447-452.

Schagatay, E. and B. Holm. 1996. Effects of water and ambient air temperatures on human diving bradycardia. *European Journal of Applied Physiology* 73: 1-6.

Schagatay, E., M. van Kampen, S. Emanuelsson, and B. Holm. 2000. Effects of physical and apnea training on apneic time and the diving response in humans. *European Journal of Applied Physiology* 82: 161-169.

Tipton, M.J. 1995. Immersion fatalities: hazardous responses and dangerous discrepancies. *Journal of the Royal Naval Medical Service* 81: 101-107.

Tipton, M.J., C.M. Eglin, and F.S. Golden. 1998a. Habituation of the initial responses to cold water immersion in humans: a central or peripheral mechanism? *Journal of Physiology (London)* 512: 621-628.

Tipton, M.J., C.M. Franks, B.A. Sage, and P.J. Redman. 1997. An examination of two emergency breathing aids for use during helicopter underwater escape. *Aviation, Space, and Environmental Medicine* 68: 907-914.

Tipton, M.J., F.S. Golden, C. Higenbottam, I.B. Mekjavic, and C.M. Eglin. 1998b. Temperature dependence of habituation of the initial responses to cold-water immersion. *European Journal of Applied Physiology* 78: 253-257.

Tipton, M.J., I.B. Mekjavic, and C.M. Eglin. 2000. Permanence of the habituation of the initial responses to cold-water immersion in humans. *European Journal of Applied Physiology* 83: 17-21.

Tipton, M.J., D.A. Stubbs, and D.H. Elliott. 1991. Human initial responses to immersion in cold water at three temperatures and after hyperventilation. *Journal of Applied Physiology* 70: 317-322.

Wallingford, R., M.B. Ducharme, and E. Pommier. 2000. Factors limiting cold-water swimming distance while wearing personal floatation devices. *European Journal of Applied Physiology* 82: 24-29.

Chapter 7

Behnke, A.R., R.M. Thomson, and E.P. Motley. 1935. The psychologic effects from breathing air at 4 atmospheres pressure. *American Journal of Physiology* 112: 554-558.

Bove, A.A. 1998. Risk of decompression sickness with patent foramen ovale. *Undersea and Hyperbaric Medicine* 25: 175-178.

Brubakk, A.O. and T.S. Neuman. 2002. *Bennett and Elliott's physiology and medicine of diving.* Amsterdam: Saunders.

Brubakk, A.O., S. Tonjum, B. Holand, R.E. Peterson, R.W. Hamilton, E. Morild, and J. Onarheim. 1982. Heat loss and tolerance time during cold exposure in heliox atmosphere at 16 ATA. *Undersea Biomedical Research* 9: 81-90.

Butcher, S.J., R.L. Jones, N.D. Eves, and S.R. Petersen. 2006. Work of breathing is increased during exercise with the self-contained breathing apparatus regulator. *Applied Physiology, Nutrition, and Metabolism* 31: 693-701.

Butcher, S.J., R.L. Jones, J.R. Mayne, T.C. Hartley, and S.R. Petersen. 2007. Impaired exercise ventilatory mechanics with the self-contained breathing apparatus are improved with heliox. *European Journal of Applied Physiology* 101: 659-669.

Cain, J.B., S.D. Livingstone, R.W. Nolan, and A.A. Keefe. 1990. Respiratory heat loss during work at various ambient temperatures. *Respiration Physiology* 79: 145-150.

Carter, R.C. 1976. Evaluation of JIM: a one-atmosphere diving suit. 5-76: Department of the Navy Experimental Diving Unit, Panama City 1-16. http://archive.rubicon-foundation.org/4790.

Carturan, D., A. Boussuges, P. Burnet, J. Fondarai, P. Vanuxem, and B. Gardette. 1999. Circulating venous bubbles in recreational diving: relationships with age, weight, maximal oxygen uptake and body fat percentage. *International Journal of Sports Medicine* 20: 410-414.

Cheung, S.S. and I.B. Mekjavic. 1995. Human temperature regulation during subanesthetic levels of nitrous oxide-induced narcosis. *Journal of Applied Physiology* 78: 2301-2308.

Clarkson, D.P., C.L. Schatte, and J.P. Jordan. 1972. Thermal neutral temperature of rats in helium-oxygen, argon-oxygen, and air. *American Journal of Physiology* 222: 1494-1498.

Daniels, S. 2008. Cellular and neurophysiological effects of high ambient pressure. *Undersea and Hyperbaric Medicine* 35: 11-19.

DeGorordo, A., F. Vallejo-Manzur, K. Chanin, and J. Varon. 2003. Diving emergencies. *Resuscitation* 59: 171-180.

Fowler, B., K.N. Ackles, and G. Porlier. 1985. Effects of inert gas narcosis on behavior—a critical review. *Undersea Biomedical Research* 12: 369-402.

Fowler, B., P.L. White, G.R. Wright, and K.N. Ackles. 1980. Narcotic effects of nitrous oxide and compressed air on memory and auditory perception. *Undersea Biomedical Research* 7: 35-46.

Gempp, E., J.E. Blatteau, E. Stephant, and P. Louge. 2009. Relation between right-to-left shunts and spinal cord decompression sickness in divers. *International Journal of Sports Medicine* 30: 150-153.

Lun, V., J. Sun, T. Passias, and I.B. Mekjavic. 1993. Effects of prolonged CO2 inhalation on shivering thermogenesis during cold-water immersion. *Undersea and Hyperbaric Medicine* 20: 215-224.

McArdle, W.D., F.I. Katch, V.L. Katch. 2007. *Exercise Physiology: Energy, Nutrition, and Human Performance.* 6th ed. New York: Lippincott Williams & Wilkins.

Mekjavic, I.B., S.A. Savic, and O. Eiken. 1995. Nitrogen narcosis attenuates shivering thermogenesis. *Journal of Applied Physiology* 78: 2241-2244.

Mekjavic, P.J. and I.B. Mekjavic. 2007. Decompression-induced ocular tear film bubbles reflect the process of denitrogenation. *Investigative Ophthalmology and Visual Science* 48: 3756-3760.

Nakayama, H. 1978. Body thermal drain under hyperbaric dry heliox environment with undersea excursion dive. *Journal of Human Ergology* 7: 177-183.

Passias, T.C., I.B. Mekjavic, and O. Eiken. 1992. The effect of 30% nitrous oxide on thermoregulatory responses in humans during hypothermia. *Anesthesiology* 76: 550-559.

Sessler, D.I. 1993. Perianesthetic thermoregulation and heat balance in humans. *FASEB Journal* 7: 638-644.

Spitler, D.L., S.M. Horvath, K. Kobayashi, and J.A. Wagner. 1980. Work performance breathing normoxic nitrogen or helium gas mixtures. *European Journal of Applied Physiology* 43: 157-166.

Talpalar, A.E. and Y. Grossman. 2006. CNS manifestations of HPNS: revisited. *Undersea and Hyperbaric Medicine* 33: 205-210.

Yildiz, S., H. Ay, A. Gunay, S. Yaygili, and S. Aktas. 2004. Submarine escape from depths of 30 and 60 feet: 41,183 training ascents without serious injury. *Aviation, Space, and Environmental Medicine* 75: 269-271.

Chapter 8

Adams, W.C., E.M. Bernauer, D.B. Dill, and J.B. Bomar Jr. 1975. Effects of equivalent sea-level and altitude training on VO_{2max} and running performance. *Journal of Applied Physiology* 39: 262-266.

Ameln, H., T. Gustafsson, C.J. Sundberg, K. Oka-moto, E. Jansson, L. Poellinger, and Y. Makino. 2005. Physiological activation of hypoxia inducible factor-1 in human skeletal muscle. *FASEB Journal* 19: 1009-1011.

Bartsch, P., B. Saltin, J. Dvorak, and Fédération Internationale de Football Association. 2008. Consensus statement on playing football at different altitude. *Scandinavian Journal of Medicine and Science in Sports* 18 Suppl. 1: 96-99.

Bassett, D.R. Jr., C.R. Kyle, L. Passfield, J.P. Broker, and E.R. Burke. 1999. Comparing cycling world hour records, 1967-1996: modeling with empirical data. *Medicine and Science in Sports and Exercise* 31: 1665-1676.

Berglund, B. and B. Ekblom. 1991. Effect of recombinant human erythropoietin treatment on blood pressure and some haematological parameters in healthy men. *Journal of Internal Medicine* 229: 125-130.

Buick, F.J., N. Gledhill, A.B. Froese, L. Spriet, and E.C. Meyers. 1980. Effect of induced erythrocythemia on aerobic work capacity. *Journal of Applied Physiology* 48: 636-642.

Burtscher, M., M. Faulhaber, M. Flatz, R. Likar, and W. Nachbauer. 2006. Effects of short-term acclimatization to altitude (3200 m) on aerobic and anaerobic exercise performance. *International Journal of Sports Medicine* 27: 629-635.

Calbet, J.A., G. Radegran, R. Boushel, H. Sonder-gaard, B. Saltin, and P.D. Wagner. 2002. Effect of blood haemoglobin concentration on VO_{2max} and cardiovascular function in lowlanders acclimatised to 5260 m. *Journal of Physiology* 545: 715-728.

Chapman, R.F., J. Stray-Gundersen, and B.D. Levine. 1998. Individual variation in response to altitude training. *Journal of Applied Physiology* 85: 1448-1456.

Coyle, E.F. 2005. Improved muscular efficiency displayed as Tour de France champion matures. *Journal of Applied Physiology* 98: 2191-2196.

Daniels, J. and N. Oldridge. 1970. The effects of alternate exposure to altitude and sea level on world-class middle-distance runners. *Medicine and Science in Sports* 2: 107-112.

Dufour, S.P., E. Ponsot, J. Zoll, S. Doutreleau, E. Lonsdorfer-Wolf, B. Geny, E. Lampert, M. Fluck, H. Hoppeler, V. Billat, B. Mettauer, R. Richard, and J. Lonsdorfer. 2006. Exercise training in normobaric hypoxia in endurance runners. I. Improvement in aerobic performance capacity. *Journal of Applied Physiology* 100: 1238-1248.

Favier, R., H. Spielvogel, D. Desplanches, G. Ferretti, B. Kayser, A. Grunenfelder, M. Leuenberger, L. Tuscher, E. Caceres, and H. Hoppeler. 1995. Training in hypoxia vs. training in normoxia in high-altitude natives. *Journal of Applied Physiology* 78: 2286-2293.

Ge, R.L., S. Witkowski, Y. Zhang, C. Alfrey, M. Sivieri, T. Karlsen, G.K. Resaland, M. Harber, J. Stray-Gundersen, and B.D. Levine. 2002. Determinants of erythropoietin release in response to short-term hypobaric hypoxia. *Journal of Applied Physiology* 92: 2361-2367.

Gore, C.J., S.A. Clark, and P.U. Saunders. 2007. Nonhematological mechanisms of improved sea-level performance after hypoxic exposure. *Medicine and Science in Sports and Exercise* 39: 1600-1609.

Gore, C.J., A.G. Hahn, R.J. Aughey, D.T. Martin, M.J. Ashenden, S.A. Clark, A.P. Garnham, A.D. Roberts, G.J. Slater, and M.J. McKenna. 2001. Live high:train low increases muscle buffer capacity and submaximal cycling efficiency. *Acta Physiologica Scandinavica* 173: 275-286.

Gore, C.J., F.A. Rodriguez, M.J. Truijens, N.E. Townsend, J. Stray-Gundersen, and B.D. Levine. 2006. Increased serum erythropoietin but not red cell production after 4 wk of intermittent hypobaric hypoxia (4,000-5,500 m). *Journal of Applied Physiology* 101: 1386-1393.

Hahn, A.G., C.J. Gore, D.T. Martin, M.J. Ashenden, A.D. Roberts, and P.A. Logan. 2001. An evaluation of the concept of living at moderate altitude and training at sea level. *Comparative Biochemistry and Physiology Part A: Molecular and Integrative Physiology* 128: 777-789.

Hendriksen, I.J. and T. Meeuwsen. 2003. The effect of intermittent training in hypobaric hypoxia on sea-level exercise: a cross-over study in humans. *European Journal of Applied Physiology* 88: 396-403.

Jedlickova, K., D.W. Stockton, H. Chen, J. Stray-Gundersen, S. Witkowski, G. Ri-Li, J. Jelinek, B.D. Levine, and J.T. Prchal. 2003. Search for genetic determinants of individual variability of the erythropoietin response to high altitude. *Blood Cells, Molecules and Diseases* 31: 175-182.

Levine, B.D. and J. Stray-Gundersen. 1997. "Living high-training low": effect of moderate-altitude acclimatization with low-altitude training on performance. *Journal of Applied Physiology* 83: 102-112.

Levine, B.D., J. Stray-Gundersen, C.J. Gore, and W.G. Hopkins. 2005. Point:counterpoint: positive effects of intermittent hypoxia (live high:train low) on exercise performance are/are not mediated primarily by augmented red cell volume. *Journal of Applied Physiology* 99: 2053-2055.

Loland, S. and A. Caplan. 2008. Ethics of technologically constructed hypoxic environments in sport. *Scandinavian Journal of Medicine and Science in Sports* 18 Suppl. 1: 70-75.

Lundby, C., J.A. Calbet, M. Sander, G. van Hall, R.S. Mazzeo, J. Stray-Gundersen, J.M. Stager, R.F. Chapman, B. Saltin, and B.D. Levine. 2007. Exercise economy does not change after acclimatization to moderate to very high altitude. *Scandinavian Journal of Medicine and Science in Sports* 17: 281-291.

Mackenzie, R.W., P.W. Watt, and N.S. Maxwell. 2008. Acute normobaric hypoxia stimulates erythropoietin release. *High Altitude Medicine and Biology* 9: 28-37.

Moore, L.G. 2001. Human genetic adaptation to high altitude. *High Altitude Medicine and Biology* 2: 257-279.

Muza, S.R. 2007. Military applications of hypoxic training for high-altitude operations. *Medicine and Science in Sports and Exercise* 39: 1625-1631.

Neya, M., T. Enoki, Y. Kumai, T. Sugoh, and T. Kawahara. 2007. The effects of nightly normobaric hypoxia and high intensity training under intermittent normobaric hypoxia on running economy and hemoglobin mass. *Journal of Applied Physiology* 103: 828-834.

Onywera, V.O., R.A. Scott, M.K. Boit, and Y.P. Pitsiladis. 2006. Demographic characteristics of elite Kenyan endurance runners. *Journal of Sports Sciences* 24: 415-422.

Peronnet, F., G. Thibault, and D.L. Cousineau. 1991. A theoretical analysis of the effect of altitude on running performance. *Journal of Applied Physiology* 70: 399-404.

Perry, C.G., J. Reid, W. Perry, and B.A. Wilson. 2005. Effects of hyperoxic training on performance and cardiorespiratory response to exercise. *Medicine and Science in Sports and Exercise* 37: 1175-1179.

Perry, C.G., J.L. Talanian, G.J. Heigenhauser, and L.L. Spriet. 2007. The effects of training in hyperoxia vs. normoxia on skeletal muscle enzyme activities and exercise performance. *Journal of Applied Physiology* 102: 1022-1027.

Pitsiladis, Y.P. and R. Scott. 2005. Essay: the makings of the perfect athlete. *Lancet* 366 Suppl. 1: S16-17.

Ponsot, E., S.P. Dufour, J. Zoll, S. Doutreleau, B. N'Guessan, B. Geny, H. Hoppeler, E. Lampert, B. Mettauer, R. Ventura-Clapier, and R. Richard. 2006. Exercise training in normobaric hypoxia in endurance runners. II. Improvement of mitochondrial properties in skeletal muscle. *Journal of Applied Physiology* 100: 1249-1257.

Roberts, A.D., S.A. Clark, N.E. Townsend, M.E. Anderson, C.J. Gore, and A.G. Hahn. 2003. Changes in performance, maximal oxygen uptake and maximal accumulated oxygen deficit after 5, 10 and 15 days of live high:train low altitude exposure. *European Journal of Applied Physiology* 88: 390-395.

Rupert, J.L. and P.W. Hochachka. 2001. The evidence for hereditary factors contributing to high altitude adaptation in Andean natives: a review. *High Altitude Medicine and Biology* 2: 235-256.

Saunders, P.U., D.B. Pyne, R.D. Telford, and J.A. Hawley. 2004a. Factors affecting running economy in trained distance runners. *Sports Medicine* 34: 465-485.

Saunders, P.U., R.D. Telford, D.B. Pyne, R.B. Cunningham, C.J. Gore, A.G. Hahn, and J.A. Hawley. 2004b. Improved running economy in elite runners after 20 days of simulated moderate-altitude exposure. *Journal of Applied Physiology* 96: 931-937.

Schmidt, W., K. Heinicke, J. Rojas, J. Manuel Gomez, M. Serrato, M. Mora, B. Wolfarth, A. Schmid, and J. Keul. 2002. Blood volume and hemoglobin mass in endurance athletes from moderate altitude. *Medicine and Science in Sports and Exercise* 34: 1934-1940.

Scott, R.A. and Y.P. Pitsiladis. 2007. Genotypes and distance running: clues from Africa. *Sports Medicine* 37: 424-427.

Semenza, G.L. 2000. HIF-1: mediator of physiological and pathophysiological responses to hypoxia. *Journal of Applied Physiology* 88: 1474-1480.

Semenza, G.L. 2006. Regulation of physiological responses to continuous and intermittent hypoxia by hypoxia-inducible factor 1. *Experimental Physiology* 91: 803-806.

Semenza, G.L. 2008. Hypoxia-inducible factor 1 and cancer pathogenesis. *IUBMB Life* 60: 591-597.

Stellingwerff, T., L. Glazier, M.J. Watt, P.J. LeBlanc, G.J. Heigenhauser, and L.L. Spriet. 2005. Effects of hyperoxia on skeletal muscle carbohydrate metabolism during transient and steady-state exercise. *Journal of Applied Physiology* 98: 250-256.

Stray-Gundersen, J., R.F. Chapman, and B.D. Levine. 2001. "Living high-training low" altitude training improves sea level performance in male and female elite runners. *Journal of Applied Physiology* 91: 1113-1120.

Terrados, N., J. Melichna, C. Sylven, E. Jansson, and L. Kaijser. 1988. Effects of training at simulated altitude on performance and muscle metabolic capacity in competitive road cyclists. *European Journal of Applied Physiology* 57: 203-209.

Tucker, R., B. Kayser, E. Rae, L. Rauch, A. Bosch, and T. Noakes. 2007. Hyperoxia improves 20 km cycling time trial performance by increasing muscle activation levels while perceived exertion stays the same. *European Journal of Applied Physiology* 101: 771-781.

Ventura, N., H. Hoppeler, R. Seiler, A. Binggeli, P. Mullis, and M. Vogt. 2003. The response of trained athletes to six weeks of endurance training in hypoxia or normoxia. *International Journal of Sports Medicine* 24: 166-172.

Weston, A.R., G. Mackenzie, M.A. Tufts, and M. Mars. 2001. Optimal time of arrival for performance at moderate altitude (1700 m). *Medicine and Science in Sports and Exercise* 33: 298-302.

Wilber, R.L. 2004. *Altitude training and athletic performance.* Champaign, IL: Human Kinetics.

Wilber, R.L. 2007. Application of altitude/hypoxic training by elite athletes. *Medicine and Science in Sports and Exercise* 39: 1610-1624.

Chapter 9

Bartsch, P. and B. Saltin. 2008. General introduction to altitude adaptation and mountain sickness. *Scandinavian Journal of Medicine and Science in Sports* 18 Suppl. 1: 1-10.

Basnyat, B. and D.R. Murdoch. 2003. High-altitude illness. *Lancet* 361: 1967-1974.

Baumgartner, R.W., P. Bartsch, M. Maggiorini, U. Waber, and O. Oelz. 1994. Enhanced cerebral blood flow in acute mountain sickness. *Aviation, Space, and Environmental Medicine* 65: 726-729.

Boyer, S.J. and F.D. Blume. 1984. Weight loss and changes in body composition at high altitude. *Journal of Applied Physiology* 57: 1580-1585.

Catron, T.F., F.L. Powell, and J.B. West. 2006. A strategy for determining arterial blood gases on the summit of Mt. Everest. *BMC Physiology* 6: 3-8.

Coote, J.H. 1995. Medicine and mechanisms in altitude sickness: recommendations. *Sports Medicine* 20: 148-159.

Cumbo, T.A., B. Basnyat, J. Graham, A.G. Lescano, and S. Gambert. 2002. Acute mountain sickness, dehydration, and bicarbonate clearance: preliminary field data from the Nepal Himalaya. *Aviation, Space, and Environmental Medicine* 73: 898-901.

Dehnert, C., M.M. Berger, H. Mairbaurl, and P. Bartsch. 2007. High altitude pulmonary edema: a pressure-induced leak. *Respiratory Physiology and Neurobiology* 158: 266-273.

Dehnert, C., J. Weymann, H.E. Montgomery, D. Woods, M. Maggiorini, U. Scherrer, J.S. Gibbs, and P. Bartsch. 2002. No association between high-altitude tolerance and the ACE I/D gene polymorphism. *Medicine and Science in Sports and Exercise* 34: 1928-1933.

Eldridge, M.W., A. Podolsky, R.S. Richardson, D.H. Johnson, D.R. Knight, E.C. Johnson, S.R. Hopkins, H. Michimata, B. Grassi, J. Feiner, S.S. Kurdak, P.E. Bickler, P.D. Wagner, and J.W. Severinghaus. 1996. Pulmonary hemodynamic response to exercise in subjects with prior high-altitude pulmonary edema. *Journal of Applied Physiology* 81: 911-921.

Firth, P.G., H. Zheng, J.S. Windsor, A.I. Sutherland, C.H. Imray, G.W. Moore, J.L. Semple, R.C. Roach, and R.A. Salisbury. 2008. Mortality on Mount Everest, 1921-2006: descriptive study. *British Medical Journal (Clinical Research Ed.)* 337: a2654.

Garner, S.H., J.R. Sutton, R.L. Burse, A.J. McComas, A. Cymerman, and C.S. Houston. 1990. Operation Everest II: neuromuscular performance under conditions of extreme simulated altitude. *Journal of Applied Physiology* 68: 1167-1172.

Green, H.J., J.R. Sutton, A. Cymerman, P.M. Young, and C.S. Houston. 1989. Operation Everest II: adaptations in human skeletal muscle. *Journal of Applied Physiology* 66: 2454-2461.

Hackett, P.H. and R.C. Roach. 2004. High altitude cerebral edema. *High Altitude Medicine and Biology* 5: 136-146.

Hochachka, P.W., C.L. Beatty, Y. Burelle, M.E. Trump, D.C. McKenzie, and G.O. Matheson. 2002. The lactate paradox in human high-altitude physiological performance. *News in Physiological Sciences* 17: 122-126.

Houston, C.S. and J. Dickinson. 1975. Cerebral form of high-altitude illness. *Lancet* 2: 758-761.

Kallenberg, K., C. Dehnert, A. Dorfler, P.D. Schellinger, D.M. Bailey, M. Knauth, and P.D. Bartsch. 2008. Microhemorrhages in nonfatal high-altitude cerebral edema. *Journal of Cerebral Blood Flow and Metabolism* 28: 1635-1642.

Kasic, J.F., H.M. Smith, and R.I. Gamow. 1989. A self-contained life support system designed for use with a portable hyperbaric chamber. *Biomedical Sciences Instrumentation* 25: 79-81.

Krasney, J.A. 1994. A neurogenic basis for acute altitude illness. *Medicine and Science in Sports and Exercise* 26: 195-208.

Leaf, D.E. and D.S. Goldfarb. 2007. Mechanisms of action of acetazolamide in the prophylaxis and treatment of acute mountain sickness. *Journal of Applied Physiology* 102: 1313-1322.

Leon-Velarde, F. and O. Mejia. 2008. Gene expression in chronic high altitude diseases. *High Altitude Medicine and Biology* 9: 130-139.

Leon-Velarde, F., M. Maggiorini, J.T. Reeves, A. Aldashev, I. Asmus, L. Bernardi, R.L. Ge, P. Hackett, T. Kobayashi, L.G. Moore, D. Penaloza, J.P. Richalet, R. Roach, T. Wu, E. Vargas, G. Zubieta-Castillo, and G. Zubieta-Calleja. 2005. Consensus statement on chronic and subacute high altitude diseases. *High Altitude Medicine and Biology* 6: 147-157.

Luks, A.M. 2008. Do we have a "best practice" for treating high altitude pulmonary edema? *High Altitude Medicine and Biology* 9: 111-114.

Lundby, C., B. Saltin, and G. van Hall. 2000. The "lactate paradox," evidence for a transient change in the course of acclimatization to severe hypoxia in lowlanders. *Acta Physiologica Scandinavica* 170: 265-269.

Moore, L.G. 2000. Comparative human ventilatory adaptation to high altitude. *Respiration Physiology* 121: 257-276.

Oelz, O., H. Howald, P.E. Di Prampero, H. Hoppeler, H. Claassen, R. Jenni, A. Buhlmann, G. Ferretti, J.C. Bruckner, and A. Veicsteinas. 1986. Physiological profile of world-class high-altitude climbers. *Journal of Applied Physiology* 60: 1734-1742.

Okazaki, S., Y. Tamura, T. Hatano, and N. Matsui. 1984. Hormonal disturbances of fluid-electrolyte metabolism under altitude exposure in man. *Aviation, Space, and Environmental Medicine* 55: 200-205.

Podolsky, A., M.W. Eldridge, R.S. Richardson, D.R. Knight, E.C. Johnson, S.R. Hopkins, D.H. Johnson, H. Michimata, B. Grassi, J. Feiner, S.S. Kurdak, P.E. Bickler, J.W. Severinghaus, and P.D. Wagner. 1996. Exercise-induced VA/Q inequality in subjects with prior high-altitude pulmonary edema. *Journal of Applied Physiology* 81: 922-932.

Pugh, L.G., M.B. Gill, S. Lahiri, J.S. Milledge, M.P. Ward, and J.B. West. 1964. Muscular exercise at great altitudes. *Journal of Applied Physiology* 19: 431-440.

Roach, R.C., P. Bartsch, O. Oelz, and P.H. Hackett. 1993. The Lake Louise acute mountain sickness scoring system. In *Hypoxia and molecular medicine*, ed: Sutton, J.R., C.S. Houston, and G. Coates, 272-274. Burlington: Queen City Press.

Rodway, G.W., L.A. Hoffman, and M.H. Sanders. 2003. High-altitude-related disorders—part I: pathophysiology, differential diagnosis, and treatment. *Heart & Lung: The Journal of Critical Care* 32: 353-359.

Rodway, G.W., L.A. Hoffman, and M.H. Sanders. 2004. High-altitude-related disorders—part II: prevention, special populations, and chronic medical conditions. *Heart & Lung: The Journal of Critical Care* 33: 3-12.

Schneider, M., D. Bernasch, J. Weymann, R. Holle, and P. Bartsch. 2002. Acute mountain sickness: influence of susceptibility, preexposure, and ascent rate. *Medicine and Science in Sports and Exercise* 34: 1886-1891.

Tsianos, G., K.I. Eleftheriou, E. Hawe, L. Woolrich, M. Watt, I. Watt, A. Peacock, H. Montgomery, and S. Grant. 2005. Performance at altitude and angiotensin I-converting enzyme genotype. *European Journal of Applied Physiology* 93: 630-633.

Viesturs, E. and D. Roberts. 2006. *No shortcuts to the top: climbing the world's 14 highest peaks.* New York: Broadway Books.

West, J.B. 1999. Barometric pressures on Mt. Everest: new data and physiological significance. *Journal of Applied Physiology* 86: 1062-1066.

West, J.B. 2002. Highest permanent human habitation. *High Altitude Medicine and Biology* 3: 401-407.

West, J.B. 2006. Human responses to extreme altitudes. *Integrative and Comparative Biology* 46: 25.

West, J.B., S.J. Boyer, D.J. Graber, P.H. Hackett, K.H. Maret, J.S. Milledge, R.M. Peters Jr., C.J. Pizzo, M. Samaja, and F.H. Sarnquist. 1983a. Maximal exercise at extreme altitudes on Mount Everest. *Journal of Applied Physiology* 55: 688-698.

West, J.B., S. Lahiri, M.B. Gill, J.S. Milledge, L.G. Pugh, and M.P. Ward. 1962. Arterial oxygen saturation during exercise at high altitude. *Journal of Applied Physiology* 17: 617-621.

West, J.B., S. Lahiri, K.H. Maret, R.M. Peters Jr., and C.J. Pizzo. 1983b. Barometric pressures at extreme altitudes on Mt. Everest: physiological significance. *Journal of Applied Physiology* 54: 1188-1194.

Wright, A., S. Brearey, and C. Imray. 2008. High hopes at high altitudes: pharmacotherapy for acute mountain sickness and high-altitude cerebral and pulmonary oedema. *Expert Opinion on Pharmacotherapy* 9: 119-127.

Chapter 10

Adams, G.R., V.J. Caiozzo, and K.M. Baldwin. 2003. Skeletal muscle unweighting: spaceflight and ground-based models. *Journal of Applied Physiology* 95: 2185-2201.

Alfrey, C.P., M.M. Udden, C.L. Huntoon, and T. Driscoll. 1996a. Destruction of newly released red blood cells in space flight. *Medicine and Science in Sports and Exercise* 28: S42-44.

Alfrey, C.P., M.M. Udden, C. Leach-Huntoon, T. Driscoll, and M.H. Pickett. 1996b. Control of red blood cell mass in spaceflight. *Journal of Applied Physiology* 81: 98-104.

Aubert, A.E., F. Beckers, and B. Verheyden. 2005. Cardiovascular function and basics of physiology in microgravity. *Acta Cardiologica* 60: 129-151.

Bagian, J.P. and D.F. Ward. 1994. A retrospective study of promethazine and its failure to produce the expected incidence of sedation during space flight. *Journal of Clinical Pharmacology* 34: 649-651.

Barratt, M. 1999. Medical support for the International Space Station. *Aviation, Space, and Environmental Medicine* 70: 155-161.

Beavers, K.R., D.K. Greaves, P. Arbeille, and R.L. Hughson. 2007. WISE-2005: orthostatic tolerance is poorly predicted by acute changes in cardiovascular variables. *Journal of Gravitational Physiology* 14: P63-64.

Bogomolov, V.V., F. Castrucci, J.M. Comtois, V. Damann, J.R. Davis, J.M. Duncan, S.L. Johnston, G.W. Gray, A.I. Grigoriev, Y. Koike, P. Kuklinski, V.P. Matveyev, V.V. Morgun, V.I. Pochuev, A.E. Sargsyan, K. Shimada, U. Straube, S. Tachibana, Y.V. Voronkov, and R.S. Williams. 2007. International Space Station medical standards and certification for space flight participants. *Aviation, Space, and Environmental Medicine* 78: 1162-1169.

Buckey, J.C. 2006. *Space physiology.* New York: Oxford University Press.

Buckey, J.C. Jr., F.A. Gaffney, L.D. Lane, B.D. Levine, D.E. Watenpaugh, S.J. Wright, C.W. Yancy Jr., D.M. Meyer, and C.G. Blomqvist. 1996a. Central venous pressure in space. *Journal of Applied Physiology* 81: 19-25.

Buckey, J.C. Jr., L.D. Lane, B.D. Levine, D.E. Watenpaugh, S.J. Wright, W.E. Moore, F.A. Gaffney, and C.G. Blomqvist. 1996b. Orthostatic intolerance after spaceflight. *Journal of Applied Physiology* 81: 7-18.

Caillot-Augusseau, A., M.H. Lafage-Proust, C. Soler, J. Pernod, F. Dubois, and C. Alexandre. 1998. Bone formation and resorption biological markers in cosmonauts during and after a 180-day space flight (Euromir 95). *Clinical Chemistry* 44: 578-585.

Cavanagh, P.R., A.A. Licata, and A.J. Rice. 2005. Exercise and pharmacological countermeasures for bone loss during long-duration space flight. *Gravitational and Space Biology Bulletin* 18: 39-58.

Clément, G. 2007, 2005. *Fundamentals of space medicine.* Dordrecht, Netherlands: Springer.

Coblentz, A., E. Fossier, G. Ignazi, and R. Mollard. 1988. Habitability design of European spacecraft Hermes—ergonomic aspects. *Acta Astronautica* 17: 223-225.

Convertino, V.A. and H. Sandler. 1995. Exercise countermeasures for spaceflight. *Acta Astronautica* 35: 253-270.

Cowell, S.A., J.M. Stocks, D.G. Evans, S.R. Simonson, and J.E. Greenleaf. 2002. The exercise and environmental physiology of extravehicular activity. *Aviation, Space, and Environmental Medicine* 73: 54-67.

Davis, J.R., J.M. Vanderploeg, P.A. Santy, R.T. Jennings, and D.F. Stewart. 1988. Space motion sickness during 24 flights of the space shuttle. *Aviation, Space, and Environmental Medicine* 59: 1185-1189.

Dervay, J.P., M.R. Powell, B. Butler, and C.E. Fife. 2002. The effect of exercise and rest duration on the generation of venous gas bubbles at altitude. *Aviation, Space, and Environmental Medicine* 73: 22-27.

Drummer, C., V. Friedel, A. Borger, I. Stormer, S. Wolter, A. Zittermann, G. Wolfram, and M. Heer. 1998. Effects of elevated carbon dioxide environment on calcium metabolism in humans.

Aviation, Space, and Environmental Medicine 69: 291-298.

Ferrando, A.A., K.D. Tipton, M.M. Bamman, and R.R. Wolfe. 1997. Resistance exercise maintains skeletal muscle protein synthesis during bed rest. *Journal of Applied Physiology* 82: 807-810.

Flouris, A.D. and S.S. Cheung. 2006. Design and control optimization of microclimate liquid cooling systems underneath protective clothing. *Annals of Biomedical Engineering* 34: 359-372.

Gibala, M.J., J.P. Little, M. van Essen, G.P. Wilkin, K.A. Burgomaster, A. Safdar, S. Raha, and M.A. Tarnopolsky. 2006. Short-term sprint interval versus traditional endurance training: similar initial adaptations in human skeletal muscle and exercise performance. *Journal of Physiology* 575: 901-911.

Gontcharov, I.B., I.V. Kovachevich, S.L. Pool, O.L. Navinkov, M.R. Barratt, V.V. Bogomolov, and N. House. 2005. In-flight medical incidents in the NASA-Mir program. *Aviation, Space, and Environmental Medicine* 76: 692-696.

Katuntsev, V.P., Y.Y. Osipov, A.S. Barer, N.K. Gnoevaya, and G.G. Tarasenkov. 2004. The main results of EVA medical support on the Mir Space Station. *Acta Astronautica* 54: 577-583.

Kawaguchi, M., S. Inoue, T. Sakamoto, Y. Kawaraguchi, H. Furuya, and T. Sakaki. 1999. The effects of prostaglandin E1 on intraoperative temperature changes and the incidence of postoperative shivering during deliberate mild hypothermia for neurosurgical procedures. *Anesthesia and Analgesia* 88: 446-451.

Kirby, C.R., M.J. Ryan, and F.W. Booth. 1992. Eccentric exercise training as a countermeasure to non-weight-bearing soleus muscle atrophy. *Journal of Applied Physiology* 73: 1894-1899.

Kreitenberg, A., K.M. Baldwin, J.P. Bagian, S. Cotten, J. Witmer, and V.J. Caiozzo. 1998. The Space Cycle self powered human centrifuge: a proposed countermeasure for prolonged human spaceflight. *Aviation, Space, and Environmental Medicine* 69: 66-72.

Lackner, J.R. and P. Dizio. 2006. Space motion sickness. *Experimental Brain Research* 175: 377-399.

Lang, T., A. LeBlanc, H. Evans, Y. Lu, H. Genant, and A. Yu. 2004. Cortical and trabecular bone mineral loss from the spine and hip in long-duration spaceflight. *Journal of Bone and Mineral Research* 19: 1006-1012.

Lang, T.F., A.D. Leblanc, H.J. Evans, and Y. Lu. 2006. Adaptation of the proximal femur to skeletal reloading after long-duration spaceflight. *Journal of Bone and Mineral Research* 21: 1224-1230.

Ma, O.J., J.G. Norvell, and S. Subramanian. 2007. Ultrasound applications in mass casualties and extreme environments. *Critical Care Medicine* 35: S275-279.

Nicogossian, A.E., C.L. Huntoon, and S.L. Pool, eds. 1994. *Space physiology and medicine.* Philadelphia: Lea & Febiger.

Norsk, P. 2005. Cardiovascular and fluid volume control in humans in space. *Current Pharmaceutical Biotechnology* 6: 325-330.

Oman, C.M., B.K. Lichtenberg, K.E. Money, and R.K. McCoy. 1986. M.I.T./Canadian vestibular experiments on the Spacelab-1 mission: 4. Space motion sickness: symptoms, stimuli, and predictability. *Experimental Brain Research* 64: 316-334.

Payne, M.W., D.R. Williams, and G. Trudel. 2007. Space flight rehabilitation. *American Journal of Physical Medicine and Rehabilitation* 86: 583-591.

Perhonen, M.A., F. Franco, L.D. Lane, J.C. Buckey, C.G. Blomqvist, J.E. Zerwekh, R.M. Peshock, P.T. Weatherall, and B.D. Levine. 2001. Cardiac atrophy after bed rest and spaceflight. *Journal of Applied Physiology* 91: 645-653.

Putcha, L., K.L. Berens, T.H. Marshburn, H.J. Ortega, and R.D. Billica. 1999. Pharmaceutical use by U.S. astronauts on space shuttle missions. *Aviation, Space, and Environmental Medicine* 70: 705-708.

Rice, G.M., C.A. Vacchiano, J.L. Moore Jr., and D.W. Anderson. 2003. Incidence of decompression sickness in hypoxia training with and without 30-min O_2 prebreathe. *Aviation, Space, and Environmental Medicine* 74: 56-61.

Sayson, J.V. and A.R. Hargens. 2008. Pathophysiology of low back pain during exposure to microgravity. *Aviation, Space, and Environmental Medicine* 79: 365-373.

Smith, S.M. and M. Heer. 2002. Calcium and bone metabolism during space flight. *Nutrition* 18: 849-852.

Smith, S.M., S.R. Zwart, M. Heer, S.M. Lee, N. Baecker, S. Meuche, B.R. Macias, L.C. Shackelford, S. Schneider, and A.R. Hargens. 2008. WISE-2005: supine treadmill exercise within lower body negative pressure and flywheel resistive exercise as a countermeasure to bed rest-induced bone loss in women during 60-day simulated microgravity. *Bone* 42: 572-581.

Sonnenfeld, G. 2003. Animal models for the study of the effects of spaceflight on the immune system. *Advances in Space Research* 32: 1473-1476.

Watenpaugh, D.E., J.C. Buckey, L.D. Lane, F.A. Gaffney, B.D. Levine, W.E. Moore, S.J. Wright, and C.G. Blomqvist. 2001. Effects of spaceflight on human calf hemodynamics. *Journal of Applied Physiology* 90: 1552-1558.

Whedon, G.D., L. Lutwak, P. Rambaut, M. Whittle, C. Leach, J. Reid, and M. Smith. 1976. Effect of weightlessness on mineral metabolism; metabolic studies on Skylab orbital space flights. *Calcified Tissue Research* 21 Suppl.: 423-430.

Wichman, H.A. and S.I. Donaldson. 1996. Remote ergonomic research in space: spacelab findings and a proposal. *Aviation, Space, and Environmental Medicine* 67: 171-175.

Wilson, T.E., J. Cui, R. Zhang, S. Witkowski, and C.G. Crandall. 2002. Skin cooling maintains cerebral blood flow velocity and orthostatic tolerance during tilting in heated humans. *Journal of Applied Physiology* 93: 85-91.

Yang, Y., A. Kaplan, M. Pierre, G. Adams, P. Cavanagh, C. Takahashi, A. Kreitenberg, J. Hicks, J. Keyak, and V. Caiozzo. 2007. Space cycle: a human-powered centrifuge that can be used for hypergravity resistance training. *Aviation, Space, and Environmental Medicine* 78: 2-9.

Chapter 11

Adams, W.C. 1987. Effects of ozone exposure at ambient air pollution episode levels on exercise performance. *Sports Medicine* 4: 395-424.

Adams, W.C. 2003. Comparison of chamber and face mask 6.6-hour exposure to 0.08 ppm ozone via square-wave and triangular profiles on pulmonary responses. *Inhalation Toxicology* 15: 265-281.

Adams, W.C., W.M. Savin, and A.E. Christo. 1981. Detection of ozone toxicity during continuous exercise via the effective dose concept. *Journal of Applied Physiology* 51: 415-422.

Campbell, M.E., Q. Li, S.E. Gingrich, R.G. Macfarlane, and S. Cheng. 2005. Should people be physically active outdoors on smog alert days? *Canadian Journal of Public Health* 96: 24-28.

Devlin, R.B., L.J. Folinsbee, F. Biscardi, G. Hatch, S. Becker, M.C. Madden, M. Robbins, and H.S. Koren. 1997. Inflammation and cell damage induced by repeated exposure of humans to ozone. *Inhalation Toxicology* 9: 211-235.

Flouris, A.D. 2006. Modelling atmospheric pollution during the games of the XXVIII Olympiad: effects on elite competitors. *International Journal of Sports Medicine* 27: 137-142.

Flouris, A.D., G.S. Metsios, A.Z. Jamurtas, and Y. Koutedakis. 2008. Sexual dimorphism in the acute effects of secondhand smoke on thyroid hormone secretion, inflammatory markers and vascular function. *American Journal of Physiology* 294: E456-462.

Folinsbee, L.J. 1993. Human health effects of air pollution. *Environmental Health Perspectives* 100: 45-56.

Foster, W.M., M. Wills-Karp, C.G. Tankersley, X. Chen, and N.C. Paquette. 1996. Bloodborne markers in humans during multiday exposure to ozone. *Journal of Applied Physiology* 81: 794-800.

Foxcroft, W.J. and W.C. Adams. 1986. Effects of ozone exposure on four consecutive days on work performance and VO_{2max}. *Journal of Applied Physiology* 61: 960-966.

Gliner, J.A., S.M. Horvath, and L.J. Folinsbee. 1983. Preexposure to low ozone concentrations does not diminish the pulmonary function response on exposure to higher ozone concentrations. *American Review of Respiratory Disease* 127: 51-55.

Grievink, L., S.M. Jansen, P. van't Veer, and B. Brunekreef. 1998. Acute effects of ozone on pulmonary function of cyclists receiving antioxidant supplements. *Occupational and Environmental Medicine* 55: 13-17.

Grievink, L., A.G. Zijlstra, X. Ke, and B. Brunekreef. 1999. Double-blind intervention trial on modulation of ozone effects on pulmonary function by antioxidant supplements. *American Journal of Epidemiology* 149: 306-314.

Hazucha, M.J. 1987. Relationship between ozone exposure and pulmonary function changes. *Journal of Applied Physiology* 62: 1671-1680.

Hazucha, M.J., D.V. Bates, and P.A. Bromberg. 1989. Mechanism of action of ozone on the human lung. *Journal of Applied Physiology* 67: 1535-1541.

Hazucha, M.J., L.J. Folinsbee, and E. Seal Jr. 1992. Effects of steady-state and variable ozone concentration profiles on pulmonary function. *American Review of Respiratory Disease* 146: 1487-1493.

Kahan, E.S., U.J. Martin, S. Spungen, D. Ciccolella, and G.J. Criner. 2007. Chronic cough and dyspnea in ice hockey players after an acute exposure to combustion products of a faulty ice resurfacer. *Lung* 185: 47-54.

Lam, T.H., S.Y. Ho, A.J. Hedley, K.H. Mak, and G.M. Leung. 2004. Leisure time physical activity and mortality in Hong Kong: case-control study of all adult deaths in 1998. *Annals of Epidemiology* 14: 391-398.

Li, J., Y. Lu, K. Huang, C. Wang, J. Lu, C. Zhang, and N. Zhong. 2008. Chinese response to allergy and asthma in Olympic athletes. *Allergy* 63: 962-968.

Linn, W.S., D.A. Medway, U.T. Anzar, L.M. Valencia, C.E. Spier, F.S. Tsao, D.A. Fischer, and J.D. Hackney. 1982. Persistence of adaptation to ozone in volunteers exposed repeatedly for six weeks. *American Review of Respiratory Disease* 125: 491-495.

Lippi, G., G.C. Guidi, and N. Maffulli. 2008. Air pollution and sports performance in Beijing. *International Journal of Sports Medicine* 29: 696-698.

Metsios, G.S., A.D. Flouris, A.Z. Jamurtas, A.E. Carrillo, D. Kouretas, A.E. Germenis, K. Gourgoulianis, T. Kiropoulos, M.N. Tzatzarakis, A.M. Tsatsakis, and Y. Koutedakis. 2007. A brief exposure to moderate passive smoke increases metabolism and thyroid hormone secretion. *Journal of Clinical Endocrinology and Metabolism* 92: 208-211.

Otsuka, R., H. Watanabe, K. Hirata, K. Tokai, T. Muro, M. Yoshiyama, K. Takeuchi, and J. Yoshikawa. 2001. Acute effects of passive smoking on the coronary circulation in healthy young adults. *Journal of the American Medical Association* 286: 436-441.

Pierson, W.E. 1989. Impact of air pollutants on athletic performance. *Allergy Proceedings* 10: 209-214.

Qi, J., L. Yang, and W. Wang. 2007. Environmental degradation and health risks in Beijing, China. *Archives of Environmental and Occupational Health* 62: 33-37.

Romieu, I., F. Meneses, M. Ramirez, S. Ruiz, R. Perez Padilla, J.J. Sienra, M. Gerber, L. Grievink, R. Dekker, I. Walda, and B. Brunekreef. 1998. Antioxidant supplementation and respiratory functions among workers exposed to high levels of ozone. *American Journal of Respiratory and Critical Care Medicine* 158: 226-232.

Rundell, K.W. 2003. High levels of airborne ultrafine and fine particulate matter in indoor ice arenas. *Inhalation Toxicology* 15: 237-250.

Sharman, J.E., J.R. Cockcroft, and J.S. Coombes. 2004. Cardiovascular implications of exposure to traffic air pollution during exercise. *QJM: Monthly Journal of the Association of Physicians* 97: 637-643.

Wang, L., J. Hao, K. He, S. Wang, J. Li, Q. Zhang, D.G. Streets, J.S. Fu, C.J. Jang, H. Takekawa, and S. Chatani. 2008. A modeling study of coarse particulate matter pollution in Beijing: regional source contributions and control implications for the 2008 Summer Olympics. *Journal of the Air and Waste Management Association* 58: 1057-1069.

Wayne, W.S., P.F. Wehrle, and R.E. Carroll. 1967. Oxidant air pollution and athletic performance. *Journal of the American Medical Association* 199: 901-904.

Wong, C.M., S. Ma, A.J. Hedley, and T.H. Lam. 2001. Effect of air pollution on daily mortality in Hong Kong. *Environmental Health Perspectives* 109: 335-340.

Wong, C.M., C.Q. Ou, T.Q. Thach, Y.K. Chau, K.P. Chan, S.Y. Ho, R.Y. Chung, T.H. Lam, and A.J. Hedley. 2007. Does regular exercise protect against air pollution-associated mortality? *Preventive Medicine* 44: 386-392.

Yip, F.Y., W.D. Flanders, A. Wolkin, D. Engelthaler, W. Humble, A. Neri, L. Lewis, L. Backer, and C. Rubin. 2008. The impact of excess heat events in Maricopa County, Arizona: 2000-2005. *International Journal of Biometeorology* 52: 765-772.

Chapter 12

Akerstedt, T. 2005. Shift work and sleep disorders. *Sleep* 28: 9-11.

Akerstedt, T., B. Peters, A. Anund, and G. Kecklund. 2005. Impaired alertness and performance driving home from the night shift: a driving simulator study. *Journal of Sleep Research* 14: 17-20.

Anglem, N., S.J. Lucas, E.A. Rose, and J.D. Cotter. 2008. Mood, illness and injury responses and recovery with adventure racing. *Wilderness and Environmental Medicine* 19: 30-38.

Atkinson, G., D. Barr, N. Chester, B. Drust, W. Gregson, T. Reilly, and J. Waterhouse. 2008. Bright light and thermoregulatory responses to exercise. *International Journal of Sports Medicine* 29: 188-193.

Atkinson, G., P. Buckley, B. Edwards, T. Reilly, and J. Waterhouse. 2001. Are there hangover-effects on physical performance when melatonin is ingested by athletes before nocturnal sleep? *International Journal of Sports Medicine* 22: 232-234.

Atkinson, G., B. Drust, T. Reilly, and J. Waterhouse. 2003. The relevance of melatonin to sports medicine and science. *Sports Medicine* 33: 809-831.

Atkinson, G., B. Edwards, T. Reilly, and J. Waterhouse. 2007. Exercise as a synchroniser of human circadian rhythms: an update and discussion of the methodological problems. *European Journal of Applied Physiology* 99: 331-341.

Atkinson, G., H. Jones, B.J. Edwards, and J.M. Waterhouse. 2005. Effects of daytime ingestion of melatonin on short-term athletic performance. *Ergonomics* 48: 1512-1522.

Balkin, T.J., P.D. Bliese, G. Belenky, H. Sing, D.R. Thorne, M. Thomas, D.P. Redmond, M. Russo, and N.J. Wesensten. 2004. Comparative utility of instruments for monitoring sleepiness-related performance decrements in the operational environment. *Journal of Sleep Research* 13: 219-227.

Bambaeichi, E., T. Reilly, N.T. Cable, and M. Giacomoni. 2005. Influence of time of day and partial sleep loss on muscle strength in eumenorrheic females. *Ergonomics* 48: 1499-1511.

Barger, L.K., K.P. Wright Jr., R.J. Hughes, and C.A. Czeisler. 2004. Daily exercise facilitates phase delays of circadian melatonin rhythm in very dim light. *American Journal of Physiology* 286: R1077-1084.

Barrett, J., L. Lack, and M. Morris. 1993. The sleep-evoked decrease of body temperature. *Sleep* 16: 93-99.

Belenky, G., N.J. Wesensten, D.R. Thorne, M.L. Thomas, H.C. Sing, D.P. Redmond, M.B. Russo, and T.J. Balkin. 2003. Patterns of performance degradation and restoration during sleep restriction and subsequent recovery: a sleep dose-response study. *Journal of Sleep Research* 12: 1-12.

Brisswalter, J., F. Bieuzen, M. Giacomoni, V. Tricot, and G. Falgairette. 2007. Morning-to-evening differences in oxygen uptake kinetics in short-duration cycling exercise. *Chronobiology International* 24: 495-506.

Cain, S.W., D.W. Rimmer, J.F. Duffy, and C.A. Czeisler. 2007. Exercise distributed across day and night does not alter circadian period in humans. *Journal of Biological Rhythms* 22: 534-541.

Cajochen, C. 2007. Alerting effects of light. *Sleep Medicine Reviews* 11: 453-464.

Castellani, J.W., A.J. Young, J.E. Kain, and M.N. Sawka. 1999. Thermoregulatory responses to cold water at different times of day. *Journal of Applied Physiology* 87: 243-246.

Dawson, D., L. Lack, and M. Morris. 1993. Phase resetting of the human circadian pacemaker with use of a single pulse of bright light. *Chronobiology International* 10: 94-102.

Demerouti, E., S.A. Geurts, A.B. Bakker, and M. Euwema. 2004. The impact of shiftwork on work—home conflict, job attitudes and health. *Ergonomics* 47: 987-1002.

Driscoll, T.R., R.R. Grunstein, and N.L. Rogers. 2007. A systematic review of the neurobehavioural and physiological effects of shiftwork systems. *Sleep Medicine Reviews* 11: 179-194.

Drust, B., J. Waterhouse, G. Atkinson, B. Edwards, and T. Reilly. 2005. Circadian rhythms in sports performance—an update. *Chronobiology International* 22: 21-44.

Edwards, B.J., G. Atkinson, J. Waterhouse, T. Reilly, R. Godfrey, and R. Budgett. 2000. Use of melatonin in recovery from jet-lag following an eastward flight across 10 time-zones. *Ergonomics* 43: 1501-1513.

Edwards, B.J., W. Edwards, J. Waterhouse, G. Atkinson, and T. Reilly. 2005a. Can cycling performance in an early morning, laboratory-based cycle time-trial be improved by morning exercise the day before? *International Journal of Sports Medicine* 26: 651-656.

Edwards, B.J., K. Lindsay, and J. Waterhouse. 2005b. Effect of time of day on the accuracy and consistency of the badminton serve. *Ergonomics* 48: 1488-1498.

Edwards, B., J. Waterhouse, G. Atkinson, and T. Reilly. 2007. Effects of time of day and distance upon accuracy and consistency of throwing darts. *Journal of Sports Sciences* 25: 1531-1538.

Folkard, S. and D.A. Lombardi. 2006. Modeling the impact of the components of long work hours on injuries and "accidents." *American Journal of Industrial Medicine* 49: 953-963.

Glenville, M., R. Broughton, A.M. Wing, and R.T. Wilkinson. 1978. Effects of sleep deprivation on short duration performance measures compared to the Wilkinson auditory vigilance task. *Sleep* 1: 169-176.

Healy, D., D.S. Minors, and J.M. Waterhouse. 1993. Shiftwork, helplessness and depression. *Journal of Affective Disorders* 29: 17-25.

Hennig, J., P. Kieferdorf, C. Moritz, S. Huwe, and P. Netter. 1998. Changes in cortisol secretion during shiftwork: implications for tolerance to shiftwork? *Ergonomics* 41: 610-621.

Krauchi, K., C. Cajochen, M. Pache, J. Flammer, and A. Wirz-Justice. 2006. Thermoregulatory effects of melatonin in relation to sleepiness. *Chronobiology International* 23: 475-484.

Landrigan, C.P., J.M. Rothschild, J.W. Cronin, R. Kaushal, E. Burdick, J.T. Katz, C.M. Lilly, P.H. Stone, S.W. Lockley, D.W. Bates, and C.A. Czeisler. 2004. Effect of reducing interns' work hours on serious medical errors in intensive care units. *New England Journal of Medicine* 351: 1838-1848.

Lemmer, B. 2007. The sleep-wake cycle and sleeping pills. *Physiology and Behavior* 90: 285-293.

Lucas, S.J., N. Anglem, W.S. Roberts, J.G. Anson, C.D. Palmer, R.J. Walker, C.J. Cook, and J.D. Cotter. 2008. Intensity and physiological strain of competitive ultra-endurance exercise in humans. *Journal of Sports Sciences* 26: 477-489.

McLellan, T.M., I.F. Smith, G.A. Gannon, and J. Zamecnik. 2000. Melatonin has no effect on tolerance to uncompensable heat stress in man. *European Journal of Applied Physiology* 83: 336-343.

Meney, I., J. Waterhouse, G. Atkinson, T. Reilly, and D. Davenne. 1998. The effect of one night's sleep deprivation on temperature, mood, and physical performance in subjects with different amounts of habitual physical activity. *Chronobiology International* 15: 349-363.

Mitler, M.M., M.A. Carskadon, C.A. Czeisler, W.C. Dement, D.F. Dinges, and R.C. Graeber. 1988. Catastrophes, sleep, and public policy: consensus report. *Sleep* 11: 100-109.

Mitler, M.M., R.M. Hajdukovic, R. Shafor, P.M. Hahn, and D.F. Kripke. 1987. When people die. Cause of death versus time of death. *American Journal of Medicine* 82: 266-274.

Nindl, B.C., C.D. Leone, W.J. Tharion, R.F. Johnson, J.W. Castellani, J.F. Patton, and S.J. Montain. 2002. Physical performance responses during 72 h of military operational stress. *Medicine and Science in Sports and Exercise* 34: 1814-1822.

O'Brien, P.M. and P.J. O'Connor. 2000. Effect of bright light on cycling performance. *Medicine and Science in Sports and Exercise* 32: 439-447.

Ohkuwa, T., H. Itoh, T. Yamamoto, H. Yanagi, Y. Yamazaki, and T. Akimaru. 2001. Effect of varying light intensity on maximal power production and selected metabolic variables. *Archives of Physiology and Biochemistry* 109: 430-434.

Reilly, T., J. Waterhouse, and B. Edwards. 2005. Jet lag and air travel: implications for performance. *Clinics in Sports Medicine* 24: 367-80, xii.

Rouch, I., P. Wild, D. Ansiau, and J.C. Marquie. 2005. Shiftwork experience, age and cognitive performance. *Ergonomics* 48: 1282-1293.

Samuels, C. 2009. Sleep, recovery, and performance: the new frontier in high-performance athletes. *Physical Medicine and Rehabilitation Clinics of North America* 20: 149-159.

Schreijer, A.J., S.C. Cannegieter, C.J. Doggen, and F.R. Rosendaal. 2009. The effect of flight-related behaviour on the risk of venous thrombosis after air travel. *British Journal of Haematology* 144: 425-429.

Souissi, N., N. Bessot, K. Chamari, A. Gauthier, B. Sesboue, and D. Davenne. 2007. Effect of time of day on aerobic contribution to the 30-s Wingate test performance. *Chronobiology International* 24: 739-748.

Stephenson, L.A., C.B. Wenger, B.H. O'Donovan, and E.R. Nadel. 1984. Circadian rhythm in sweating and cutaneous blood flow. *American Journal of Physiology* 246: R321-324.

Van Dongen, H.P., G. Maislin, J.M. Mullington, and D.F. Dinges. 2003. The cumulative cost of additional wakefulness: dose-response effects on neurobehavioral functions and sleep physiology from chronic sleep restriction and total sleep deprivation. *Sleep* 26: 117-126.

Waterhouse, J., S. Aizawa, A. Nevill, B. Edwards, D. Weinert, G. Atkinson, and T. Reilly. 2007a. Rectal temperature, distal sweat rate, and forearm blood flow following mild exercise at two phases of the circadian cycle. *Chronobiology International* 24: 63-85.

Waterhouse, J., B. Edwards, A. Nevill, S. Carvalho, G. Atkinson, P. Buckley, T. Reilly, R. Godfrey, and R. Ramsay. 2002. Identifying some determinants of "jet lag" and its symptoms: a study of athletes and other travellers. *British Journal of Sports Medicine* 36: 54-60.

Waterhouse, J., T. Reilly, G. Atkinson, and B. Edwards. 2007b. Jet lag: trends and coping strategies. *Lancet* 369: 1117-1129.

Young, A.J., J.W. Castellani, C. O'Brien, R.L. Shippee, P. Tikuisis, L.G. Meyer, L.A. Blanchard, J.E. Kain, B.S. Cadarette, and M.N. Sawka. 1998. Exertional fatigue, sleep loss, and negative energy balance increase susceptibility to hypothermia. *Journal of Applied Physiology* 85: 1210-1217.

Zhang, P. and H. Tokura. 1999. Thermoregulatory responses in humans during exercise after exposure to two different light intensities. *European Journal of Applied Physiology* 79: 285-289.

Index

Note: The italicized *f* and *t* following page numbers refer to figures and tables, respectively.

About the Author

Stephen S. Cheung, PhD, is the Canada research chair in environmental ergonomics in the department of physical education and kinesiology at Brock University in St. Catharines, Ontario. Dr. Cheung has published more than 46 papers on topics that span the chapters in this book, including extensive publication on hyperthermia and its effect on exercise capacity and fatigue as well as the effects of cold on manual function and marine survival.

Dr. Cheung is a member of many professional organizations, including the Canadian Society for Exercise Physiology, the American College of Sports Medicine, and the Aerospace Medical Association. He serves on the executive committee of the International Conference on Environmental Ergonomics, and he graduated with honors from the prestigious International Space University in Barcelona. He received his PhD in exercise science from the University of Toronto.

In his leisure time, Dr. Cheung enjoys bicycling and racing, playing squash, and reading.